Firewood Crops

Shrub and Tree Species for Energy Production

Report of an Ad Hoc Panel of the
Advisory Committee on Technology Innovation
Board on Science and Technology for International Development
Commission on International Relations

NATIONAL ACADEMY OF SCIENCES
Washington, D.C. 1980

NOTICE: The project that is the subject of this report was approved by the Governing Board of the National Research Council, whose members are drawn from the Councils of the National Academy of Sciences, the National Academy of Engineering, and the Institute of Medicine. The members of the committee responsible for the report were chosen for their special competences and with regard for appropriate balance.

This report has been reviewed by a group other than the authors according to procedures approved by a Report Review Committee consisting of members of the National Academy of Sciences, the National Academy of Engineering, and the Institute of Medicine.

The National Research Council was established by the National Academy of Sciences in 1916 to associate the broad community of science and technology with the Academy's purposes of furthering knowledge and of advising the federal government. The Council operates in accordance with general policies determined by the Academy under the authority of its congressional charter of 1863, which establishes the Academy as a private, nonprofit, self-governing membership corporation. The Council has become the principal operating agency of both the National Academy of Sciences and the National Academy of Engineering in the conduct of their services to the government, the public, and the scientific and engineering communities. It is administered jointly by both Academies and the Institute of Medicine. The National Academy of Engineering and the Institute of Medicine were established in 1964 and 1970 respectively, under the charter of the National Academy of Sciences.

The Board on Science and Technology for International Development (BOSTID) of the Commission on International Relations addresses a range of issues arising from the ways in which science and technology in developing countries can stimulate and complement the complex processes of social and economic development. It oversees a broad program of bilateral workshops with scientific organizations in developing countries and conducts special studies. BOSTID's Advisory Committee on Technology Innovation publishes topical reviews of technical processes and biological resources of potential importance to developing countries.

This report has been prepared by an ad hoc advisory panel of the Advisory Committee on Technology Innovation, Board on Science and Technology for International Development, Commission on International Relations, National Academy of Sciences–National Research Council, for the Office of Science and Technology, Bureau for Development Support, Agency for International Development, Washington, D.C., under Contract No. AID/csd-2584, Task Order No. 1.

Second Printing, August 1981

ISBN: 978-1-60322-056-9

Library of Congress Catalog Number: 80-83796

PANEL ON FIREWOOD CROPS

EDWARD S. AYENSU, Director, Office of Biological Conservation, Smithsonian Institution, Washington, D.C., USA, *Chairman*

JOHN G. BENE, Senior Advisor to the President, International Development Research Centre, Ottawa, Canada

JAMES S. BETHEL, Dean, School of Forest Resources, University of Washington, Seattle, Washington, USA

LOUTFY BOULOS, Professor, National Research Centre, Dokki, Cairo, Egypt

GERARDO BUDOWSKI, Head, Natural Renewable Resources Program, Centro Agronomico Tropical de Investigacion y Ensenanza, Turrialba, Costa Rica

JEFFERY BURLEY, University Lecturer, Forestry Department, Commonwealth Forestry Institute, University of Oxford, Oxford, England

ERIK ECKHOLM, Senior Researcher, Worldwatch Institute, Washington, D.C., USA

HANS M. GREGERSEN, Professor, College of Forestry, University of Minnesota, Minneapolis, Minnesota, USA

ANTHONY V. HALL, Senior Lecturer and Assistant Curator, Bolus Herbarium, University of Cape Town, Rondebosch, South Africa

R. D. JOHNSTON, Principal Research Scientist, Division of Forest Research, Commonwealth Scientific and Industrial Research Organization, Canberra, Australia

STANLEY L. KRUGMAN, Principal Research Forest Geneticist, Timber Management Research, United States Department of Agriculture-Forest Service, Washington, D.C., USA

BOHUMIL F. KUKACHKA, Wood Technologist, Pioneer Research Unit, Forest Products Laboratory, Madison, Wisconsin, USA

HANS-JURGEN VON MAYDELL, Senior Research Scientist, Federal Research Centre for Forestry and Forest Products, Hamburg-Reinbeck, Germany

FRANÇOIS MERGEN, Professor, School of Forestry and Environmental Studies, Yale University, New Haven, Connecticut, USA

JULIA MORTON, Director, Morton Collectanea, University of Miami, Coral Gables, Florida, USA

H. Y. MOHAN RAM, Head, Department of Botany, University of Delhi, Delhi, India

JOZEF SWIDERSKI, Chief, Mechanical Wood Products Branch, Forest Industries and Trade Division, Food and Agriculture Organization of the United Nations, Rome, Italy

FRANK H. WADSWORTH, Director, Institute of Tropical Forestry, United States Department of Agriculture-Forest Service, Rio Piedras, Puerto Rico

K. FREERK WIERSUM, Staff Member, Department of Silviculture, Agricultural University, Wageningen, The Netherlands

NOEL D. VIETMEYER, Professional Associate, National Academy of Sciences, Washington, D.C., *Staff Study Director*

MARY JANE ENGQUIST, Staff Assistant, National Academy of Sciences, Washington, D.C., *Staff*

GREGORY MALINOWSKI, Research Associate, National Academy of Sciences, Washington, D.C., *Staff* www.KnowledgePublications.com

Foreword

Among the many transformations that mankind has wrought on this earth, none is so vivid and bewildering as those made in the landscape. To meet their daily requirements for food, our ancestors hunted the animals of the wild, slashed and burned parts of their surroundings to bare the ground for cultivation, and cut down the forests for fuel and shelter necessary for survival.

Our ancestors may have believed that the land had the power to regenerate, especially if the intervals between their periodic depredations were long enough for the natural ecosystem to reestablish itself. But during the last century, there have been rapid and extensive inroads into hitherto untouched and unspoiled forests in the name of economic development and, in some instances, for pleasure.

In the less-affluent nations of the world, the last half-century has seen a steady increase in human population. This has affected the availability of food, shelter, and fuel. Small- and large-scale farming, as well as excessive exploitation of forests, is already affecting the supply of fuelwood upon which many of the less fortunate depend for their source of energy.

More than one-third of the world's population depends on wood for cooking and heating. Eighty-six percent of all the wood consumed annually in the developing countries is used for fuel, and of this total at least half is used for cooking. The situation is growing so desperate that wood is poached from forest reserves; hedges planted around homes are stolen at night; and even scaffolding is stolen from building sites to meet shortages in firewood supplies.

In the face of global concern over the dwindling supply of fuelwood, the rate of forest decimation to provide basic human necessities in developing countries is alarming. We must look upon woody plants as renewable resources that, if effectively managed, could alleviate the problem not only for the present, but for posterity.

We hope this report will prove useful to developing countries by suggesting potentially significant fuelwood candidates for introduction to suitable environments. This report should stimulate initiatives for restoring our renewable resources and increasing the world's supply of fuelwood.

On behalf of the members of the panel I would like to thank the National Academy of Sciences for recognizing this subject as one of the most important problems facing mankind today and for giving the panel the opportunity to help bring this information to the attention of those who need it most.

EDWARD S. AYENSU, *Chairman*
Panel on Firewood Crops

Preface

No less than one and a half billion people in developing countries derive at least 90 percent of their energy requirements from wood and charcoal. Another billion people meet at least 50 percent of their energy needs this way. Indeed, it has been estimated that at least half the timber cut in the world still serves its original role for mankind: as fuel for cooking and heating.*

This essential resource, however, is seriously threatened. The developing world is facing a critical firewood shortage as serious as the petroleum crisis. The growth in human population is far outpacing the growth of new trees—not surprising when the average user burns as much as a ton of firewood a year. The results are soaring prices for wood, a growing drain on incomes and physical energy expended to satisfy basic fuel needs, the wasteful burning of animal manures to cook food rather than produce it, and an ecologically disastrous and potentially irreversible spread of treeless landscapes.

If the pace of tree planting around the world is not greatly accelerated, a recent paper from the World Bank says, "By the turn of the century, at least a further 250 million people will be without wood fuel for their minimum cooking and heating needs and will be forced to burn dried animal dung and agricultural crop residues, thereby further decreasing food crop yields."†

This report does not suggest a solution to the whole firewood crisis. It examines but one part of the solution: the selection of species suitable for deliberate cultivation as firewood crops in developing countries.

The panel that produced the report met at Airlie, Virginia, in July 1977. Prior to the meeting, an inquiry was sent to several hundred plant scientists and foresters asking their suggestions for species that might become important sources of firewood in the developing world. About 150 responses were received, containing nominations for over 1,200 species. Of these, about 700 were given top ranking (see Appendix 4), signifying that they are potentially valuable firewood crops that deserve increased recognition and research. Using them as a guide, the panel selected the species included in this book.

Primary emphasis here is on species suitable for growing firewood for individual family needs. However, species suited to plantation cultivation for

*Eckholm, Erik. 1975. *The Other Energy Crisis: Firewood.* Worldwatch Paper 1. Worldwatch Institute, Washington, D.C.

†Spears, J. S. 1978. *Wood as an Energy Source: The Situation in the Developing World.* Presented to the 103rd Annual Meeting of the American Forestry Association, Hot Springs, Arkansas, October 8, 1978.

fueling small industrial factories, electric generators, and crop driers are also considered. Most of the plants are little known in traditional forest production. Some are woody shrubs rather than forest trees, but even these many-branched, crooked, sometimes short-lived species may meet many requirements for small-scale village use.

The panel particularly looked for:

- Multiple-purpose plants that have uses in addition to providing fuel;
- Plants that adapt well to different sites, that establish easily, and that require little care;
- Plants for problem environments such as steep hillslopes, low-nutrient or toxic soils, arid zones, and tropical highlands; and
- Plants not consumed by goats and wildlife.

Special consideration was given to such characteristics as:

- Nitrogen-fixing ability;
- Rapid growth;
- Ability to coppice;
- Ability to produce wood of high calorific value that burns without sparks or toxic smoke; and
- Ability to grow successfully in a wide range of environments, including different altitudes, soil types, rainfall regimes, amounts of sunlight, and terrain.

In producing this report the panel had in mind those sites now causing the greatest concern: where fuelwood has virtually disappeared. The report therefore concentrates on new plantings. There are, however, large areas where tree resources still exist that, with careful husbandry, could be maintained as fuel resources. Moreover, there is scope for harvesting firewood as a by-product of industrial logging and of conventional plantation forestry. The potential for producing fuel from these conditions is also important for readers to consider.

An alternative and complementary approach to the firewood crisis is the substitution of well-designed stoves, kilns, or boilers for the generally inefficient devices now widely used. Some promising fuel-efficient devices are described in Appendix 1.

This is one in a series of reports that identifies unconventional scientific research with promise for developing countries. In each study, the experience and knowledge of distinguished scientists is incorporated in a book that provides new ideas for decision makers. Publications that contain information on some exceptionally promising firewood species and related technologies* are:

*For information on how to order these and other reports, see page 234.

- *Leucaena: Promising Forage and Tree Crop for the Tropics*
- *Tropical Legumes: Resources for the Future.*
- *Producer Gas: A Little-Known Fuel for Motor Transport**
- *Sowing Forests from the Air†*

Projects now being planned to complement this one on shrubs and trees for firewood include studies of fast-growing tropical grasses as fuel crops, detailed studies of *Calliandra calothyrsus* and *Acacia mangium*, two little-known but apparently promising tropical tree crops, and another study on shrubs and trees that will describe species not covered in this report.

The present report was prepared under a contract with the Office of Science and Technology, Development Support Bureau, Agency for International Development. Travel funds for John Bene were paid by the International Development Research Centre, Canada.

The final text was edited and prepared for publication by F. R. Ruskin. The cover art was prepared by Debbie Hanson, and Wendy D. White edited the bibliography.

Most of the plants described in this report are so little known that information about their requirements and performance in fuelwood plantations is sparse. The panel would greatly appreciate hearing from readers having useful additional details for inclusion in subsequent editions of this report. Suggestions and information from readers about species not covered in this volume will also be welcome. Comments should be sent to the Staff Officer, Dr. Noel Vietmeyer, National Academy of Sciences, 2101 Constitution Avenue, Washington, D.C. 20418, USA.

WARNING

If misunderstood, this book is potentially dangerous. Because of the severity of the firewood crisis, the panel has selected trees and shrubs that are aggressive and quick-growing. These seem appropriate for cultivation in areas of extreme fuel shortage, particularly where climates and soil conditions are harsh. But in more equable environments and where no fuelwood shortages exist, such potentially invasive plants should be introduced only with great care. The threat of their weediness is too great. **In any trials of fuelwood plantations local species should always be given first priority.**

*This report, now in preparation, examines the use of charcoal- and wood-fueled producer gas units for powering vehicles.

†Also in preparation, this report describes the experiences, notably in Canada, the United States, Australia, and New Zealand, with establishing forests by broadcasting seed from aircraft.

Contents

APPENDIXES

(J. Ling, UNESCO)

Introduction

ERIK ECKHOLM

For more than a third of the world's population, the real energy crisis is a daily scramble to find wood to cook meals. Diplomats, economists, and the media have given little attention to the scarcity of firewood, but the problem is enormous, if less dramatic than the scarcities of food or oil. As one Indian official has expressed the problem, which gets worse each year, "Even if we somehow grow enough food for our people in the year 2000, how in the world will they cook it?"

Most people in the industrialized nations have no idea how important firewood is to the less-developed countries. In most poor countries today, 90 percent of the people depend on firewood as their chief source of fuel, and each year the average user burns anywhere from a fifth of a ton (in extremely poor, wood-short areas such as India) to well over a ton (in parts of Africa and Southeast Asia). An American may use far more than that in an open fireplace, but a ton per user is a huge amount when multiplied by hundreds of millions. The firewood "crisis" is the total of thousands of local and regional scarcities that are becoming steadily more serious.

Firewood scarcity is probably most acute today in the countries of the Indian subcontinent and in the semiarid stretches of Africa below the Sahara, although it affects many other areas as well. In Latin America the scarcity of wood and charcoal is a problem throughout most of the Andean region, Central America, and the Caribbean. One of the obvious results of this scarcity is that as firewood prices rise, so does the economic burden on the poor. In some cities the price of firewood has risen by more than 300 percent in 2 years. Government agencies have collected few statistics on this trend, but firewood prices seem to be rising beyond the reach of many city dwellers.

In some cases, the rise in the price of fuelwood has been prompted by the rising cost of imported kerosene, the principal alternative energy source for the poor. But firewood prices have often risen *faster* than kerosene prices, a fact that reflects the growing difficulty with which wood is procured. Gathering firewood is now an entire day's task in some mountain villages of Nepal; a generation ago the same expedition would have taken an hour or two.

Erik Eckholm is a member of the NAS Firewood Panel. This introduction is derived from some of his recent magazine articles.

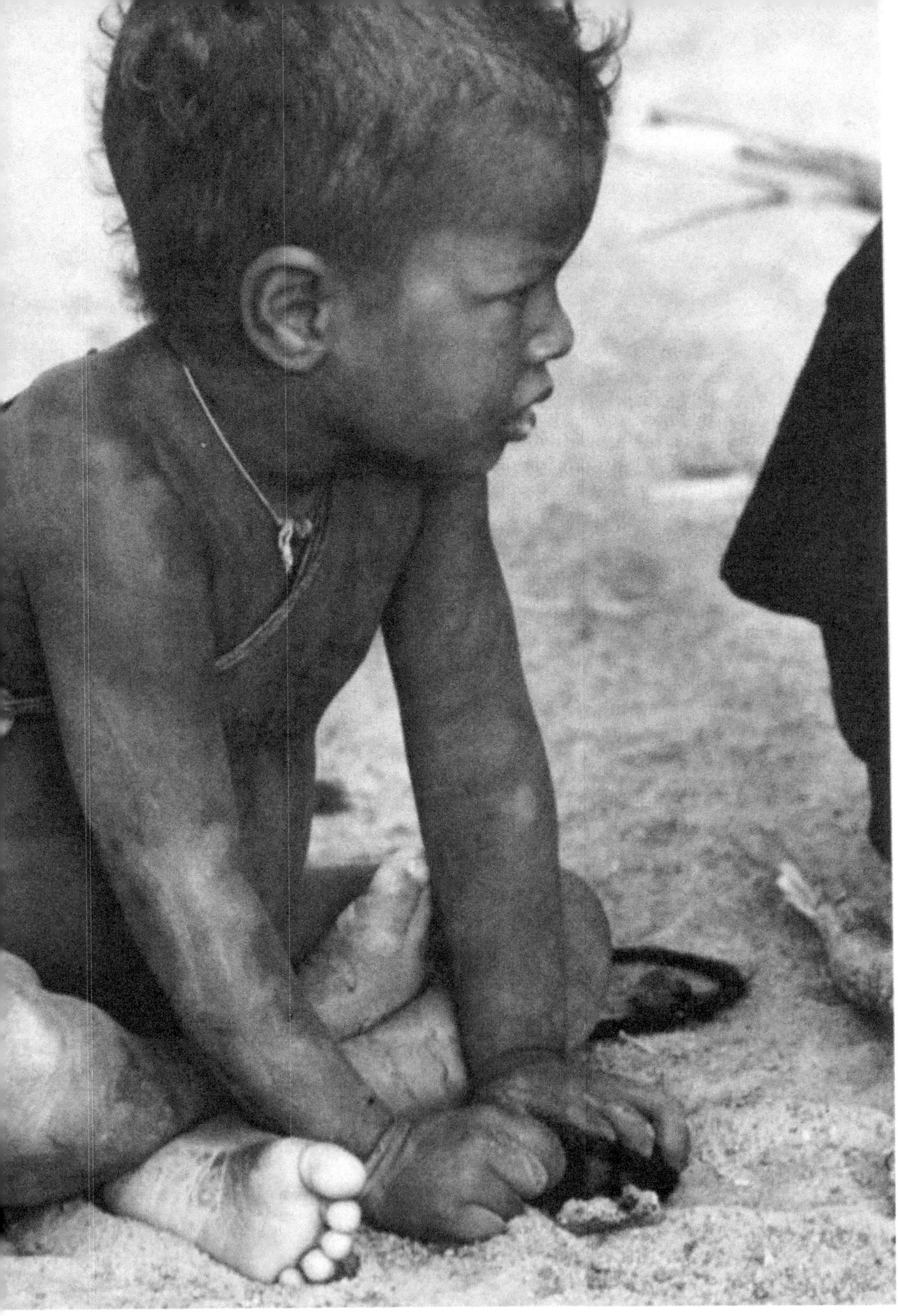

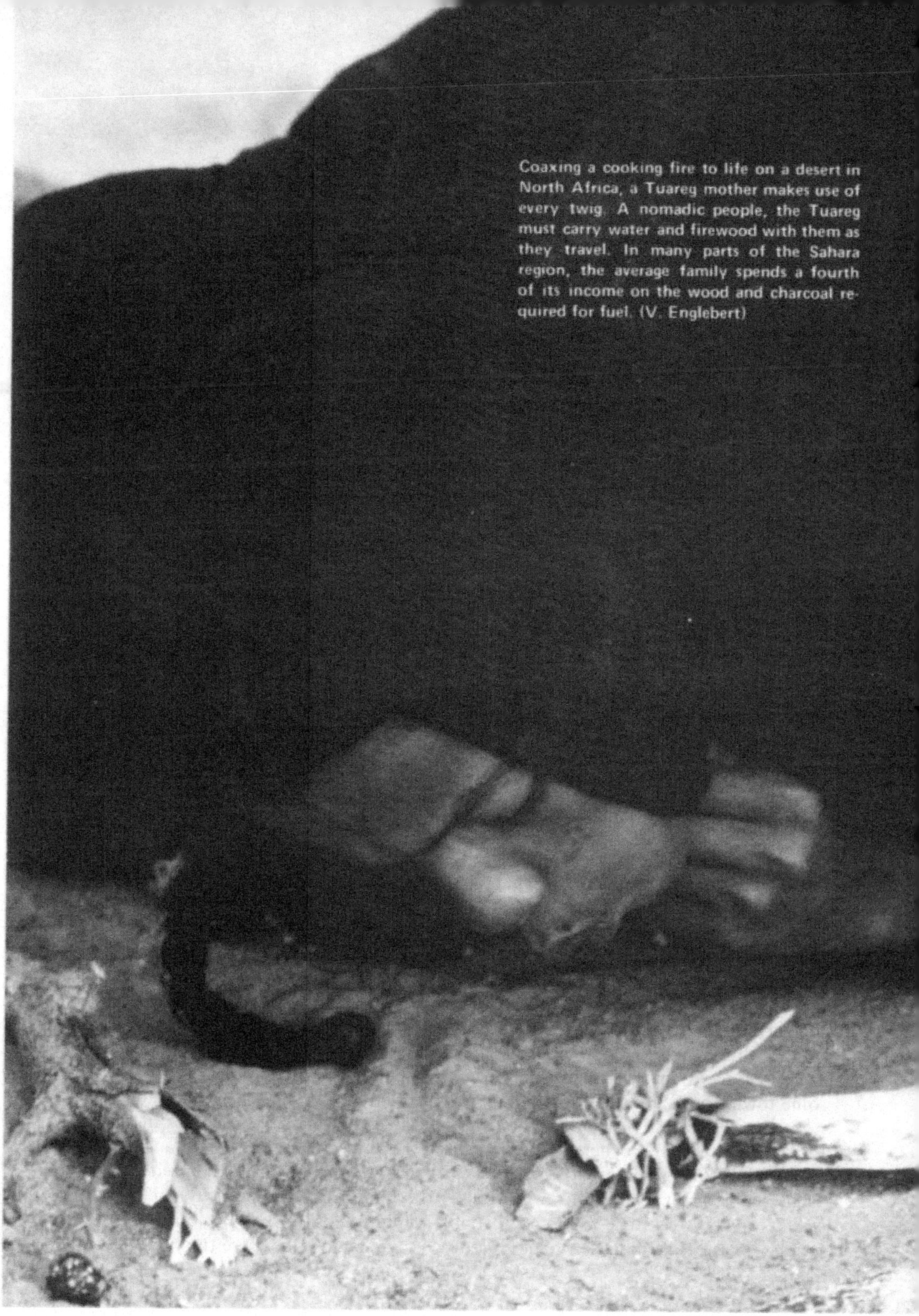

Coaxing a cooking fire to life on a desert in North Africa, a Tuareg mother makes use of every twig. A nomadic people, the Tuareg must carry water and firewood with them as they travel. In many parts of the Sahara region, the average family spends a fourth of its income on the wood and charcoal required for fuel. (V. Englebert)

Sahara. Tuareg girl collecting camel dung. In the absence of wood in many developing countries, manure is used for fuel and the land is robbed of sorely needed nutrients. (V. Englebert)

Throughout most of Asia, Africa, and Latin America, those who can possibly pay the price for wood and charcoal do so, spending much of their income for fuel, and thereby foregoing consumption of other essential goods. Wood is simply accepted as one of the major expenses of living. In Niamey, Niger, the average manual laborer's family now spends one-fourth of its income on firewood and charcoal, which merchants bring into town by donkey cart and truck. Those who cannot pay may send their children (or hike into the surrounding countryside themselves) to forage fuel—if there are enough trees within a reasonable walking distance. Otherwise, they may scrounge about the town for twigs, garbage, or anything else that burns, including, in some towns, bark from the ornamental and shade trees that line the streets.

In the past, most firewood was burned in villages. But as wood prices in the towns go up, landowners naturally find an advantage in carting their available timber to the nearest town to sell instead of giving it to rural laborers. This commercialization of firewood raises the hope that entrepreneurs will start to plant trees to develop a profitable, labor-intensive business, but so far the usual result has been depletion of woodlands. In either case the rural poor, with little or no cash to spare, are in deep trouble.

Scarcity of firewood creates further problems. Once the farmland trees and the scrubby woodlands of unfarmed areas begin to disappear, both the needy and the entrepreneurs tend to poach for fuelwood in legally protected national forest preserves. These preserves are essential to the economy and the ecology of a country, and in India special mobile guard squads and even mobile courts have been formed to stop poachers. Such measures have been largely ineffectual; the problem is too widespread and deep-rooted.

Perhaps one reason that the firewood scarcity has not provoked much world attention is that the shortage appears essentially local and seems limited in its consequences to the actual users of the wood. But the problem is spreading into larger areas, increasing in severity, and exacerbating other problems. In that sense it is like the oil crisis. If dwindling oil reserves threaten the productivity of our industrial system, the deforestation that results in part from gathering firewood threatens a significant portion of the world's agricultural system. Much deforestation is intentional, carried out to clear more land for agriculture. Some clearing and a great deal of woodland depletion, however, occur because wood is needed for fuel, and the frequent result is reduced productivity of the land because of accelerated soil erosion, increasingly severe flooding, and creeping deserts.

Throughout the sub-Saharan fringe of Africa, from Senegal to Ethiopia, people living in towns like Niamey are inadvertently contributing to the creation of desert-like conditions in a wide band below the desert's edge whenever they buy firewood. Unlike oil, wood is not shipped thousands of miles,

Cutting firewood, Kilmi, Chad. In many areas, large-diameter trees are less valuable than small "stove-sized" ones because villagers lack the saws, wedges, and other tools needed to handle them easily. (F. Mattioli, WFP photo)

but in some areas it is used to make charcoal, which is often transported hundreds of miles. Virtually all trees accessible for road transport within 40 kilometers of Ouagadougou, Upper Volta, have been consumed as fuel by the city's inhabitants, and the circle of land that has been "strip mined" for firewood without being reclaimed is still expanding.

In the absence of suitable alternative energy sources and all-out efforts to disseminate more-efficient cooking stoves, future firewood needs in developing countries will be determined largely by population growth. The populations in many of the countries facing the most severe wood shortages may double over the next 25 years, putting unbearable pressures on their remaining woodlands. It should be obvious that the rise in energy demands, including demands for firewood, could be partially curbed by family planning.

The scarcity of firewood has damaged some countries in another way besides through deforestation. Throughout much of India, Pakistan, and Bangladesh, one often sees pyramids of hand-molded dung patties drying in the sun. In many areas these dung cakes have been the only source of fuel for generations, but as population increases and supplies of firewood decrease, farmers who once returned all or part of the available dung to the soil have been forced to use it for cooking. Between 300 and 400 million tons of wet dung—which shrink to 60–80 millions tons when dried—is annually burned as fuel in India, robbing farmland of nutrients and organic matter. Looking only at this direct economic cost, it is easy to see why the country's National Commission on Agriculture recently declared that "the use of cow dung as a source of noncommercial fuel is virtually a crime."

Fortunately, trees, when properly managed, are a renewable resource. The immediate logical response to the firewood shortage, one that will have many incidental ecological benefits, is to plant more trees in plantations, on farms, along roads, in shelterbelts, and on unused land throughout the rural areas of poor countries. For many regions, fast-growing varieties of trees are available that can be culled for firewood in less than 10 years, and some species spring back without replanting.

The concept is simple; its implementation is not. The governments of nearly all the wood-short countries have had tree-planting programs—in some cases for decades. But problems have plagued these programs from the beginning. One is the sheer magnitude of the need for wood and the growth in demand. The increase in population has simply swallowed most moderate efforts at tree planting. The problem of scale is closely linked to a second obstacle: to national leaders, the increasing scarcity of firewood does not seem to be urgent. With elections to win, wars to fight, dams to build, and hungry mouths to feed, it is hard for politicians to concentrate funds and attention on the problem. Some ecologists in poor countries have been warning their governments for years about the dangers of deforestation and fuel shortages, but tree-planting campaigns do not win elections.

Even when the political will is there and the funds are allocated, reforestation campaigns are unexpectedly difficult and complex. Planting millions of trees and successfully nurturing them to maturity is quite different from such well-bounded technical tasks as building factories, and the projects frequently fail. For example, most of the regions with too few trees also have too many cattle, sheep, and goats, and where rangelands are badly overgrazed, livestock may eat the leaves off saplings. To be successful, reforestation efforts require a formidable administrative effort to protect the young trees for years until they are grown—and once they are grown there is the problem of monitoring timber harvests and of systematic replanting as the trees reach maturity.

Reforestation also requires massive popular support. In country after country, the lesson is plain. Tree-planting programs are most successful when local communities are most involved and when the people perceive clearly that success is in their self-interest. Central or state governments can provide plans, money, and advice, but unless community members understand why lands to which they have traditionally had free access for grazing and wood gathering are being demarcated into a plantation, they are apt to view the project with suspicion.

There is no single magic solution to the firewood scarcity, but some blend of fuel conservation, tree planting, and new technologies could certainly relax its stranglehold on any country. The failure of many affected countries to meet the firewood challenge does not, in the final analysis, reflect an absence of suitable technologies, but rather a failure of political systems, of social organization. Should firewood shortages continue to worsen, no dramatic event like an Arab oil embargo will flash crisis signals to the world. For the world's poor, the energy "emergency" is a constant reality, one submerged in the daily struggle to get by. A deepening firewood crisis, and the environmental degradation it entails, means a steady deterioration in their prospects for a better life.

Gathering firewood.
top left: Obokeo, Cameroon. (M. Benaissa, FAO photo)
top center: Southern China. (M.B. Bullock)
lower center: Collecting water and *Acacia mearnsii* branches for fuel, rural South Africa. (R.J. Poynton)
top right: Mountains of Afghanistan. (FAO photo)
lower right: Near Arequipa, Peru. (L. McIntyre)
bottom left: Fuelwood from a *Cassia siamea* plantation for the city of Ouagadougou, Upper Volta. (H.J. von Maydell)

I Wood as Fuel

As underground fuels become more scarce and expensive, the importance of wood is increasing dramatically. This has catapulted tree-growing into the arena of world energy production. Trees, if better managed and utilized, could rapidly increase the energy available to developing countries. And this could be done fairly inexpensively, without masses of foreign exchange or technology, and in many cases by using unskilled workers who are already available and underemployed in countries with severe energy shortages. The additional benefits in improved environment are hard to quantify but are likely to be substantial.

The major concern of most governments at present is to increase and conserve energy supplies. Since energy for rural development has become one of the more crucial issues, firewood is beginning to enter the mainstream of national and international priorities and policies. Now there is new regard for foresters and appreciation of the importance of their skills to a nation's present and to its future. Many government ministries other than forestry have suddenly become interested in funding tree-growing–for example, ministries of energy, agriculture, rural development, conservation, natural resources, and community services.

Development-assistance agencies have also shown a new awareness of the importance of trees. They realize, for example, that it is pointless to worry about producing more food if there is no fuel to cook it with, and foolish to invest in expensive engineering projects if continual deforestation by firewood gatherers will silt up the dams and irrigation canals and make mountain roads impassable because of soil slippage. "Forestry supports agriculture," said Mao Tse Tung, and agricultural scientists and planners attest to the wisdom of his words.

All over the world foresters and even national political leaders are beginning to recognize the need to integrate forestry into rural development in new ways. The U.S. Congress has directed the Agency for International Development (AID) to focus on forestry and firewood plantations in its rural development programs. The World Bank, the world's largest lending institution, in a recent forestry sector policy paper has announced its intention to multiply its support for such activities as village woodlots, farm forestry, and environmental rehabilitation. And in Nairobi, the International Council for Research in Agroforestry has been set up, aimed at coordinating and stimulating efficient land use through combining wood and food production.

Firewood–far from becoming outmoded–has become a recognized source of energy, and demand will increase enormously in the coming decades. Indeed, wood products are likely to continue as the most important universal fuel for rural areas of developing countries.

John Spears of the World Bank calculates that even with optimistic assumptions that wood-conserving stoves, biogas reactors, and solar cookers will be used where available, an additional 20–25 million hectares of trees must be planted by the year 2000. At the present rate of reforestation that is 10 times more than will be achieved.*

Regional shortfalls, of course, may be more serious. According to the Club du Sahel, tree planting in the Sahelian zone of Africa must increase fiftyfold if the firewood demands for the year 2000 are to be met.†

Wood can be grown where no fuel was produced before. It is renewable and its production can be sustained. The Indian neem tree (page 114), introduced to the Accra Plains of Ghana early this century, has been providing firewood to towns such as Kumasi (and to villages in many West African countries) ever since, and the supply seems to be keeping up with demand as the vigorous and resilient neem is reseeded by bats that eat its fruit.

Another example of sustained firewood production is the Paliparan area in the Philippines. In the early 1920s when *Leucaena leucocephala* was planted there, it was a green desert of *Imperata* grass (a scourge of the tropics, known commonly as blady grass, cogon, or alang-alang). Since then the area has been a productive energy forest, yielding 20 m^3 or more of firewood per hectare per year. Even today it is the main firewood supply for the city of Laguna and surrounding districts.

Wood can be a "self-reliant" fuel that requires little foreign exchange and is independent of the vagaries of international commerce. It does not demand a sophisticated distribution network; indeed, it is the most accessible form of energy for most rural populations in developing countries. It is a familiar fuel whose use requires few extension services. One of the cheapest sources of energy available, it can, if necessary, be used without expensive metal appliances. It is one of the few fuels that an individual can produce for himself. Near Mendi in Papua New Guinea, highland tribesmen plant *Casuarina oligodon* (page 41) for firewood around community "longhouses" and in abandoned farmland. Although contact with this remote valley was made only in the last 20 years, today it is common to see tribesmen sitting beside the road surrounded by cords of firewood for sale.

Fuelwood can also provide export income. Kenya exports charcoal to Arabian Gulf nations and Suriname regularly ships charcoal to northern Europe. This can, however, be a mixed blessing. Eastern Kenya, for instance, has been severely denuded by the demands for exports added to those for wood to fuel the area's cooking fires.

Fuelwood is not limited to household use. Large energy plantations are planned or already underway in the United States and many other countries to fuel machines such as:

*J. S. Spears. 1978.

†Club du Sahel. 1978. *Energy in the Development Strategy of the Sahel* (Paris, October 1978).

Traditional cooking practices:

top left: *Leucaena leucocephala* as firewood in traditional Filipino stove, Los Baños, Phillipines. (N.D. Vietmeyer)

top center: Nepal. (D. Mason, AFP/FAO photo)

top right: Mud brick rural stove, Sirs-el-Layyan, Manoufie, Egypt. (FAO photo)

bottom right: The traditional three-rock fireplace, Upper Volta. (F. Botts, FAO)

bottom center right: Pigeon pea stalks in use as firewood, Hyderabad, India. (D. Sharma)

bottom left: Makeshift stoves in (upper) Africa and (lower) Latin America. (VITA)

above: Ullong-do Island, South Korea, squid fisherman's wife cooks breakfast over a driftwood fire in a pot fitted into a cast-iron stove. (J.A. Bannigan)

Marketing firewood.

top left: Yemen Arab Republic. Youthful woodcutters in the stark mountain area between Hodeida and Sana with branches for sale for cooking and heating fuel. (F. Mattioli, FAO photo)

center: Limbe, Haiti. Fuel is so scarce in areas like this that even tiny piles fetch high prices. (M.D. Benge)

top right: Kenya town firewood and charcoal. (N. McKee)

center right: Scaffolding removed from construction site and sold for firewood in Takoradi, southwestern Ghana. (E.S. Ayensu)

bottom right: Syria. In many cases nomads harvest for fuel shrubs that are palatable to livestock, thus causing serious pasture deterioration. (FAO photo)

above: Hyderabad, India. Pigeon pea stalks gathered for firewood. (D. Sharma)

bottom left: Papua New Guinea. Southern Highland villagers with cords of *Casuarina oligodon* wood (see page 41) wait beside the road for passing trucks from nearby Mendi. (N.D. Vietmeyer)

- Electric generators;
- Railroad locomotives;
- Driers for fish, tobacco, lumber, grain, copra, and other agricultural products;
- Factories milling sugar, timber, or other new materials;
- Pottery, brick, charcoal, and limestone kilns; and
- Metal smelters.

The lumber and paper industries have long used bark and waste to generate steam and electricity. Indeed, 8 percent of Sweden's energy and 15 percent of Finland's energy is generated from wood; in the United States in 1974 more energy was generated from wood than from nuclear power. An electric generator designed to consume about 25,000 dry tons of wood fuel each month has recently been installed at the Jari Project in the Brazilian Amazon to power a pulp mill, a saw mill, and a township of over 10,000 people.

The government of the Philippines plans to support rural electrification with several wood-fueled power plants. Early in 1978 it planted an energy forest of *Leucaena leucocephala* (page 50), which will be used to fuel a 75-megawatt steam power plant. A study that considers hauling distance, plantation and cooling-water requirements, and available government reforestation areas has identified 25 potential energy plantation sites with a total area of approximately 320,000 hectares. If fully planted, these could support 19 wood-fired power plants with an overall generating capacity of 1,425 megawatts. According to the official report, a wood-fired electrical generating plant could not only compete economically with an oil-fired plant but could also generate an estimated net foreign exchange savings of approximately $146 million in the first 10 years of operation.*

Firewood production can be good for economic development in rural areas. Growing trees for firewood can be successfully combined with the production of posts, poles, and timber. The production and sale of wood to nearby urban centers can provide many jobs and much rural cash income. It has been estimated that some 6,000 families are involved in supplying wood and charcoal to the city of Maputo in Mozambique, for example. Few other energy sources can provide and maintain such high employment.

FIREWOOD PLANTATIONS

A logical first response to the firewood situation is to plant more trees. In theory, probably all countries have the physical resources to meet their most urgent rural firewood needs. It has been estimated that at least 75 percent of tropical land is inherently unsuited to sustained conventional agriculture, but 35 percent of the population of the tropics lives on this land. About half of

*J. A. Semana, P. V. Bawagan, F. R. Siriban, and V. B. Mendoza. 1977. *A Feasibility Study of the Utilization of Man-Made Forests for Generating Electricity*. Forest Products Research and Industries Development Commission, National Science Development Board, College, Laguna, Philippines.

it will not support vegetation of any kind. The best use of the remainder is tree cultivation (or the simultaneous or sequential cultivation of trees with annual crops or with the pasturing of animals). Virtually everywhere, villages have unused or misused areas on which woodlots can be planted.

But there is little modern experience with growing trees explicitly for firewood, even by most foresters. Fuel production has long been considered the lowest use of wood, and foresters have traditionally cultivated trees primarily for other purposes, such as for timber and pulpwood. For these products, the species they choose to grow are not those that would be grown purely for fuel. Moreover, the yield measurements reported, the management techniques devised, and the varieties (provenances) selected almost never reflect potentials for firewood production.

Firewood plantings can use species with short boles, crooked trunks, or wood that warps or splits as it dries. These features are not as detrimental to fuelwood use as to timber production. Nor is stem size. In simple cookstoves, for example, branches as small as one or two centimeters in diameter may be ideal. Thus a shrub may prove satisfactory for village fuelwood silviculture if it grows fast and produces a dense wood that burns with intense heat. In practice, fuelwood may come both as a primary crop from fuelwood forests and as a secondary crop from timber forests.

In 1936, horticulturists transported seed of *Calliandra calothyrsus*,* a small Central American shrub, from Guatemala to Indonesia. They were interested in it as a shade tree and ornamental, for like some other *Calliandra* species, it has flowers that are showy crimson powderpuffs. But Indonesians took up *Calliandra calothyrsus* as a firewood crop instead. Indeed, for 25 years steadily expanding fuelwood plantations of *Calliandra* have been established, until they now cover more than 30,000 hectares in Java (page 36).

The wood, too small for lumber, is dense, burns well, and is ideally sized for domestic cooking. It is also useful for fueling brick, tile, and lime kilns and for copra and tobacco driers. Indonesian villagers now cultivate it widely on their own land, often intercropping with food crops. The plant's value is dramatically exemplified by the village of Toyomarto in East Java. There, land that was once grossly denuded and erosion-pocked is now covered with a *Calliandra* "forest" and is fertile once more. Today the villagers make a good living selling the firewood, actually earning more from it than from their food crops.

Another example of shrubs grown successfully for firewood is found in South Korea. Known as "miracle plants," the leguminous bushes *Lespedeza bicolor* and *L. thunbergii* rapidly cover bare ground, and their long, shallow roots bind and hold down soil. Prolific root nodules provide nitrogen, and their protein-rich foliage is fed to livestock. Foresters plant *Lespedeza* species around pine seedlings to protect the soil, to provide plant nutrients, and to produce forage and firewood until the forest is established. The dense wood of the small stems is eagerly sought for cooking fuel, and it is a well-established Korean practice to plant *Lespedeza* species along ridges as a

*They identified it, probably correctly, as *Calliandra calothyrsus;* the previous NAS report *Tropical Legumes: Resources for the Future* misspelled the specific epithet as *callothyrsus*.

In addition to domestic use, firewood is important to local industries; e.g., tobacco, brickmaking, bakeries, blacksmiths. Shortages can be the limiting factor to the continuation and expansion of these industries.

top left: Baking bread, Herat, Afghanistan. (R.A. Bannigan)
top center: Drying tobacco, Philippines. (M.D. Benge)
top right: Blacksmith forging sickles. Kirkuk, Iraq. (H.A. Hopfen, FAO photo)
bottom right: Brickmaking, Setwane, Botswana. (J. Murphy, WFP photo)
bottom left: Brewing tea, Kabul, Afghanistan. (R.A. Bannigan)
center left: Distilling mint oil, Chiang Mai, Thailand. (Narong Chomchalow)

firewood crop. It is harvested a year after planting, yielding 2.9 t (dry weight) of firewood per hectare and 6.1 t after 2 years. The flowers are a source of honey that has become a specialty throughout the country.

In addition to fuel, woodlots in and around villages and cities can provide stable and pleasant surroundings. They provide shade, shelter, beautification, and habitats for wildlife, which in many areas are a valued food source. The plantations also reduce wind erosion, beneficially influence local temperature and humidity, and in many cases replenish and redistribute essential soil nutrients.

A covering of plants slows rainfall runoff, which generally allows for greater groundwater recharge and so helps maintain year-round stream flow. Thus it decreases the likelihood of floods and the buildup of silt in reservoirs, thereby increasing the capacity and life of hydroelectric and irrigation facilities. Although most efforts to combat flooding and siltation (as well as snowslides and landslides) have entailed engineering measures—dams, embankments, and dredging, for example—these address symptoms, not causes. Reforesting denuded areas is a preventive measure that decreases the severity of flooding and decreases the load of sediment entering waterways. For decades floods, drought, and famine were recurring calamities of Fukien Province in China, but in the 5 years since forests were reestablished in the hills and watersheds, the inhabitants have lived free from all three. Growing trees or shrubs within watersheds also improves the water quality of rivers and lakes and the growth of fish and other aquatic foods.

Indiscriminate firewood collection is currently one of the principal destroyers of native forests. Intensively cultivated woodlots on accessible sites can help relieve this pressure on natural forests by supplying a large share of the needed firewood more conveniently.

Woody plants can provide more than fuel calories. They can also be sources of:

- Vegetable oil and fruits and nuts for food;
- Edible leaves and shoots for sauces, curries, salads, and beverages;
- Honey;
- Forage for livestock and silkworms;
- Green manure for fertilizing soil;
- Tanbark for the tannin used in leather making;
- Medicines and pharmaceuticals;
- Extractives such as resins, rubber, gums, and dyes;
- Timber, lumber, posts, poles, and pulp for paper, cardboard, and construction boards; and
- Shade for pastures or plantation crops such as coffee and cacao.

Some of these uses are inimical to fuel production, but they give the tree owner more flexibility; he has the option of using some specimens perhaps for forage, while keeping others for fuel. In times of hardship, he may sacrifice some tree growth to feed his family or animals with the foliage. In some cases, dense forests can produce a great deal of burnable waste material without a living tree being felled. In others, the owner may sell the best-formed trees for timber or pulp and use the remainder as fuel. Having such options is important to a rural farmer, and in this report we note the main alternative uses for the species selected, even if they conflict with firewood use.

FUELWOOD MANAGEMENT

Today, international aid agencies and foresters in the Third World are receptive to new notions about the purposes and practices of forestry. In essence, they recognize the modern necessity of taking forestry outside the forests—of involving people throughout the countryside in growing trees to meet their own requirements as well as to protect the land off which they and their livestock live.

Firewood production is particularly appropriate to this philosophy. It is less dependent on silvicultural expertise than sawtimber is and therefore can be done by nonprofessionals who learn the basic techniques for their own use. Firewood can best be produced like a farm crop without government intervention. Thousands of Brazilian farmers in the state of São Paulo, Minas Gerais, and Paraná already grow eucalyptus woodlots, partly for firewood, on land generally unsuitable for agriculture.

However, although the cultivation of firewood species does not demand continuous professional supervision, a forest service may be needed to provide seed or planting stock and advice for getting the trees established. Further, silvicultural practices (such as weeding and pest control) can greatly increase yields. What is sorely needed is the greater involvement of trained forestry experts in firewood production at all levels from the village woodlot to the national forest.

Trees for firewood can be planted in "nonforest" areas: along roadsides, in shelterbelts, on farms, on unused land, and in schoolyards, cemeteries, churchyards, market squares, parks, and home gardens. Fuelwood trees can be cultivated in small woodlots, even as individual specimens around a house or village. In some areas, such as Java and the People's Republic of China, home gardens already supply a good share of family firewood needs. Correct spacing of planted trees is important to production, but it is unnecessary to assure geometric precision as is required where mechanized equipment must pass between them.

Rural areas can probably supply their own fuelwood from small, local plantings, but urban areas can best be supplied from concentrated large plantations, strategically located and possibly government administered.

Firewood plantations, if carefully managed and protected from fire, animals, and "poachers," can be self-renewing. They are usually managed on rotations of about 10 years (much less in some moist tropical regions). The timing varies with the quality of the soil, species used, temperature, moisture available, and intensity of cultivation. Rotations of less than 5 years seem feasible in many areas, especially for those species that regenerate by sprouts (coppice).

Wood fuel for transportation.
top left: In 1949 the Royal Siam railway used 700,000 m^3 of fuelwood. (FAO photo)
top right: Wood-fueled steam tractor, used to haul logs in Maine, USA, early this century. (Maine State Museum)
bottom right: Car powered by wood gas that in 1979 traveled coast to coast across the United States on about 1,400 kg of wood. (A report on this technology, widely used during World War II but forgotten since, is in preparation. For ordering information see page 234). (B. Russell)
bottom center: Wood-powered American riverboats about 1855. (Detail of print, courtesy National Gallery of Canada)
bottom left: Locomotive taking on fuelwood, Chiang Mai Division, Thailand. (FAO photo)

FUELWOOD SPECIES

Trees most likely to prove useful for fuelwood plantations are those termed "pioneers," which in nature colonize deforested areas. These withstand degraded soils, exposure to wind, and drought. Many are rapid-growing legumes that fix atmospheric nitrogen. The very nature of a pioneer species endows it with adaptability, aggressiveness, and hardiness on hostile terrain. The most promising species are often the last natives to survive repeated grazing, cutting, fire, and soil degradation. These are usually well known, available nearby, obviously adapted, and tough. Some of these species may not require nursery production and laborious transplanting. Some pioneer species can be seeded directly on site; large stands of *Leucaena leucocephala* and *Sesbania grandiflora* have even been established in Southeast Asia merely by scattering seed from the air. (In addition, almost half a million hectares of pine forests have been established this way in the United States and Canada, and in Australia, plantations of eucalyptus are routinely established from the air.*)

At the same time, the vigor and adaptability of some of these species makes them potential weeds.

The ability to coppice or grow rapidly from root suckers is exceptionally important in a firewood species. The stumps of coppicing plants do not die; instead, dormant or adventitious buds regenerate new shoots. This allows repeated harvest without the cost and effort of replanting seedlings each time. Moreover, the living roots continue to bind the soil, and a canopy of new foliage quickly develops to shield the soil surface from rain and wind and helps suppress weeds. Coppice sprouts usually grow vigorously because they are served by roots big enough to feed the former tree. The continual cropping of coppice regrowth was an established technique in Europe in Roman times, and "coppice farms" were widespread in medieval England. Today eucalyptus pulpwood plantations, in Brazil and elsewhere, are based on coppice regrowth. The first crop (grown from seedlings) is usually harvested after 7 or 8 years; subsequent harvests (the result of coppice growth) are on 5- to 6-year rotations. In Europe and the United States, large tracts of redwood, aspen, maple, beech, and ash are grown on coppice rotations.

Most woods burn, but there are properties that differentiate their relative value for fuel. Density is the most general gauge of a wood's burning quality. The heavier the wood (when dry), the greater its calorific value. The heat given off is enhanced (by about 20 percent) by the resins found in conifer woods and by the oils and gums sometimes found in hardwoods. Green wood has a lower heating value than dry wood because energy is used in evaporating the moisture. The loss may reach 20 percent, which represents the wastage of 1 growing year in every 5. Air drying the wood yields more calories per unit of time than does tree growth in the forest. Air drying is thus a critical firewood conservation measure.

Some woods burn well even when green, a particularly important advantage

*See forthcoming companion report: *Sowing Forests from the Air*. To order see page 234.

Coppice farming, England. Left: Hornbeam (*Carpinus betulus*) flanking an old Roman road in Hertfordshire. The trees on the left are growing in pasture land and are cut (pollarded) above the reach of browsing animals. In the background is a regular coppice woodland.

Bottom left. Freshly coppiced ash. This tree has been cut and allowed to resprout on a regular rotation probably for at least 500 years. Bradfield Woods, Suffolk (W. H. Palmer)

Right. Woodland that has been continuously coppiced at least since the thirteenth century. The trees, mainly ash (*Fraxinus excelsior*) and hazel (*Corylus avellana*), are used for fuel and rural handicrafts. The ones left standing ("standards") are mainly oak (*Quercus robur*) grown for carpentry and joinery timber. Bradfield Woods, Suffolk (W. H. Palmer)

At least 1,000 years ago, Europeans learned to run their woods as a self-renewing resource. By coppice farming natural woodlands "woodmen" produced poles, small wood for fences and light construction, and firewood without destroying the trees. On the other hand, "foresters" managed plantations of large trees to produce planks and beams. Traditionally the woodman's profession was the more important of the two, but in Great Britain it declined around 1850. Since the late 1960s, however, the production of firewood by the old woodmen's coppicing techniques has increased steadily. This remarkably trouble-free coppice farming, practiced since the Stone Age in northwestern Europe, is so gentle to the land that medieval earthworks and even Ice Age glaciation can sometimes be detected among the trees. It is also "gentle" on the trees: some have been coppiced every 10 or 11 years for centuries.

That wood fuels can be produced in large plantations is evidenced by the eucalypt plantations (*Eucalyptus grandis* and other species) of the Companhia Siderurgica Belgo-Mineira in Brazil. Above: Part of the 200,000 ha of plantations near Monlevade that are harvested on 5- to 7-year cycles. Below: The wood is charcoaled for use as a carbon source in blast furnaces producing carbon steel wire rod. (Companhia Agrícola e Florestal Santa Bárbara, Belo Horizonte, Minas Gerais, Brazil)

New Zealand's largest dairy company has reportedly cut its fuel bill 80 percent by replacing its oil-fired process burners with a facility that burns sawdust and waste-wood chips. The plant raises steam for the various processes in a dairy for making milk powder, butter, and buttermilk.

The Hikurangi Cooperative Dairy Company chose a West German-designed boiler with a heat output of 50 GJ that would run on waste from a local sawmill. The dairy has also ensured future fuel supplies by planting 80 ha of *Pinus radiata* pine woods. It expects that the facility will repay the original investment within 4 years by burning an annual 14,000 t of sawdust, chips, and bark, instead of 3,000 t of fuel oil. (Hikurangi Cooperative Dairy Company) where fuel is so urgently needed that the users are unlikely to take the time to dry their firewood.

Rotted wood gives less heat than sound wood so that a species's natural durability and resistance to fungal decay may be important. (In any location where fuelwood is in high demand, however, the wood is not likely to be left long enough to decay.)

Some woods are poor fuels for use in unsophisticated stoves because they spark excessively or because their smoke is odorous, toxic, or irritating. The species catalogued in this report generally do not have these characteristics.

For the foreseeable future in developing countries, fuelwood will be hand harvested. Therefore, when choosing a species for firewood plantations, one consideration should be the ease with which it can be transported. Crooked and thorny species are difficult or bulky to transport, though for local village use this may be unimportant.

FUEL-EFFICIENT STOVES

In 1740, the threat of a fuelwood shortage around Philadelphia moved Benjamin Franklin to design his "Pennsylvania Fire-place," a cast-iron heating and cooking stove that greatly reduced heat lost up the chimney. In Franklin's words: "By the help of this saving invention our wood may grow as fast as we consume it, and our posterity may warm themselves at a moderate rate, without being obliged to fetch their fuel over the Atlantic."

Today, Franklin's approach is even more relevant. The open fire typically used to cook food in many developing countries wastes 90 percent of the fuel's heat. Even simple stoves with adjustable air intake to control the burning rate can double the efficiency of cooking with wood. Further gains can be made by designing the firebox to extract the most heat from the exhaust gases. These stoves need not be cast iron or expensive. Appendix 1 depicts some innovative approaches to the design of efficient stoves and kilns.

An open fire may extract only 10 percent of the energy potential of wood; an improved stove can use 20 percent or more.* If such stoves were universally adopted in areas where open kitchen fires are used, the demand for fuelwood would drop significantly.

Efficiency can also be increased by designing better cooking pots, by improving insulation and house-heating systems, and by finding better techniques for air drying fuelwood.

The issues to be considered when advocating use of more-efficient stoves are much broader than simply energy efficiency or even fuel costs. The kitchen, in effect, is a miniature food-processing plant where the time and effort involved in preparing the food and fire and in the cooking and clean-up are all important factors in the efficiency calculations. Moreover, the ease, and the prestige, of using a gas stove should not be underestimated.†

*S. Draper. 1977.
†Information supplied by D. Donovan.

Tradition remains the most serious barrier to the use of new stoves. Considerable extension work and demonstration may be needed to convert people to the use of new stoves. In some cases unexpected problems may arise. For example, some Nepalese villagers recently abandoned highly successful and widely adopted stoves when their roof thatch, no longer infiltrated by smoke, became infested with termites.

CHARCOAL

Nearly everywhere, reliance on charcoal as a source of fuel is increasing. In Tanzania, for example, the charcoal share of the wood fuel burned, which was 3 percent in 1970, is expected to rise to 25 percent by the year 2000. In principle this is discouraging, because in preparing the charcoal more than half the wood's energy is wastefully burned away. But charcoal makes wood energy easier and cheaper to transport, and the growing reliance on it is a result of the increasing distance from harvest site to the user. Also, charcoal is preferred because of its steady and concentrated heat, its smokeless burning, and because it can easily be extinguished when the fire is no longer needed. Charcoal also can substitute for fossil fuels, which in some places is an urgent need. Regardless of overall inefficiency, it seems clear that more meals will be cooked over charcoal in the future.*

IMPLEMENTATION

Planting fuelwood on a large scale demands social and political commitment. Decision makers in many countries, including Nepal, South Korea, the Philippines, and the Sahelian countries, are already funding national tree-growing programs, but elsewhere the pending crisis is generally not attracting the necessary financial commitment. This neglect is likely to disrupt energy supplies and lead to severe hardship in rural areas. As the pressures for fuel increase, it may lead to civil unrest and the devastation of new plantations. Research on appropriate species and management methods should start before the crisis is reached. The forest services have a crucial role to play in this race against time.

What is needed is a change in priorities in the use of trained foresters *and* agronomists. Forests for fuel can be treated as just one more farm crop. This makes firewood production more suitable for developing countries with few foresters. It seems possible that agronomists, rather than foresters, will be responsible for much of the small-scale firewood production in the future.

The existing forests are too important and too vulnerable to be abandoned by foresters in favor of village woodlots in the farm lands. In addition to making fuelwood production an agricultural responsibility, the suggestion has been made† that what is needed are "barefoot foresters" to persuade

*One hope for the future is that better charcoal kilns (see Appendix 1) will permit recovery of liquids of economic value; some of these liquids may even prove to be useful petroleum substitutes.

†E. Eckholm. 1979.

small farmers (whose economic horizons usually extend only to the next harvest) to plant trees for the future, to teach how to do it, and to introduce cook stoves that conserve firewood. Such extension services are particularly important in the case of individual and village-level fuelwood projects. Even with finances available, poor management and inadequate extension work are often critical bottlenecks. The accomplishments in South Korea (see Appendix 3) have demonstrated the importance of having such trained people working at the village level.

To be successful, fuelwood programs should fit into the social, cultural, economic, religious, political, and legal framework of the local area where plantations are to be established. Projects are often doomed from the outset by religious or cultural taboos, resistance to change, attitudes toward property ownership, or government intervention. Without the commitment of local residents to the orderly management and protection of the fuelwood plantings, they are likely to fail. Projects that demand basic changes in lifestyle are more likely to fail than those that adapt to existing lifestyles and outlooks.

Further, as a practical matter, fuelwood plantings are more likely to succeed if they are considered in the context of local land-use planning and local development objectives. The integration of fuelwood production with other agricultural activities can help ensure and sustain local interest; it can also cut costs and balance the development of local communities by providing them with an energy supply.

Local communities must be motivated to assume responsibility for the management of all vegetation, including trees, in their area. Forestry departments in some African countries, for example, have been producing millions of tree seedlings for distribution to villages at subsidized prices, or even free of charge, hoping to encourage individuals and communities to plant more trees. They have met with major success only in Ethiopia, where an estimated 50,000-100,000 hectares of eucalyptus have been established—almost all of it before Ethiopia had a professional forestry service—and all planted by illiterate but self-motivated peasants (see Appendix 2).

RECOMMENDATIONS

To alleviate the growing shortage of wood fuel is one of mankind's major challenges. To this end, firewood research is vital, requiring the combined efforts of government, industry, landowners, villagers, researchers, philanthropic institutions, and development-assistance agencies. Some activities to be undertaken (sequentially or concurrently) include:

1. Searching out and reducing the wasteful use of available fuel. For example:
   - Ensuring that existing wood resources are harvested and used without waste;
   - Testing and developing fuel-efficient stoves, particularly ones that villagers can make themselves; and

• Instituting policies and programs that encourage the use of alternative energy sources such as biogas and solar heat.

2. Conserving existing fuelwood sources. For example:
   • Controlling harvesting intensity to preserve forest productivity.
3. Identifying available production areas such as those in existing forests, wastelands, and farmlands.
4. Inventorying the tree species found locally, noting especially the species traditionally preferred for fuel.
5. Testing tree growth in all the available production areas. Tests should include:
   • Optimizing growth of volunteer trees of acceptable fuelwood species;
   • Test planting the best native fuelwoods; and
   • Test planting selected exotic species.

In all trials of exotic species, local species must be included for comparison.

Characteristics of the trees that should be assessed include growth rates by volume and dry weight (of stem and branch wood), wood and chemical properties, and resistance to pests. The importance of selecting and using the best seed source within a species cannot be overemphasized.

For species that show promise, adaptability trials should be set up to determine the responses of various species to different soils, growing conditions, altitudes, latitudes, temperatures, moisture conditions, and pests. This will indicate the relative advantages and limitations of each and will provide the technical backup for pilot-sized fuelwood plantations that can serve as demonstration plots. It will also enable researchers and officials to become familiar with the species tested. These trials might also provide the germplasm to start local woodlots and firewood plantations.

Comparison of experiences from such pilot plantations will provide the foundation for decisions establishing and utilizing firewood plantations and allow predictions of economic success.

Most of the species selected are likely to be grown mainly in woodlots and backyards or planted to double as living fences, or shade trees in pastures, or in plantations of coffee, cacao, or other crops. Testing should therefore be done over a range of such conditions.

II Fuelwood Species for Humid Tropics

While firewood shortages may not have reached crisis point in the lowland tropics, trees are being destroyed there on a massive scale.

For example, it is anticipated that the lowland forests of the Philippines, the Malay Peninsula, much of Indonesia, western Africa, Madagascar, Central America, and the West Indies will be converted to other uses within the next 10 years. Firewood harvests will contribute to this loss. Energy plantations could help slow deforestation caused by firewood harvesting.

It is thought that perhaps one billion people live in the humid tropics, with about 200 million of them living within or on the fringes of forests. Each is thought to burn an average of 0.5–1.3 m^3 of firewood per year. Moreover, they ship substantial amounts of fuelwood and charcoal to distant urban markets. The wood taken out of tropical moist forests and used for fuel each year has been estimated at some 150 million m^3.*

Insufficient information is available to judge whether the tropical moist forests can sustain an annual harvest of this size, but as human populations expand and petroleum prices rise, it is inevitable that firewood collection will contribute increasingly to deforestation.

In some tropical countries the deliberate cultivation of trees for firewood is already financially profitable. The rising costs of other fuels is likely to make this increasingly attractive. Trees grow well in the humid tropics. The heat and humidity ensure high rates of photosynthesis and short harvest cycles, thereby making these areas suitable for supplying appreciable amounts of wood energy on a sustained basis.

This section describes species worth testing as fuelwood crops in the humid tropics. Others worth considering include:

Ailanthus altissima	page 74	*Eucalyptus citriodora*	page 128
Albizia lebbek	page 110	*Eucalyptus globulus*	page 82
Alnus acuminata	page 76	*Eucalyptus grandis*	page 84
Cajanus cajan	page 118	*Grevillea robusta*	page 86
Cassia siamea	page 120	*Inga* spp.	page 88
Eucalyptus camaldulensis	page 126	*Pithecellobium dulce*	page 144

*Myers, N. 1980. *Conversion of Tropical Moist Forests.* National Academy of Sciences, Washington, D. C. 205 pp.

(FAO Photo)

Acacia auriculiformis

Botanic Name *Acacia auriculiformis* A. Cunn. ex Benth. The specific epithet is also spelled *auriculaeformis*

Family Leguminosae (Mimosoideae)

Main Attributes *Acacia auriculiformis* merits large-scale testing as a fuelwood species. It has the capacity to produce good fuelwood on poor soils, even in areas with extended dry seasons.

Description It is a resilient, vigorously growing, small tree with a generally crooked trunk up to 60 cm in diameter. It can reach a height of 30 m.

Distribution The species is native to the savannas of Papua New Guinea, the islands of the Torres Strait, and the northern areas of Australia. Because of its ability to grow on very poor soils, it has been introduced into countries such as Indonesia, Malaysia, India, Tanzania, and Nigeria.

Use as Firewood The tree is already established on large-scale private fuelwood plantations and in national forests in Indonesia. It is also grown in Bihar and West Bengal in India. The wood is well suited for fuelwood, with a high specific gravity (0.6-0.75) and a calorific value of 4,800-4,900 kcal per kg. The wood also yields excellent charcoal that glows well and burns without smoke or sparks. The trees coppice poorly, although in Indonesian experiments coppicing was found to be possible if stems were cut at least 50 cm above ground. When the trees are felled, however, a crop of seedlings shoots up so rapidly that cutover stands regenerate readily.

Yield The tree's growth rate under optimal conditions is unknown, but it grows very fast, even on marginal land. In Papua New Guinea, on a site abandoned by farmers as too infertile, trees grew to 6 m in 2 years, with a diameter of 5 cm and had reached 17 m in 8 years. In Indonesia and Malaysia, annual wood production of 17-20 m^3 per ha, with rotations of 10-12 years is achieved. Even on poor soils, the production reaches 10 m^3 per ha per year in moist conditions. However, in semiarid West Bengal on shallow soils, yields were only 5 m^3 per ha per year at the 15th year.

Other Uses

- Pulp. *Acacia auriculiformis* shows promise as a source of wood pulp. Recent tests conducted in Australia have shown that 10-year-old trees grown in a Papua New Guinea plantation can be pulped readily by the sulfate process to give high yield of pulp with very good strength properties. High-quality pulps were also produced by the neutral sulfite semichemical (NSSC) process. It is being used regularly for making pulp by papermills in West Bengal.
- Ornamental and shade tree. With its dense foliage—which remains through the hot season—*Acacia auriculiformis* makes a useful shade tree and soil-cover crop. An attractive ornamental that withstands city heat better than most broad-leaved trees, it requires little attention. It is widely planted to shade and beautify streets in Indonesia and Malaysia.
- Tannin. The bark contains around 13 percent water-soluble tannin that produces a good-quality leather but is inclined to redden on exposure to sunlight.

Environmental Requirements

- Temperature. This species can grow under humid, tropical conditions and thrives where mean annual temperatures range from 26° to over 30° C. However, it is also able to survive in dry savanna conditions, as its thick, leathery "leaf" (actually a flattened and expanded leaf stalk) withstands heat and desiccation.
- Altitude. It is suitable for a forest-plantation crop at altitudes up to about 600 m.
- Rainfall. Although well adapted to drought, *Acacia auriculiformis* grows most quickly in humid climates. Its natural habitat has an average annual rainfall varying from 1,500 to 1,800 mm and a dry season of 6 months.
- Soil. *Acacia auriculiformis* will grow in a wide range of deep or shallow soils including sand dunes, mica schist, clay, limestone, podsols, laterite, and lateritic soils. These problem soils are often poor in nutrients, but the plant produces profuse bundles of nodules and can often survive on land very low in nitrogen and organic matter where eucalypts and other species fail.

At Rum Jungle in northern Australia, the tree grows on alkaline sand dunes (pH 9.0) as well as on acid spoil (pH 3.0) from uranium mining. It is the only native woody plant adapt-

Acacia auriculiformis, 10-year-old plantation at Kunjingini, East Sepik District, Papua New Guinea. (A.F.J. Logan)

able enough to colonize these uranium spoil heaps, and even on 20-year-old heaps it is the only tree to be found. In Malaysia, *Acacia auriculiformis* has grown well on the spoil heaps left after tin mining. In Indonesia, it has been successfully planted on steep, unstable slopes for erosion control and is recommended for planting on the poorest soils in the national forest estates. Recent trials in India have shown that this species grows quite well in papermill sludge, which is highly alkaline (pH about 9.5).

Establishment This species can be easily established by direct seeding or from nursery-raised seedlings. It flowers early and profusely, and seeds are usually plentiful. The plant has a taproot and can withstand root competition from nearby trees. It adapts well to plantation cultivation. The seedlings are hardy; plantations require no extensive site preparation other than clearing vegetation and trash and weed control during the early years.

- Seed treatment. Seeds should be pretreated before sowing by immersing them in boiling water and leaving them to cool and soak for 24 hours.
- Ability to compete with weeds. Young seedlings can be easily smothered by weeds, as the tree is shade intolerant. But *Acacia auriculiformis* is inherently a colonizing species and, once seedlings are well established, they outstrip any competition. Because of the susceptibility of the seedlings to weed competition, it is advisable to raise them in the nursery.

Pests and Diseases In Indonesia, where the plant is used widely, no pests or diseases have been recorded. In Zanzibar, seedlings have been attacked by insects and nematodes.

Limitations *Acacia auriculiformis* can grow very large lateral branches, which often begin low on the trunk. It is, however, easily pruned. Branches break easily in storm winds. It is less fire resistant than most eucalypts. Although drought resistant, *Acacia auriculiformis* cannot withstand drought conditions as severe as hybrid eucalypts can, probably because its roots are closer to the soil surface.

Calliandra calothyrsus

Botanic Name *Calliandra calothyrsus* Meissn.*

Synonym *Calliandra confusa* Sprague Riley

Common name Calliandra

Family Leguminosae (Mimosoideae)

Main Attributes This small bush is unusually promising as a firewood source because of its excellent coppicing ability and very quick growth. In Indonesia it has been cut for fuel after only a year's growth and harvested annually for the next 15-20 years. Even when harvested on such short rotations, it produces a sizable yield of branch wood that makes good household fuel.

Description *Calliandra calothyrsus* is a leguminous shrub that rarely reaches more than 10 m tall, with a maximum diameter of 20 cm.

Distribution The plant is native to Central America, but seeds were introduced from Guatemala to Indonesia in 1936. Calliandra proved so successful as a plantation crop that in 1950 the Indonesian State Forest Enterprise (Perum Perhutani) began planting it on a large scale, so that by early 1979 about 30,000 ha in Central, East, and West Java were under cultivation.

Use as Firewood In many parts of Java, *Calliandra calothyrsus* has become a favorite fuelwood. (In one instance, an experimental plantation of 0.5 ha was established in 1963; by 1975, over 250 ha of firewood plantations had been independently established on nearby privately owned farms and home lots.) The wood has a specific gravity of 0.51-0.78, its calorific value is 4,500-4,750 kcal per kg, and its ash content is 1.8 percent. It is used for cooking as well as in small industries; for example, those making lime, tiles, or bricks.

Yield Trial plots in Indonesia showed initial growth of 2.5-3.5 m in only 6-9 months. After 1 year's growth, calliandra can be cut at about 50 cm above the ground, reportedly yielding about 5-20 m^3 per ha. Afterwards, yearly cuttings are possible, producing between 35 and 65 m^3 of small-sized fuelwood per ha.

Other Uses

- Erosion control. The species grows very quickly, its dense foliage provides ground cover, and its extensive and deep root system binds soil, thereby making *Calliandra calothyrsus* particularly suitable for erosion control on slopes and for rejuvenating degraded soils. Extensive use is planned for stream-bank protection in Java.
- Soil improvement. By its nitrogen fixation and litter production, calliandra improves soil quality and productivity. Because of this, farmers in East Java sometimes rotate agricultural crops with calliandra plantations.
- Fodder. Livestock relish the leaves and the plant is a good fodder crop. In Indonesia, annual yields of 7-10 t of dry fodder (22 percent crude protein) per ha have been recorded. It has been grown together with elephant grass for fodder in large areas previously unable to support any crop.
- Ornamental. The bush is an exciting ornamental, producing beautiful red "powderpuff" flowers. It forms attractive hedges.
- Firebreaks. It is planted in strips on Indonesian state forest lands to protect the forest against fire (as well as illegal woodcutting).
- Bee forage. Honey produced by bees that forage on calliandra flowers has bittersweet flavor.

Environmental Requirements

- Temperature. Unknown.
- Altitude. On Java, the plant grows at altitudes between 150 and 1,500 m.
- Rainfall. The plant grows where rainfall is over 1,000 mm per year, though it can withstand drought periods lasting several months.
- Soil. It can grow on many different soils, including infertile ones, and even grows on heavily compacted clay-type soils with poor aeration.

Establishment Plantations are easily established by direct seeding or by seedlings. Seeds or seedlings are usually planted at the beginning of the wet season. Seedlings are transplanted from the nurseries at about 4-6 months at spacings of 2 m x 2 m or 1 m x 1 m.

*A recent study by Wiersum and Breteler has shown the type specimens of *C. confusa* (the name most commonly used in Central America) and *C. calothyrsus* (the oldest name) are identical. In the NAS report *Tropical Legumes: Resources for the Future*, the name was misspelled *C. callothyrsus.*

Some Indonesian villagers now cultivate *Calliandra calothyrsus* widely on their own land, often intercropping it with food crops. The plant's value is dramatically exemplified by the village of Toyomarto in East Java. There, land that was once grossly denuded and erosion-pocked is now covered with *Calliandra* forest and is fertile once more. Today the villagers make a good living selling the firewood, actually earning more from it than from their food crops. (Perum Perhutani)

• Seed treatment. Seeds are treated with hot water and then soaked in cold water for 24 hours.

• Ability to compete with weeds. Because it grows so rapidly and densely, calliandra suppresses competing plants very quickly.

Pests and Diseases Unreported.

Limitations There is little information on performance of this species on different sites. The plant is so hardy and reproduces so easily that it may become a weed of sorts and may be difficult to keep in check.

Casuarina equisetifolia

Botanic Name *Casuarina equisetifolia* L.

Synonym *Casuarina littoralis* Salisb.

Common Names Casuarina, she-oak, horsetail oak, Australian beefwood, Australian pine, ironwood, whistling pine, agoho (Philippines), ru (Malaysia), filao, nokonoko (Fijian)

Family Casuarinaceae

Main Attributes Almost all of the approximately 35 *Casuarina* species produce top-quality firewood. There are rapid-growing, carefree species for sites and climates as varied as coastal sand dunes, high mountain slopes, the hot humid tropics, and semiarid regions. They tend to be salt tolerant, wind resistant, and adaptable to moderately poor soils. Although they are not legumes, they do have the ability to form root nodules and fix atmospheric nitrogen.

Casuarina equisetifolia is, perhaps, the most widely used so far, but other *Casuarina* species deserve more attention and testing.

Description Viewed from a distance, casuarinas look like somber pine trees with long, drooping, gray-green needles and small cones. They are medium to lofty evergreens with open, feathery crowns that appear to be leafless. The leaves are actually reduced to small sheaths on the needle-like branchlets. By dispensing with leaves, the plants have reduced the surface exposed to the elements and this makes them adaptable to dry sites and salt spray.

Casuarina equisetifolia can attain heights of up to 50 m, with diameters of up to 1 m. However, it is generally only 15-25 m tall.

Distribution *Casuarina* species are native to the Southern Hemisphere from tropical India to Polynesia. Most are native to Australia where they occur in subtropical and tropical coastal regions as well as in the arid central areas.

Casuarina equisetifolia is indigenous to north and northeast Australia, some Pacific islands, and from Indonesia and Malaysia to India and Sri Lanka. It has been introduced for firewood, beautification, and other purposes to India; Pakistan; East, Central, and West Africa; and the West Indies, as well as to Florida and the Gulf of Mexico area in the United States.

Use as Firewood The wood of *Casuarina equisetifolia* burns with great heat and has been called the best firewood in the world. It is used for both domestic and industrial fuel. In India, it fuels some railroad locomotives and in the State of Karnataka it is the major species planted for firewood. It burns readily, even when green, and the ashes retain heat for a long time. It makes exceptionally fine charcoal. It has a specific gravity of 0.8-1.2 and calorific value of about 4,950 kcal per kg.

Yield On good sites in Malaysia and the Philippines the tree commonly grows 2-3 m a year. In general, the yield per ha varies from 75-200 t, on a rotation of 7-10 years, with a spacing between plants of about 2 m. In the Philippines, even higher yields have been noted.

Other Uses

- Wood. Timber is generally dark colored, fissile, strong, heavy, and very tough. It is used for house posts, rafters, electric poles, mine props, roofing shingles, tool handles, oars, yokes, and wagon wheels.
- Erosion control. Because it is salt tolerant and can grow and reproduce in sand, *Casuarina equisetifolia* is used to control erosion along coastlines, estuaries, riverbanks, and waterways.
- Windbreak. An abundance of switchy twigs absorb wind energy amazingly well. A wind strong enough to blow hats off can be stripped of its force by a belt of casuarinas two or three deep, leaving the leeward air heavy and still.
- Dye, tanning. The bark contains 6-18 percent tannin and has been used extensively in Madagascar for tanning purposes. It penetrates the hide quickly and furnishes a fairly plump, pliant, and soft leather of pale reddish-brown color.
- Pulp. The wood has been found to make a useful pulp by use of the neutral sulfite semichemical process.

Environmental Requirements

- Temperature. This is a species for warm to hot subtropical and tropical climates. The monthly mean maximum temperature in its native areas is 10°-33°C, but it is adapted to a wide range of temperatures. It is not frost hardy, although some other *Casuarina* species are.
- Altitude. This is a lowland tree that can be planted from sea level up to 1,500 m.
- Rainfall. In its natural habitat, annual

Casuarina equisetifolia, planted for fuelwood and sand-dune stabilization at Wu Yang People's Commune, Guangdong Province, China. With 44 km of beach front along the South China Sea, this commune formerly suffered greatly from sand dunes invading agricultural land. Some 1,500 ha of *Casuarina* shelterbelt has now been planted along the beach front. Small branches and leaf litter are collected for fuel as often as once every 2 weeks (see rake marks). These 13-year-old trees average over 13 cm diameter and 17 m height. They are harvested for fuel and wood, but only in small clear-cut patches so as to retain shelter and sand stabilization. (T.M. Catterson, FAO Forestry Department)

Rangsit, Thailand. *Casuarina junghuniana* growing on acid soil poorly suited to most other tree crops. Grown from cuttings, the trees are cut for pilings, posts, and firewood after 5 years. (Narong Chomchalow)

rainfall is from 700 to 2,000 mm, often with a dry season of 6-8 months. However, it has been planted successfully in areas with annual rainfall as little as 200-300 mm or as much as 5,000 mm.

• Soil. *Casuarina equisetifolia* trees have root nodules containing nitrogen-fixing actinomycete microorganisms. They are therefore not dependent on soil nitrogen for good growth. The species tolerates calcareous and slightly saline soils, but it grows poorly on heavy soils such as clays. It can withstand partial waterlogging for a time.

Establishment The trees produce seed plentifully. Normally, seedlings are raised in a nursery for transplanting to the field. Transplanting is carried out at the onset of the rainy season, 4-18 months after sowing. In dry sites, irrigation may be needed immediately after transplanting and as long as the first 3 years.

• Seed treatment. When seeds are planted outside their natural range, the soil may need inoculating with crushed nodules from natural stands. The seeds should also be treated to repel ants.

• Ability to compete with weeds. Initially poor, especially in dense grass cover.

Pests and Diseases Seedlings are vulnerable to attack by ants, crickets, and other insect pests. They are also susceptible to root rot.

Limitations *Casuarina equisetifolia* can exhaust the moisture in the soil, lower the water table of the site, and restrict growth of a healthy understory, leaving the soil exposed. The tree is fire sensitive and can be browsed only lightly without being damaged. Although other *Casuarina* species coppice readily, *C. equisetifolia* does not.

In some cases casuarinas tend to be aggressive plants. *C. equisetifolia* has invaded South

Casuarina oligodon near Mendi, Papua New Guinea. This *Casuarina* species grows well at this altitude of 2,100 m. Villagers plant for fuel in fencelines and in abandoned cropland (see also picture page 14). (N.D. Vietmeyer)

Florida, displacing much native vegetation, and proving nearly impossible to control.

Related Species Other *Casuarina* species worth testing as firewood crops include:

Australian Species	Adaptation
• *Casuarina cristata* (*C. lepidophloia*)	Arid and semiarid zones; prefers moderately heavy soils; tolerates alkalinity
• *C. cunninghamiana*	Coastal tablelands and higher elevations
• *C. decaisneana*	Arid sandy soils
• *C. glauca*	Saline and swampy ground; tidal flats; heavy soils; warm temperate and tropical coasts
• *C. leuhmannii*	Arid and semiarid zones; saline, clay soils
• *C. littoralis*	Infertile, tropical, and temperate acid soils
• *C. stricta*	Warm temperate zones; poor coastal sands and clays; calcareous and saline soils
• *C. torulosa*	Tropical and subtropical coastal tableland and high-altitude regions

Papua New Guinea Species	
• *Casuarina oligodon*	High elevations, including ultrabasic rocks
• *C. papuana*	

Indonesian Species	
• *Casuarina junghuniana* (*C. montana*)	High elevations
• *C. sumatrana*	Lowland podsols

Fijian Species	
• *Casuarina nodiflora*	Wetter areas (2,000–3,000 mm)

Philippine Species	
• *Casuarina rhumphiana*	A high-elevation species grown at 200–1,000 m altitude

Derris indica

Botanic Name *Derris indica* Bennet

Synonyms *Pongamia glabra* Vent., *P. pinnata* Merr., *P. pinnata* (L.) Pierre

Common Names Pongam, ponga, oil tree, kona, kanji, pari-pari, karanda, karanja, Indian beech

Family Leguminosae (Papilionoideae)

Main Attributes *Derris indica* provides two sources of energy: in nearly every country where it grows, its wood is burned for cooking fuel, and in India the thick oil from its seeds is burned in lamps. The tree adapts well to extremes of temperature, soil, and moisture. It can be grown in the shade of other trees, and its spreading roots make it valuable for checking erosion and for binding shifting sand dunes.

Description A medium-sized, deciduous, thornless tree with spreading or drooping branches, the pongam reaches a height of about 8 m. When its leaves first develop, it turns a vivid lime green that later deepens to dark green. The pale-pink flowers bloom in great numbers just after the leaves have formed. Its pods, each usually containing a single, oblong seed, are produced in huge numbers.

Distribution Native to the Indian subcontinent, the pongam is now widely distributed throughout the world's humid lowland tropics. It can be found in the Philippines, Malaysia, Australia, Oceania, the Seychelles, Florida, and Hawaii, usually in coastal forests near the sea and alongside tidal streams and rivers. It is being planted for afforestation in the drier parts of the Indian peninsula. A handsome tree, it is planted for shade and ornament, especially along roadsides.

Use as Firewood Pongam wood is yellowish-white and fairly hard. It burns well and is commonly employed for fuel in India. The calorific value is 4,600 kcal per kg.

Yield Classed as a fast-growing species, the trees often reach adult height in 4 or 5 years. They coppice well.

Other Uses

- Wood. The wood is coarse textured. While it is somewhat difficult to work, it is beautifully grained and often employed for cabinetwork. It is also used for making cart wheels and posts.
- Fiber. The bark fiber is made into string and rope.
- Fodder. The leaves are a valuable fodder for livestock, especially in arid regions. Since pasture grasses grow well in its shade, it could be a valuable tree fodder. Its foliage is probably rich in protein.
- Oil. The seeds contain a yellow or red-brown oil (30-40 percent). It is not edible, but it can be used as a lubricant or—because it burns well—as a fuel in kerosene lamps. It is also used as a leather dressing by the Indian tanning industry and in the manufacture of soap, varnish, and paints. It is reputed to have some antiseptic value in treating human and animal skin diseases.
- Pest control. When dried, pongam leaves are used to protect stored grains because their odor repels insects. They are often ploughed into the soil as a green manure and are thought to reduce nematode infestations. The presscake left after oil has been extracted from the seeds also has insecticidal and pesticidal properties and is sought after by Indian farmers. Researchers in India have shown recently that the presscake makes a useful poultry feed.

Environmental Requirements

- Temperature. Mature trees withstand temperatures from slightly below 0° to 50°C.
- Altitude. Thrives up to 1,200 m above sea level.
- Rainfall. Requires from 500 to 2,500 mm per year.
- Soil. The tree grows wild on sandy and rocky soils (including oolitic limestone). It will grow in most soil types; it is so highly tolerant of salinity that it can survive even with its roots in salt water.

Establishment The pongam is easily raised from seed and cuttings; even branches stuck into moist ground develop roots readily. It tolerates shade well and can be close-planted.

- Seed treatment. None required. Seeds remain viable a long time. Direct sowing is usually successful; nursery production is therefore often not needed, although seedlings transplant easily.

Derris Indica, a medium-sized specimen. Nadia, West Bengal, India. (L. Mandal)

• Ability to compete with weeds. Unreported.

Pests and Diseases The tree hosts a large number of insects and some fungi.

Limitations The pongam has a wide-spreading, aggressive surface root system. It sheds leaves, flowers, and pods heavily and is therefore classed as a "trashy" tree. Moreover, because of its spontaneous seedlings and root suckers it may run wild and create serious weed problems.

Gliricidia sepium

Botanic Name *Gliricidia sepium* (Jacq.) Steud.

Synonym *Gliricidia maculata* (H.B.K.) Steud.

Common Names Madre de cacao, mother of cocoa, mata-raton, kakauati (Philippines), Mexican lilac, madera negra

Family Leguminosae (Papilionoideae)

Main Attributes This fast-growing tree is good for cultivation in populated areas; for example, in villages, farms, backyards, and along fence lines, paddy bunds, and the edges of roads and paths. It produces good fuelwood. It fixes nitrogen efficiently and grows well in, and enriches, poor soils. During dry (or cold) seasons it drops its heavy mantle of leaves and so conserves precious groundwater.

Description *Gliricidia sepium* is a small, thornless tree that grows up to 10 m high. It has an open crown and an often contorted trunk that is 30 cm or less in diameter. It is one of the most common and best-known trees of Mexico, Central America, and northern South America.

Distribution *Gliricidia sepium* has been introduced to the West Indies, where it is becoming naturalized. It has also been introduced to Africa and Asia and has become naturalized in the Philippines. It has been planted in southern Florida and in South America as far south as Brazil.

Use as Firewood Wherever *Gliricidia sepium* grows, its hard, heavy wood is used for fuel. Although not tall, the tree produces much branch wood and coppices easily. Its calorific value is 4,900 kcal per kg.

Yield Unreported.

Other Uses

- Timber. The wood finishes smoothly and is suitable for furniture, small articles, agricultural implements, and tool handles. Highly resistant to termites and decay, it is also used for posts and heavy construction.
- Living fence. *Gliricidia sepium* is easily propagated by cuttings, provided there is ample soil moisture. Even large branches will sprout roots and grow when they are stuck in the ground. A row of these makes a very effective living fence or windbreak that will last for many years without maintenance. Trimming these "fences" every month or 2 during the rainy season assures large amounts of foliage for green manure or ruminant feed.
- Ornamental. The tree produces dense masses of attractive white or pink flowers.
- Shade and green manure. The tree's long, leafy branches make it ideal as a shade tree. It is widely used to shade cacao, coffee, vanilla, and tea. The foliage is rich in nitrogen and the falling leaves enrich the soil beneath the trees. The foliage can also be cut and used to fertilize nearby crops.
- Fodder. The leaves contain over 20 percent crude protein and are nutritious for cattle. (They are, however, toxic to most other animals, including horses.)
- Honey. The flowers are a good source of forage for bees.

Environmental Requirements

- Temperature. 22°-30°C.
- Altitude. It is found growing on the plains and foothills of Central America extending up to about 1,600 m elevation, mainly below 500 m.
- Rainfall. 1,500-2,300 mm per year and more.
- Soil. It does well in moist or dry soil, even with heavy concentrations of limestone.

Establishment The plant is propagated easily by seed; however, the natural regeneration of large cuttings (often nearly 2 m long) is a simple method of getting large specimens quickly.

- Seed treatment. Soak in hot water, cool off during the night, sow the next morning.
- Ability to compete with weeds. Unreported.

Pests and Diseases The tree is losing popularity in Puerto Rico because the foliage is often attacked by aphids that secrete a sweet honeydew that attracts ants and causes the leaves to fall.

Limitations The roots, bark, and seeds are poisonous. The leaves may also be toxic to humans, although they are eaten in some parts of the tropics. Perhaps cooking inactivates the toxin.

Above: David, Panama. A living fence of *Gliricidia sepium*, which provides fuel and forage. (G. Budowski)
Right: East Java, Indonesia. This fast-growing species out-competes the weedy scourge *Imperata cylindrica* and is widely grown for fuel in Indonesia. It is often grown in fire belts around the borders of forests and farm fields because the living trees are "fireproof." (Perum Perhutani)

Gmelina arborea

Botanic Name *Gmelina arborea* Roxb.

Common Names Gmelina, yemane (Burmese name, also used internationally), gumhar (India), gamar (Bangladesh)

Family Verbenaceae

Main Attributes Gmelina (pronounced mel-eye-na) is especially promising as a fuelwood crop for plantations in humid lowlands because it can be established easily and more cheaply than other species and regenerates well from both sprouts and seed. Its silviculture is relatively well known, it is already found in many countries, its seeds are readily available because the trees produce flowers and fruit copiously, it is easy to handle and propagate, it adapts to a wide range of soils and climates, and it can be extremely fast growing.

Description Gmelina is a medium to large deciduous tree. In well-thinned plantations under optimum conditions it may reach heights of about 30 m and stem diameter over 60 cm, but it is more usually about 20 m with a clear bole of 6-9 m. In the open, however, it develops low, heavy branches, a wide crown, and a strongly tapered stem. Clusters of brown and yellow flowers appear when the trees are more or less leafless, but some trees can be found in flower and fruit in most locations throughout the year.

Distribution Gmelina is native to moist forests of India, Bangladesh, Sri Lanka, Burma, and much of Southeast Asia and southern China. It occurs in some comparatively dry areas in central India. It is now on trial in many tropical countries and commercial plantations are found principally in Brazil, Gambia, Sierra Leone, Ivory Coast, Nigeria, Malawi, Malaysia, and the Philippines.

Use as Firewood Gmelina is already being planted for fuel in Malawi (especially for tobacco curing), Sierra Leone, and Nigeria (especially Lyo and Kano states). The wood is relatively light (specific gravity, 0.42-0.64); calorific value of sapwood is about 4,800 kcal per kg. It burns quickly. Charcoal from gmelina wood burns well and without smoke, but it leaves a lot of ash.

Yield On the best sites very high annual increments, probably exceeding 30 m^3 per ha, are possible.* Some trees have reached 3 m a year after planting and 20 m after 4.5 years. Annual average diameter increases of 5 cm have been recorded. Rotations of 5-8 years seem most commonly used and may produce 20-35 m^3 per ha per year or more. Coppice rotations are usually at 5-year intervals for fuel.

The trees coppice well and coppice growth is also rapid.

Other Uses

- Wood. Straw-colored gmelina wood is one of the best utility timbers of the tropics—useful for particle board, plywood core stock, pit props, matches, and sawtimber for light construction, general carpentry, packing, and furniture.
- Pulp. Gmelina wood gives average yields of paper with properties superior to those obtainable from most hardwood pulps. In the Amazonian Rio Jari region, millions of gmelina have been planted to feed a large (750 t per day) kraft pulp mill.
- Honey. The flowers produce abundant nectar from which high-quality honey is produced. The forestry department of Gambia produces large quantities of gmelina honey in a pilot beekeeping project.

Environmental Requirements

- Temperature. Withstands up to 52°C, but can be severely injured by frost.
- Altitude. Up to 1,000 m.
- Rainfall. It grows satisfactorily on sites that receive from 750 mm to 4,500 mm annually. At least some provenances have drought tolerance: in East Africa, gmelina is grown where the dry season is 7 months long. Also, gmelina is one of the plantation species that can be grown in the northern Philippines with its 6-month dry season.
- Soil. Gmelina is fairly adaptable and survives well on a wide range of soil types: acid soils, calcareous loams, and lateritic soils. It cannot withstand waterlogging and it gets badly stunted on very thin soils, impermeable soil layers, highly leached acid soils, or dry sand. It produces best in moist, base-rich, well-drained alluvium.

*Lamb. 1968.

Gmelina arborea, 7 years old, Kuala Langsar, Malaysia. (Forest Research Institute, Malaysia)

Establishment Gmelina is easy to handle and simple to establish. Its silviculture is somewhat similar to that of teak. It can be propagated by seed, cuttings, or by budding or grafting. Tubed seedlings are widely used, but in several countries direct seeding has proved cheap and quick if rainfall is reliable. A spacing of 2 m x 2 m has been recommended for fuelwood plantations in Malaysia. Gmelina can be established among agricultural crops (the *taungya* system) such as peanuts (groundnuts), cashew, tobacco, corn, and beans. It then benefits from better manuring and care. However, it grows too rapidly for longer-term crops such as cassava.

- Seed treatment. Soak in water 24 hours (not necessary if there is irrigation or daily rainfall). Seeds lose viability within a year.
- Ability to compete with weeds. Weeding is essential during the first year; gmelina cannot compete with aggressive grasses or vines.

Pests and Diseases The tree is usually remarkably free from diseases. However, in the American tropics it is subject to constant attack by leaf-cutter ants, and in India it has been defoliated by insects. Also a bark disease ("worm disease") is known that can girdle the base of the tree, and "machete" disease (*Ceratocystis fimbriata*) may be locally severe in humid climates. Gmelina seed and foliage are eaten widely by deer and rabbits. Cattle eat foliage and bark of young trees. The trees are seldom killed by fire, but fire damage can facilitate insect and fungal attack.

Limitations The trees cast such heavy shade that nothing will grow under dense (2 m x 2 m) stands. On sloping ground, ridges may be built before planting to avoid erosion.

The carpet of dead leaves under the trees creates a mild fire hazard during extended dry seasons.

A tendency for the trees to die at a young age (10 years) has been noted in Sierra Leone.

It is often a requirement that firewood plantations be established on fallow or marginal lands near villages, and for this purpose gmelina is not too well suited.

Some plantations have been destroyed by browsing livestock. The palatability of the foliage may be a drawback for establishing gmelina woodlots near villages.

Guazuma ulmifolia

Botanic Name *Guazuma ulmifolia* Lam.

Synonym *Guazuma tomentosa* H.B.K.

Common Names Guacima or guacimo (Spain); majagua de toro (Mexico); boss d'orme, West Indian elm (Trinidad); bastard cedar; goeaazoema (Dutch West Indies)

Family Sterculiaceae

Main Attributes Guacima is a vigorous, thornless plant whose wood makes exceptional fuel. It adapts well to adverse soils, appears compatible with agricultural crops, and its foliage is not toxic to livestock. The fruit is very much sought after by cattle, and distribution and germination is favored by having the seed run through the digestive tract. Unknown as yet in organized cultivation, it is worth widespread testing in tropical energy plantations and is particularly promising for subsistence-farming areas.

Description It occurs either as a many-branched shrub or as a single-trunked tree, ranging from 2 to 30 m in height. Mature trunks may be 30-40 cm in diameter. The trees are usually evergreen; leaves are shed only after prolonged drought. The warty fruits contain a little sweet pulp and many hard seeds.

Distribution The tree is popular and valued as a street and shade tree throughout tropical America and the Caribbean. It is little known elsewhere. It is particularly abundant along cattle trails.

Use as Firewood The wood is moderately hard (specific gravity, 0.55-0.58), with the consistency of elm wood, and is regarded as excellent fuel. In colonial days it was the main source of charcoal for the gunpowder used to protect the city of Santo Domingo (Dominican Republic) from invasion. It was used to make gunpowder in Puerto Rico and Guatemala as well. It is still exploited for fuelwood and charcoal in the West Indies.

Yield Although it grows rapidly, actual rates are yet to be determined.

Other Uses

- Wood. Sapwood is light brown and heartwood pinkish brown. The wood, easily worked, is used chiefly for posts in Puerto Rico. Elsewhere, it is used for general carpentry, interior construction, and furniture making, and for barrel staves, boxes and crates, tool handles, and gunstocks.
- Forage. Young foliage and fruits are browsed by horses, cattle, and deer in the dry periods; the fruits are sought by hogs and cattle. (The fruits may cause intestinal obstruction in cattle if eaten to excess.)
- Food. The mucilaginous green and black fruits are edible when fresh, dried, or cooked; the native Indians often drink a beverage made by crushing the fruits in water. The flowers attract bees and are a source of good-quality honey.
- Shade. The tree is often grown as a shade tree for pastures and streets.

Environmental Requirements *Guazuma ulmifolia* is particularly characteristic of second-growth forests but is also common in dry and moist thicket areas, open clearings, stream banks, pastures, and lower mountain slopes.

- Temperature. Tropical.
- Altitude. Grows from sea level to about 1,200 m elevation.
- Rainfall. The tree grows best where annual rainfall is 700-1,500 mm; this corresponds usually to a 4- to 7-month dry season in tropical regions. The tree adapts to drier situations as well.
- Soil. It adapts to a wide range of soils and does not usually require fertilization.

Establishment Seeds are plentiful and readily available in the American tropics. They germinate easily, though propagation is largely untested.

- Seed treatment. Unreported.
- Ability to compete with weeds. Unreported.

Pests and Diseases Unreported.

Limitations The seeds are believed to be distributed by birds and mammals, including cattle and possibly horses, and although the tree is

Guazuma ulmifolia, Puerto Rico. (T. H. Schubert)

not now classed as a pest, it could well become a weed tree when introduced to new areas.

The wood is not durable and is very susceptible to attack by dry-wood termites.

Leucaena leucocephala

Botanic Name *Leucaena leucocephala* (Lam.) de Wit

Synonym *Leucaena glauca* Benth.

Common Names Leucaena, ipil-ipil (Philippines), lamtora (Indonesia), guaje, yaje, uaxin (Latin America), leadtree

Family Leguminosae (Mimosoideae)

Main Attributes Of all tropical legumes, leucaena probably offers the widest assortment of uses. Through its many varieties, leucaena can produce nutritious forage, firewood, timber, and rich organic fertilizer. Its diverse uses include revegetating tropical hillslopes and providing windbreaks, firebreaks, shade, and ornamentation. Individual leucaena trees have yielded extraordinary amounts of wood–indeed, among the highest annual total yields ever recorded.

Description Depending on variety, leucaena is either a tall, slender tree that may grow up to 20 m, or a rounded, many-branched shrub less than 5 m high. It has feathery leaves, bunches of long, brown pods–often almost translucent–and small, white "powderpuff" flowers.

Distribution Leucaena originated in the midlands of southern Mexico and was introduced to the Pacific islands, the Philippines, Indonesia, Papua New Guinea, Malaysia, and East and West Africa, so that now it is truly pantropical.

Use as Firewood Leucaena wood makes excellent firewood and charcoal. It has long been used for these purposes in the Philippines. New varieties are so productive that they are already being planted to provide fuel for electric generators, factories, and agricultural-processing facilities. The wood has uncommonly high density and calorific value for a fast-growing tree, and because the stumps readily coppice, the plant could become a renewable fuel resource in areas suited to its agronomic requirements. The calorific value is 4,200-4,600 kcal per kg.

Yield In the Philippines, dense leucaena plantations have yielded higher annual quantities of wood than any species yet measured. Annual leucaena increments have been measured from 24 to over 100 m^3 per ha. Average annual increments, however, are expected to be between 30 and 40 m^3 per ha.

Other Uses

- Forage. Cattle feeding on leucaena foliage in Queensland, Australia, have shown some of the highest weight gains ever measured in the tropics. Suited mainly to cattle, water buffalo, and goats, leucaena forage is highly palatable, digestible, and nutritious. Both beef and dairy cattle thrive on it and can live on leucaena alone until mimosine-related toxicity occurs. This can be delayed or eliminated entirely by supplementing the diet with other forages. The plant's drought tolerance and hardiness make it a promising candidate for increasing meat and milk supplies throughout the dry tropics.
- Wood. The newly discovered arboreal leucaena varieties grow rapidly, yielding wood of useful size for lumber and timber. Leucaena wood has the potential to become a major source for pulp and paper, roundwood, and construction materials.
- Soil improvement. Leucaena is a nitrogen-fixing legume that helps to enrich soil and aid neighboring plants. Its foliage rivals manure in nitrogen content, and natural leaf-drop returns this to the soil beneath the shrubs. Its aggressive root system also breaks up impervious subsoil layers, improving moisture penetration and decreasing surface runoff.
- Reforestation. Its ability to thrive on steep slopes, in marginal soils, and in areas with extended dry seasons makes it a prime candidate for restoring forest cover to watersheds, slopes, and grasslands that have been denuded through wood cutting or fire.

Environmental Requirements

- Temperature. Leucaena is restricted to the tropics and subtropics; frost kills it.
- Altitude. This is a species for lowland areas mainly below 500 m. The plant continues growing at high elevations, but without its lowland vigor.
- Rainfall. The species grows best where annual rainfall is 600-1,700 mm. However, it is the dominant vegetation covering Honolulu's Diamond Head, where annual rainfall amounts to only 250 mm.
- Soil. Leucaena's root system allows it to tolerate a wide array of soil conditions. It is found in soils varying from rock to heavy clay to coral. Unaided, leucaena grows well only in neutral or alkaline (especially limestone) soils.

One-year-old leucaena trees in demonstration plots. *Left:* Near Batangas City, Philippines. (N.D. Vietmeyer) *Right:* Near Port-au-Prince, Haiti. (M. Benge)

It grows poorly in acidic soils, and much of the tropics has acidic latosolic soils high in alumina and often deficient in molybdenum and zinc.

Establishment Seed viability is high and the seeds can be successfully planted by hand or machine. Seedlings are slow starters. Leucaena can be reproduced by cuttings or grafts, but only with difficulty.

- Seed treatment. Eighty percent germination within 8 days can be achieved by treating the seeds with hot water (80°C) for 2-3 minutes. Further increases can be obtained by then soaking the seed for 2-3 days.
- Ability to compete with weeds. To establish leucaena, weeds must be controlled. Once rapid growth begins, the leucaena plants form a canopy of foliage that shades out weeds.

Pests and Diseases It is highly resistant to pests and diseases. Common pests are seed weevils, twig borers, and termites.

Limitations Leucaena's reputation has suffered in some areas because of a rugged, persistent variety that has become a weed. Also, its foliage contains mimosine, toxic to ruminants if consumed in excessive amounts.

Mangroves

Botanic Names There are many genera of mangroves. The most widely distributed are *Rhizophora* (Rhizophoraceae), *Avicennia* (Avicenniaceae), and *Bruguiera* (Rhizophoraceae).

Main Attributes Mangroves produce fuel in coastal and estuarine areas that are virtually unusable for other crops yet easily accessible by water. Mangroves are a unique biological phenomenon because they survive waterlogging, poor soil aeration, salinity, high humidity, and strong winds–a combination of conditions not tolerated by other plants. Moreover, mangrove wood is widely recognized as an ideal firewood.

Throughout the tropics, water projects are increasingly diverting river water for agriculture and cities. This is destroying many mangrove swamps because mangrove trees must be flushed with fresh water at least during one season a year. Further, swamps are being ravaged for fuel, tanbark, fish ponds, rice paddies, and human settlements. Their wholesale destruction imperils coastlines and in many areas they are threatened with extinction. Firewood production may provide an important reason for husbanding the vast mangrove resources.

Description Mangroves form "forests of the sea." The trees are often shrublike, but *Rhizophora mangle* and many other species can grow up to 40 m tall and 1 m or more in diameter. *Rhizophora* species, typical mangrove, are usually the first to occupy newly formed saline mud flats. Mangroves have developed remarkable mechanisms to survive the stress in such harsh environments. For example, in *Rhizophora* species, aerial roots descend from the branches, and "prop" or "stilt" roots spring out from the sides of their trunks and arch downward deep into the soft, black mud. These roots function as aerators, while the subterranean, absorptive roots, and even the lower parts of the trunks, are submerged in salt water–conditions that would completely destroy other forest trees. In *Avicennia* and, to a lesser extent, *Laguncularia* species, spongy, breathing vertical roots project above the mud, absorb air, and aerate the anchoring root system. Also, the seeds of many mangroves can float and germinate in salt water. *Rhizophora* seeds germinate while still attached to the tree; the fully formed, spindle-like seedling falls from the tree, spearing into the mud beneath, or floating until it finds lodgement, develops roots, becomes upright, and sprouts leaves at the apex.

Distribution Mangroves afforest some 45 million ha of shallow water and muddy tidal flats throughout the tropics and subtropics. They are abundant around the Indian Ocean from Mozambique to northern Australia and fringe East Asia as far north as Korea. They are found throughout Polynesia and around the coasts of Central America and northern South America, the West Indies, Bahamas, southern Florida, Bermuda, and West Africa. Conditions most favorable for mangrove development are found in quiet bays into which rivers flow gently. Mangroves are unusual among tropical forests in that only a few species are found in each and they are normally free of undergrowth except in open areas and around their borders where salt-tolerant shrubs and herbaceous plants abound.

Use as Firewood The most highly prized firewood in many countries comes from local mangrove species. In the Philippines, mangrove was the principal domestic fuel until World War II. The wood burns well, even when freshly cut. (The bark also makes excellent fuel.) It splits easily and is a very heavy wood (specific gravity, 0.7 to over 1.0, averaging 0.9 in *Rhizophora* species). Mangrove wood has a high calorific content that is similar in most species (4,000–4,300 kcal per kg). It is reported that 5 t of *Rhizophora mucronata* equals the calorific value of 2–3 t of coal. The wood burns leaving little ash.

Mangrove charcoal is exceptional. It burns steadily, giving off intense heat without sparking. In Bangkok, Thailand, it sells at twice the price of other charcoal. In Puerto Rico, mangrove stands have been a major source of charcoal.

Yield Mangrove silviculture has been attempted in some areas and is an established practice in some Asian countries, notably Thailand, Malaysia, and Bangladesh, and mangrove plantations have also been established in Puerto Rico. Most species seem to grow rapidly, and the salt flats soon become dense forests. Planting is usually not needed because natural regeneration is so successful.

Firewood being sold in Dacca, Bangladesh. Most of the firewood originates in the huge Sundarbans mangrove forest. (B. Christensen)

Rhizophora mangle, south Florida, USA. (J. Morton)

Other Uses

• Wood. Many mangrove species produce strong, attractive timber that finishes and polishes well. The wood of some species is notably durable in water. Older buildings in downtown Singapore are built on pilings of mangrove wood, mainly from *Rhizophora* species. In India, the heartwood of *Bruguiera gymnorrhiza* is prized for furniture.

• Coastal protection. Mangroves can be instrumental in developing coastal and estuarine areas. They provide natural protection against a turbulent sea, reducing typhoon and storm damage (and effecting their own repairs), and binding and building sand and soil.

• Food and wildlife production. Mangrove swamps are a "cradle of life." They are spawning and nursery grounds for many species of fish, as well as shrimp, crab, clams, oysters, and crocodiles. They are feeding and nesting grounds for many sea birds and home to other wildlife. Thus, many people indirectly draw their livelihood from mangroves. Mangrove areas are potential resources for aquaculture.

• Extractives. Tannins used to produce hard leather (such as for shoe soles) and resins used for bonding plywood are extracted from various species.

• Pulp. Each year, hundreds of thousands of t of mangrove wood chips (from *Rhizophora* and *Bruguiera* species) are exported from Indonesia, Sarawak, and Sabah for pulp and for rayon manufacture, in particular.

Environmental Requirements

• Temperature. Mangroves are found only in the tropics and subtropics. They are very sensitive to frost.

• Altitude. The trees are suited to coastal and estuarine zones.

• Rainfall. Mangroves grow best in moist, tropical areas where annual rainfall is over 1,000 mm.

• Soil. Mangroves can survive in highly saline, poorly drained soils that are irregularly inundated by the tide. They grow best, however, in areas that are regularly flushed by the tide or in areas with freshwater seepage. Regular flushing with seawater or freshwater appears to be necessary to avoid excessively high salinities, which arise through evaporation and would otherwise restrict growth. Consequently, attempts to impound mangrove areas within levees often reduce productivity and sometimes lead to death of trees. Soil in mangrove swamps is usually, deep, black mud that is almost continuously water logged, but mangroves also will colonize sand and carbonate soils.

Establishment There have been few attempts to establish mangroves in organized plantations. However, *Rhizophora mangle*, which is planted for coastal protection in Florida, was planted successfully in great numbers in Hawaii in 1922. Mangrove swamps in Malaysia have been managed for fuelwood since about 1900. The Mangrove Research Center of the Philippines (see Research Contacts) has a mangrove nursery and a working group on the silviculture of mangrove forests. Studies reveal that direct seeding methods result in about 90 percent survival using the *Rhizophora* and *Avicennia* species. Both air-layering and the planting of propagules in Florida have been successful. Air-layering has recently become particularly important in mangrove propagation.

• Seed treatment. None required.

• Ability to compete with weeds. Mangroves have little difficulty competing with weeds because few other woody species can survive the saline, swampy conditions. However, the leather fern (*Acrostichum aureum*) quickly invades clearings in cut-over mangrove forests and prevents the germination and regrowth of mangroves.

Pests and Diseases Mangroves have high resistance to attack by insects, pests, and diseases.

Limitations Mangrove swamps are often breeding places for mosquitoes; they may smell at low tide; and although they are a much-photographed curiosity, some people regard them as having an extremely sinister, repellent appearance, resent their presence, and favor their eradication. Usually, however, mosquito and other problems are caused by poor management, especially when the mangroves are enclosed by levees that prevent tidal inundation.

Members of the family Rhizophoraceae (*Rhizophora, Bruguiera, Ceriops*) grow from the tips of branches and can easily be killed by indiscriminate lopping of branches. Many other species, however, can regrow rapidly from buds beneath the bark along the trunk and branches (e.g., *Avicennia, Sonneratia*) and thus suffer little from removal of a large part of the branchwood.

Felling *Rhizophora* species, Matang, West Malaysia, after a rotation of 30 years. This mangrove forest has been managed since the beginning of the century. (B. Christensen)

Collecting mangrove firewood, Calapan, Philippines. Easy access by water is an advantage of many mangrove forests. (A.B. Velasco)

Mimosa scabrella

Botanic Name *Mimosa scabrella* Benth.

Synonym *Mimosa bracatinga* Hoehne

Common Names Bracatinga, bracaatinga, or abaracaatinga

Family Leguminosae (Mimosoideae)

Main Attributes This leguminous tree, virtually unknown outside of southern Brazil, grows rapidly, and in a few years produces fairly straight stems whose wood is an exellent source of fuel.

Description *Mimosa scabrella* is a thornless shrub or slender tree, sometimes 12 m high and 20-40 cm in diameter.

Distribution The plant is native to the Paraná region of southeastern Brazil. In recent years, small trial plots have been planted in Portugal, Zaire, Senegal, Ethiopia, Spain, Guatemala, Venezuela, El Salvador, Colombia, Argentina, Mexico, and Jamaica.

Use as Firewood Before the advent of the diesel locomotive, bracatinga wood was grown in plantations to fuel railroads in parts of Brazil.

Yield *Mimosa scabrella* is a fast-growing tree. In 14 months it grows to 5 m tall; in 2 years, 8-9 m; and in 3 years it sometimes attains a height of 15 m. Plantations have been harvested on rotations as short as 3 years.

Other Uses

- Pulp. *Mimosa scabrella* fiber is 1.2 mm long, and its pulp is good enough to use in the manufacture of printing and writing papers.
- Beautification. It makes a fine ornamental, garden, living fence, or avenue tree.
- Green manure. The tree is able to fix atmospheric nitrogen and also sheds an enormous quantity of nitrogen-rich leaves that easily decompose and form a good humus.

Environmental Requirements

- Temperature. Bracatinga is native to the cool, subtropical plains of southeastern Brazil, but it is very robust and can grow in warmer and drier areas.
- Altitude. It flourishes at an elevation of 2,400 m in Guatemala.
- Rainfall. Unreported.
- Soil. The tree grows well in many types of well-drained soil and is not exacting with regard to soil quality, though wet soils stunt its growth. Its nitrogen-fixing and humus-producing capacities aid greatly in improving soil conditions.

Establishment It is easily planted by seed and may be sown directly into the field. Generally, three or four seeds are set in a shallow depression (3-4 cm deep) at distances of 2-3 m apart. After germination, the inferior seedlings are rogued out.

- Seed treatment. Unreported.
- Ability to compete with weeds. Unreported.

Pests and Diseases Unreported.

Limitations Unreported.

Railway station, Rio Branco do Sul, Paraná, Brazil. Before World War II, *Mimosa scabrella* was grown as fuel for railroads in parts of southern Brazil. (F.C. Hoehne)

Curitiba, Paraná, Brazil. Natural regeneration of *Mimosa scabrella*, 3 years of age. (F.C. Hoehne)

Muntingia calabura

Botanic Name *Muntingia calabura* L.

Common Names Jamaica cherry (Florida), strawberry tree (Jamaica and Florida), Panama berry (Hawaii), Japanese cherry (Malaysia), buah cheri (Singapore), capulin (Mexico, Central America), pasito (Panama), chitato, majaguito (Colombia), majaguillo (Venezuela), bolina yamanaza (Peru), calabura (Brazil)

Family Elaeocarpaceae

Main Attributes *Muntingia calabura* is a small, fast-growing, short-lived evergreen tree often used for firewood in its native habitats.

Description With a dense, spreading crown and drooping branches, this tree reaches 8-13 m tall, with a trunk 8.5-20 cm in diameter. It often begins to bear within 1.5-2 years from seed; thereafter it flowers and fruits continuously. Its small white flowers closely resemble those of the strawberry. The fruits are 1 cm wide, round, and deep red, with a sweet light-brown pulp filled with minute seeds, thus resembling small figs internally.

Distribution The species is native to the New World from southern Mexico to Bolivia and Brazil. It is cultivated for its fruit in southern Florida and for shade and beautification in the Bahamas, Cuba, Jamaica, the Dominican Republic, and Puerto Rico. It has also been planted in Hawaii and throughout the Pacific islands. It is naturalized in Southeast Asia (Thailand, Vietnam, for example) and Malaysia and is so common in gardens and villages that the people consider it native.

Use as Firewood When thoroughly dry, the wood ignites quickly, producing a high flame with little smoke (none later); the heat is intense. In a test, small branches burned for 10 minutes and the flames then died down, leaving red, glowing embers (180°C sustained for 20 minutes and then dropping to 100°C). Residual ash and cinders represented 20 percent of the test material.

Yield Unknown.

Other Uses

- Shade and beautification. This attractive tree is commonly planted in residential areas for ornament and shade, though some people object to the musky scent of its foliage and the splatter of the juicy fruits. It seems suited for interplanting among agricultural crops and makes a good shade tree for livestock.
- Fruit. The sweet, juicy berries are sought by children and livestock, and by birds, bats, and other wildlife, but they have little flavor.
- Fiber. The bark provides a tough, silky fiber used for bark cloth and cordage. With a 43-percent yield of cellulose, the tree has been considered a potential source of paper pulp in Brazil.
- Wood. The wood is little used. In Colombia it is made into barrel staves.

Environmental Requirements

- Temperature. This tree grows best under humid, tropical conditions.
- Altitude. Up to 500 m in Puerto Rico; up to 1,300 m in Colombia.
- Rainfall. From 1,000 to 2,000 mm in Puerto Rico, an average of 1,650 mm in southern Florida.
- Soil. The tree grows so easily and widely that it seems well adapted to many different soil types. It is known to tolerate poor soil but prefers sand. In the Pacific it is recommended for planting in the sandy coral soils of low islands and atolls. It thrives on well-drained limestone in Florida and Jamaica. In Malaysia it has become established on the tailings of old tin mines.

Establishment Seed is plentiful and the seedlings establish easily. The plant is more quickly propagated by cuttings.

- Seed treatment. None required.
- Ability to compete with weeds. No plants survive under its dense shade.

Pests and Diseases In southern Florida the foliage is subject to a leaf spot and many small twigs tend to die back in limestone conditions.

Limitations The tree may become a weed. Its seeds are spread by squirrels, birds, and fruit bats. In Guatemala it colonizes abandoned agricultural land. The wide-spreading limbs tend to break in strong winds. It is recommended they be trimmed back once a year.

Muntingia calabura about 10 m tall, naturalized on wasteland, Kuala Lumpur, Malaysia. (Forest Research Institute, Kepong, Malaysia)

Muntingia calabura fruit. (K. and J. Morton)

Related Species

- *Muntingia rosea* Karst. Native to northern Venezuela, the tree has slightly larger pinkish-purple petals and larger leaves than *Muntingia calabura*. It would be more attractive as an ornamental than the white-flowered tree.

Sesbania bispinosa

Botanic Name *Sesbania bispinosa* (Jacq.) W. F. Wight

Synonym *Sesbania aculeata* Pers., *S. cannabina* (Retz.) Pers.

Common Name Dhaincha, prickly sesban

Family Leguminosae (Papilionoideae)

Main Attributes *Sesbania bispinosa* is a quick-growing shrub that can produce firewood in only 6 months. It can be used as a rotation crop to grow fuel as well as fertilize the soil in preparation for food crops. In Vietnam, in the Red River Delta, it is grown in rice fields and its stems harvested for firewood before the rice crop is planted. It matures so rapidly in tropical areas that two harvests a year are possible. All parts of the plant are useful and the crop appears easy to produce on large scale with little care or investment.

Description Normally a spreading shrub-like plant, *Sesbania bispinosa* grows tall and straight in crowded stands. Its slender stem may reach 4 m tall. It is a prickly-leaved annual with a 5- to 6-month maturing time. Because it nodulates vigorously, the plant seems to improve soil fertility.

Distribution A native of tropical and subtropical areas of the Indian subcontinent, the plant has been distributed to parts of tropical Africa, Southeast Asia, China, and the West Indies. To researchers, however, it is little known outside of two or three laboratories in India and Pakistan.

Use as Firewood The stems have a low density (specific gravity, 0.3), but can be produced in high yield in about 6 months.

Sesbania bispinosa is used as a firewood crop in northern Pakistan and in Vietnam. In the Cameroons, villagers plant a similar *Sesbania* species for firewood. It is commonly used in Pakistani villages for evaporating water from sugar.

Yield In one farmer's field in Italy, a crop of *Sesbania bispinosa* yielded 15 bone-dry t per ha.* In the tropics, where more than one crop can be harvested each year, the annual production could be even higher.

*Information supplied by L. Markila.

Other Uses

- Gum. Seeds of *Sesbania bispinosa* contain a water-soluble gum, like guar gum, that produces a smooth, light-colored, coherent, and elastic film useful for sizing textiles and paper products and for thickening and stabilizing solutions.
- Pulp. The plant is an exciting potential new source of paper products. Its fibers are very similar to those of birch, one of the best hardwoods used for pulping.
- Fiber. The stems can be processed to provide a cordage fiber with jute-like qualities, useful for items such as fishing nets, gunnysacks, and sails.
- Green manure. The living plant is used to provide windbreaks, hedges, erosion control, and shade and cover for crops. Its foliage is a choice green manure used to increase soil fertility, especially on saline and wet soils.
- Forage. The leaves reportedly make good cattle fodder.

Environmental Requirements

- Temperature. This extremely versatile plant grows well in the tropics and subtropics (e.g., Vietnam, India, Pakistan), and successful plantation trials have been carried out in temperate zones (e.g., Italy from the Rome region to Sicily).
- Altitude. Grows from sea level up to 1,200 m.
- Rainfall. It is highly resistant to drought, although it grows best in regions with rainfall ranging from 550 to 1,100 mm.
- Soil. *Sesbania bispinosa* is well adapted to difficult soils. It will grow on saline and alkaline wastelands and wet, almost waterlogged soils (such areas often remain barren for want of suitable crops).

Establishment The plant establishes easily by direct seeding.

- Seed treatment. None required.
- Ability to compete with weeds. The plant grows so quickly that it is reportedly excellent for suppressing vigorous weeds such as *Imperata cylindrica* on sites where moisture is adequate.

Sesbania bispinosa, Lucknow, Uttar Pradesh, India. (M.I.H. Farooqi)

Pests and Diseases Unreported.

Limitations The tree grows fast, seeds freely, and reportedly has become a noxious weed in rice paddies in some areas.

Related Species Other *Sesbania* species with promise as firewood include:

- *S. paludosa*
- *S. aegyptiaca* (*S. sesban*)
- *S. speciosa.*

Sesbania grandiflora

Botanic Name *Sesbania grandiflora* (L.) Pers.

Synonym *Agati grandiflora* (L.) Desv

Common Names Agati, bacule, katurai (Philippines), August flower (Guyana), West Indian pea tree, turi (Malaysia, Java), gallito, chogache (India)

Family Leguminosae (Papilionoideae)

Main Attributes This small tree produces firewood, forage, pulp and paper, food, and green manure and appears to hold promise for reforesting eroded and grassy wastelands throughout the tropics. It combines well with agriculture (agroforestry) in areas where trees are not normally grown and becomes an important fuelwood source. After the plant is harvested, shoots resprout with such vigor that they seem irrepressible. The tree's outstanding quality is its rapid growth rate, particularly during its first 3 or 4 years.

Description *Sesbania grandiflora* grows to a height of 10 m, with a trunk diameter of about 30 cm. The bole is straight and cylindrical, the wood white and soft. The bark is light grey, deeply furrowed, and corklike in texture.

Distribution Native to many Asian countries, for instance, India, Malaysia, Indonesia, and the Philippines, *Sesbania grandiflora* is commonly seen growing on the dikes between rice paddies, along roadsides, and in backyard vegetable gardens. It has been widely distributed in southern Florida and the West Indies as well as from southern Mexico through most countries of Central America down to South America. It has also been cultivated in Mauritius. A closely related species, *S. formosa*, is native to northern Australia.

Use as Firewood *Sesbania grandiflora* has long been used as firewood in Southeast Asia and has been planted in several areas in Indonesia to provide fuel and other products in "turinisation" projects (after turi, the indigenous name). However, the wood is white, soft, and has a rather low specific gravity of about 0.42, which is poor for fuelwood.

Yield In India, plantation-grown trees have reached 8 m in as little as 3 years (with average diameter not less than 10 cm). Moreover, *Sesbania grandiflora* can be planted very densely (up to 3,000 stems per ha). Wood yields of 20-25 m^3 per ha per year are commonly achieved in plantations in Indonesia. Even when planted only along the edges of agricultural fields, as in Java, yields of 3 m^3 of stacked firewood per ha from 2-year rotation periods have been recorded.

Other Uses

- Utility plantings. In Manila, the tree is often planted for beautification along roadsides, fence lines, and other boundaries. The large, handsome flowers and long pods make it a striking ornamental. With its open, spreading crown of feathery leaves, *Sesbania grandiflora* gives light shade. It also makes useful windbreaks and is often grown as a living fence.
- Food. The young leaves, tender pods, and giant flowers of *Sesbania grandiflora* are favorite Asian vegetables. They are used in curries and soups or sometimes fried, lightly steamed, or boiled. The leaves contain over 36 percent crude protein (dry weight), and with their high mineral and vitamin content, they make a nutritious, spinach-like vegetable.
- Forage. Cattle relish the feathery leaves and long pods (up to 60 cm). In parts of Java where cattle breeding is important, the tree is frequently planted as a source of forage, continually topped to keep it within the animal's reach.
- Green manure. Foliage of *Sesbania grandiflora* makes excellent green manure; in Indonesia, trees planted along dikes are often used to fertilize nearby crops. They are considered an excellent support and nurse crop for betel and pepper vines.
- Reforestation. In Taiwan, *Sesbania grandiflora* has proved useful for reforesting eroded hill regions.
- Gum and tannin. When cut, the bark of the tree exudes a clear gum that has been used as a gum arabic substitute in foods and adhesives. The bark also yields a tanning agent.
- Pulp and paper. In East Java, the tree is extensively used as a pulp source. The wood's fiber length (1.1 mm, about average for hardwoods used for paper) and chemical composition are suitable for pulping.

Sesbania grandiflora, age 4½ years, trial plot at Kimberley Research Station, Kununurra, Western Australia. (A.F.J. Logan)

Sesbania grandiflora, 7 m tall after only 1 year of growth. The seed was direct-sown in deep sandy loam. East Godavari District, Andhra Pradesh, India. (A.S. Bhat)

Environmental Requirements

- Temperature. The tree is frost sensitive and is adapted only to tropical conditions.
- Altitude. Grows up to 800 m above sea level.
- Rainfall. *Sesbania grandiflora* grows best where annual rainfall exceeds 1,000 mm and there are only a few months of dry season. It is widely grown in areas where there is extensive irrigation or flooding, such as in Asian rice paddies. On the other hand, the species grows abundantly on the semiarid Timor Islands of Indonesia.
- Soil. *Sesbania grandiflora* is able to grow in a wide range of soils, even poor ones, including black, poorly structured clay. Its extraordinary nodulation undoubtedly helps restore fertility to these soils. On Timor, the species is commonly found on abandoned swidden land. This, coupled with its rapid growth, suggests that its soil improvement qualities—though unmeasured—may be exceptional.

Establishment The species propagates easily by cuttings or seedlings. It requires little maintenance and can readily be planted on a large scale by direct seeding and even by aerial sowing.

- Seed treatment. None.
- Ability to compete with weeds. Unreported.

Pests and Diseases *Sesbania grandiflora* is very susceptible to nematodes. In northern Australia, plantations have also been damaged by birds (cockatoos) and grasshoppers.

Limitations There is no information available on general silviculture of this plant; thus, those who may wish to grow it face considerable uncertainty.

Syzygium cumini

Botanic Name *Syzygium cumini* (L.) Skeels

Synonyms *Syzygium jambolanum* DC., *Eugenia cumini* Druce, *E. jambolana* Lam.

Common Names Jambolan, Java plum, jaman, jambu, jamun, jambul, jambool, black-olum tree, Portuguese plum, Malabar plum, Indian blackberry, duhat

Family Myrtaceae

Main Attributes The fast-growing jambolan tree is a good source of fuelwood. Acclaimed for its handsome appearance, it is planted for shade along streets and roadways and as an ornamental fruit tree in home landscaping.

Description It is usually a large, erect tree with shiny, evergreen, leathery leaves, but sometimes has low branches. Mature trees are generally about 13 m tall, but may reach as high as 30 m, with trunk diameter up to 1 m.

Distribution Native to India, Burma, Sri Lanka, and the Philippines, jambolan has been widely planted and naturalized in many parts of the tropics and subtropics. It is common in southern Florida, is occasionally found in the West Indies and Central America, and in Australia is established in both New South Wales and Queensland.

Use as Firewood Jambolan wood is considered to be an excellent fuel. It is reddish gray, rough, and moderately hard, with no distinct heartwood. Its specific gravity is about 0.77 and it burns well, giving off about 4,800 kcal per kg. The plants can be coppiced and they regrow vigorously after pruning.

Yield Jambolan seedlings may reach 4 m in only 2 years.

Other Uses

- Wood. The wood is durable in water and resistant to termites. While it is difficult to work, it is often used for beams and rafters and for posts, agricultural implements, boats, oars, and masts.
- Food. The tree bears many oblong or round plum-like fruits, generally astringent and with a flavor that varies from acid to fairly sweet; the astringency can be reduced by soaking in salt water. They can be eaten raw, but more often they are made into a juice similar to grape juice. They are also made into jellies, syrup, vinegar, and wine.
- Hedges, windbreaks, and ornamental uses. The jambolan is one of the most popular avenue trees in India. Its abundant foliage produces good shade. When closely planted in rows, it makes good windbreaks and, if topped regularly, forms a dense hedge.
- Bark. The bark contains 13–19 percent tannin and is much used for tanning and dyeing in India.
- Honey. The flowers are rich in nectar that yields high-quality honey.

Environmental Requirements

- Temperature. Tropical and warm subtropical conditions are required by the jambolan.
- Altitude. In southern Asia, the tree is found growing in elevations up to 1,800 m, but in the Hawaiian Islands, it has not grown well beyond 600 m.
- Rainfall. The jambolan grows vigorously in regions of heavy rainfall (from 1,500 mm to as much as 10,000 mm per year), prospers on riverbanks, and has been known to withstand prolonged flooding. The species also grows well in higher, well-drained land, and once established, is observed to tolerate drought.
- Soil. The jambolan thrives in sand, marl, and in well-drained oolitic limestone, as well as in many other soil types.

Establishment The tree is easily grown in nurseries and easily transplanted. However, direct seeding is the most common method of propagation. Cuttings have been rooted in sand, and both air-layering and budding onto seedlings of the same species have been successful.

- Seed treatment. None required. Seeds are usually sown during the rainy season and germinate in 2–4 weeks.
- Ability to compete with weeds. It successfully competes with weeds because of rapid growth and vigor.

Pests and Diseases Some trees are susceptible to infestation by scale insects, white flies, and leaf-eating caterpillars, but these seem unlikely to cause much problem in firewood plantations.

Limitations In southern Florida the tree is considered a nuisance because it is too tall and

Syzygium cumini fruit.
(K. and J. Morton)

Syzygium cumini, northern India. (Forest Research Institute, Dehra Dun, India)

too fruitful. The fruits fall and ferment rapidly, attracting insects and staining concrete and shoes. However, heavy fruiting is not likely to be a problem in plantations designed for fuel production, and is an advantage when the tree is planted for shade in chicken yards or livestock pastures. The fallen fruits are devoured by animals.

Introduced into Hawaii for its fruit, it has run wild (because birds distribute the seeds) and is considered a noxious weed that shades out more desirable forage plants.

Terminalia catappa

Botanic Name *Terminalia catappa* L.

Common Names Indian almond, tropical almond, sea almond, West Indian almond, Malabar almond, almendra, badan (Burmese)

Family Combretaceae

Main Attributes A hardy, quick-growing, easily propagated tree, *Terminalia catappa* grows well in coastal sand dunes and other problem sites. Its wood is used for fuel in several tropical areas. The trees tolerate salt spray and drought, as well as both moderate shade or full overhead light.

Description It is a medium-sized tree, with peculiar whorls of horizontal branches stacked one above the other up the trunk, which is usually quite short and crooked. As it ages, its attractive appearance is enhanced by large, reddish-tinged, leathery leaves. It bears hard fruits, somewhat similar to greenish almonds, each containing a large, edible nut. It is primarily a coastal tree and commonly forms beach forests extending from sandy shores into the forests behind.

Distribution *Terminalia catappa* is native to Malaya and the Andamans, but is very common in India, being planted in villages and bungalow compounds and gardens for beautification as well as for its fruit. It is also found in the Philippines, Indonesia, the Seychelles, Ghana, and in the South Pacific from Polynesia to Papua New Guinea. It has been naturalized in Hawaii, Puerto Rico, and the Virgin Islands. It is planted in southern Florida and the West Indies, and can be found growing in tropical America from Mexico to Peru and Brazil.

Use as Firewood The wood is moderately heavy (specific gravity, 0.59), air-seasons quickly, and makes good firewood. It is often employed as fuel.

Yield Under favorable conditions the tree grows quite fast—sometimes reaching 1 m in the first year, 3 m in the second, and 6 m in the third. Rotations of 10-15 years seem to be average for plantation growth. The expected 10-year yield from a plantation is 22.5-36.0 t.

Other Uses

- Wood. The pale-brown to reddish-brown heartwood is attractive and hard. While machining is difficult, it is suitable for millwork, furniture, veneer, and cabinetwork. Further, it is useful for boat building, general construction, bridge timbers, cross-ties, flooring, and for boxes and crates.
- Fruit. Within the tough shell of the fruit there are one or two delightfully flavored seeds from which a bland, edible oil can be extracted. The seeds are sweet, almond-like, and widely eaten roasted or raw.
- Tannin. The bark, leaves, roots, and fruits all contain tannin and have been used for tanning purposes.
- Dune fixation. Because it is salt tolerant and grows well on sand, it has been used to arrest beach erosion.
- Shade and ornament. The tree is cultivated in gardens and along roadsides in Hawaii, Puerto Rico, and elsewhere because of its unusual and attractive branching and leaf coloration.
- Silk production. The foliage has been found suitable for feeding tasar or katkura silkworms.

Environmental Requirements

- Temperature. *Terminalia catappa* is strictly a tropical species, thriving in humid climate.
- Altitude. It grows best at low altitudes and is extensively planted in India and Sri Lanka at altitudes from sea level to 300 m; it is conspicuous in coastal towns such as Madras and Rangoon.
- Rainfall. The tree needs at least 1,000 mm of annual rainfall unless the water table is high enough for its roots to penetrate.
- Soil. *Terminalia catappa* is not particular as to soil; for example, though it grows well in sand or shingle, it also thrives in marl and permeable siliceous limestone.

Establishment The tree's natural regeneration is facile; its seeds float and are often dispersed by ocean currents. They germinate readily. Nursery beds sown with ripe seed give high yields of seedlings suitable for transplanting in the wet season the following year.

- Seed treatment. None required.
- Ability to compete with weeds. Excellent.

Terminalia catappa used as an ornamental tree, Oahu, Hawaii, USA. (Institute of Pacific Island Forestry

Pests and Diseases It is very susceptible to drywood termites and the leaves are often riddled with insects.

Limitations It is a messy tree, dropping fruit that rapidly becomes fetid.

Related species *Terminalia* is a large genus found throughout the tropics. Local species should be tested for firewood production. A few examples are:

- *Terminalia arostrata* — northern Australia
- *T. avicennioides* — tropical West Africa
- *T. bursarina* — northern Australia
- *T. glaucescens* — West African savannas
- *T. ivorensis* — tropical Africa
- *T. kaernbachii* — Papua New Guinea
- *T. macroptera* — tropical West Africa
- *T. mollis* — tropical Africa
- *T. prunioides* — East Africa
- *T. racemosa* (*T. americana*) — Mexico to northern South America and the West Indies
- *T. sericea* — East Africa
- *T. superba* — tropical Africa
- *T. tomentosa* — India, Pakistan

Trema species

Family Ulmaceae (or Urticaceae)

Main Attributes There are about 15 species of *Trema* distributed throughout the tropics. They are pioneer plants of eroded soils and volcanic ash and are generally among the first trees to come up on newly cleared land such as forest clearings, thickets, and roadsides. They will grow in poor soils and in barren environments. The species generally have many characteristics in common, offering like benefits and uses. The wood is soft and of only limited value as firewood. However, it probably can be produced easily in large quantities in appropriate locations.

Description *Trema* species are shrubs or small trees that reach heights of about 10 m and diameters of about 20 cm. They are found in open forests and have spreading crowns and evergreen leaves. Some examples are:

- *Trema orientalis*, an Asian species native to the moist parts of India where it is called "charcoal tree." It is also found throughout the Malay peninsula, extending into China, and has had some successful cultivation trials in the southern Philippines.
- *Trema politoria*, also an Asian species found in dry zones of subtropical India and throughout northern India from Gujarat in the west to Assam in the east.
- *Trema guineensis*, a small African tree known as the "charcoal tree" in Malawi. It is a pioneer species common to deciduous forests and is found distributed throughout tropical Africa to South Africa, extending into Arabia and Madagascar. (Some botanists consider this to be the same plant as *T. orientalis*.)
- *Trema micrantha*, native to Central and South America as well as to the West Indies.
- *Trema cannabina*, native to Southeast Asia.

Use as Firewood These and other *Trema* species are widely used as fuel, though their calorific value is not high (4,500 kcal per kg). *Trema orientalis* is used to make gunpowder in Java and as a charcoal for fireworks.

Yield *Trema* species are claimed to be exceedingly fast growing. They coppice well.

Other Uses

- Afforestation programs. These plants are pioneer species suitable for planting on poor lands, in afforestation of denuded and disturbed areas, and for use as general soil binders. All may prove to be soil improvers as well.
- Shade trees. *Trema orientalis* is often planted in Asia and Africa as a shade tree on plantations for coffee, cacao, and other crops.
- Pulp and wood. In combination with bamboo pulp the wood is employed in the manufacture of writing and printing papers. Bleached pulps in 46-50 percent yields can be prepared from *Trema orientalis*, with an average fiber length of 1 mm.
- Wood. In some areas (for example, in Gongola State, Nigeria), the wood is used for construction poles.
- Fodder. The leaves and branches of the plants are lopped for fodder. However, there has been some evidence, particularly in Australia, that they may contain a toxic glucoside.

Environmental Requirements

- Temperature. These are tropical and subtropical plants.
- Altitude. *Trema orientalis* is found extending up to about 2,000 m in the Himalayas, *Trema politoria* up to about 1,500 m.
- Rainfall. *Trema orientalis* needs a moist and humid climate, *Trema politoria* a dry climate, and *Trema guineensis* grows in both humid and dry environments.
- Soil. *Trema* species have no particular soil requirements. They are among the first woody plants to colonize denuded, fallow, or poor soils.

Establishment The plants regenerate very easily and can be propagated by stump cuttings.

- Seed treatment. Steeping seeds in giberellic acid in agar at 500 ppm or refrigerating at 2°C for 3-4 months breaks dormancy.
- Ability to compete with weeds. This species is sun loving and has the capacity to outgrow other species in newly cleared areas.

Pests and Diseases The trees harbor insect pests, which cause defoliation. These pests can spread to other plants of economic utility; therefore, caution is advised when these species are introduced outside their natural habitat. Two fungal diseases are known to attack them.

Limitations Unreported.

Trema orientalis natural regeneration, central Taiwan. (Taiwan Forestry Research Institute)

Parasponia root nodules which, like those of legumes, can fix atmospheric nitrogen (M.J. Trinick)

Related Species In the Malay Archipelago are found five species of *Parasponia* that are also pioneer trees and differ from *Trema* species only in tiny details of flower structure. These *Parasponia* species bear root nodules (containing *Rhizobium* bacteria) and can fix nitrogen as if they were legumes (despite previous literature reports, nitrogen-fixing nodules have not yet been found on *Trema* species*). Examples are:

- *Parasponia andersonii*
- *P. rugosa*
- *P. rigida*
- *P. parviflora.*

*Information supplied by M. J. Trinick. See Research Contacts.

III Fuelwood Species for Tropical Highlands

A major problem in the hill lands of the tropics (those areas above 1,000 m elevation) is the indiscriminate cutting and clearing of native forests. In part, this has been caused by shifting cultivators in their search for arable farm land, but the quest by impoverished people for firewood and for wood to make charcoal has greatly accelerated the process.

The unrelenting cutting of forests in some countries has led to critical problems in preserving natural resources. Haiti may be in the most serious jeopardy, but Nepal and parts of Java are also facing great difficulty.

In Nepal, where firewood is the principal source of domestic energy, people must travel at least one full day to collect wood, while a generation ago it could be collected within an hour's walk from most dwellings.

Comparable situations exist in hill and mountain regions of Central America, the Andes, Central Africa, Southeast Asia, Papua New Guinea, and Southwest Asia (Afghanistan and Iran). Population increases force people onto the forested hillsides where they clear the land for crops. As trees become more scarce, they gather their firewood from ever-higher slopes. This throws the forests into continual recession.

While 10 percent of the human population live in the highlands, another 40 percent live in the adjacent lowlands, and their future is intimately bound to developments on the slopes and plateaus above. Overgrazing and firewood collection despoils the environment and has far-reaching effects. For example, when the highlands are denuded they do not retain the rainwater. This can render fertile valleys unproductive as a result of flash flooding and siltation or streams drying up when water is urgently needed for irrigation and livestock.

Reforestation of tropical highlands is crucial to developing countries where subsistence farming is the predominant way of life. Efforts to isolate those tree species best suited for tropical highlands are underway.

This section describes cool-zone species. Some of them (e.g., *Ailanthus altissima* and *Alnus rubra*) are native to temperature latitudes; their performance in tropical latitudes is uncertain.

Almost all the species described in the humid tropics section of this report have wide adaptability and might well be grown in the highland tropics.

Collecting wood in the highlands of Nepal. The hills are almost bare due to constant scavenging for firewood, building timber, and forage. (K. Gunnar)

Species listed in other sections that deserve trials as fuelwood species in tropical highlands include:

Casuarina oligodon	page 41
Eucalyptus camaldulensis	page 126
Gliricidia sepium	page 44
Mimosa scabrella	page 56
Pinus halepensis	page 142
Sesbania bispinosa	page 60
Syzygium cumini	page 64
Terminala catappa	page 66
Trema spp.	page 68

Acacia mearnsii

Botanic Name *Acacia mearnsii* De Wild.

Synonym *Acacia mollissima* auct. mult., non De Wild.

Common Names Black or tan wattle

Family Leguminosae (Mimosoideae)

Main Attributes *Acacia mearnsii* yields excellent fuelwood and charcoal and has proved suitable for growing in many countries. It is a very good species for small-holder or village woodlots and for line plantings along roads or trails or around farm fields or villages.

Description This Australian native is a handsome tree that may grow to be 25 m tall. Solitary specimens have spreading crowns, but when crowded into plantations the trees grow erect and slender. The feathery foliage is dark green and the trunk appears black.

Distribution *Acacia mearnsii* is native to New South Wales, Queensland, South Australia, Tasmania, and Victoria. It is cultivated (often for the tannin in its bark) in New Zealand, South, Central, and East Africa, India, Sri Lanka, parts of Central America, and Indonesia. Large plantations have been established in Natal province of South Africa.

Use as Firewood This dense wood (specific gravity, 0.7-0.85) has a calorific value of 3,500-4,000 kcal per kg and ash content of about 1.5 percent. It also yields quality charcoal (specific gravity, 0.3-0.5; calorific value 6,600 kcal per kg; ash content, 0.4 percent). Oven-dry wood in Indonesia has a calorific value of 4,650 kcal per kg.

In Central Java, individual farmers plant the tree in small farm forests (generally less than 1 ha each) for their personal firewood needs and for sale. The wood is used as fuel for curing tobacco leaves, the foliage as a green manure to improve agricultural yields, and the bark for tannin extraction. In some areas the acacia forests are now almost the exclusive source of fuelwood, which relieves the pressure on native forests. Grasses and herbs that grow naturally in the acacia forests are used for grazing.

Yield In Indonesia, it has been found that short rotations of 7-10 years or less are the most economical for plantation growth. This provides an annual thickwood production ranging from 10 to 25 m^3 per ha and yielding an average bark production of 800-4,000 kg per ha, depending on the site.*

Other Uses

- Tannin. The tree's bark may contain up to 40 percent of exceptionally good tannin. *Acacia mearnsii* is the principal source of the world's tanbark, which is used in the manufacture of hard leathers for shoes, saddles, and other products.
- Erosion control. In Indonesia, Zaire, and Sri Lanka, *Acacia mearnsii* is often grown at high elevations and on poor and sloping soils that are unstable and will not support agricultural crops. Densely packed plantations have proved effective in preventing further erosion even on hillsides of up to 50-degree slope.
- Soil improvement. When densely planted and carefully managed, *Acacia mearnsii* can regenerate soils thinned by erosion or impoverished by crop growing. It is an efficient nitrogen fixer, and in Indonesia has annually yielded 21-28 t of wet leaves per ha containing 240-285 kg of nitrogen. In the Wonosobo volcanic region of Indonesia, where there are extensive *Acacia mearnsii* plantations, farmers claim that vegetable crops and tobacco grown in rotations with the trees give twice the yield because of the green manure the trees provide.†
- Pulp. Testing has shown the species to yield a good pulp having good strength characteristics for wrapping-paper.

Environmental Requirements

- Temperature. *Acacia mearnsii* is native to areas that have cool winters. Its growth slows if temperatures are too high. Therefore, in equatorial countries like Indonesia, it is grown only in the cool highlands. It is moderately frost tolerant.
- Altitude. In its Australian home, the species grows in low valleys and on hillslopes to 1,100 m elevation. In Indonesia, it is only grown at altitudes above 1,000 m. In Natal, South Africa, there are huge plantations at altitudes between 300 and 1,100 m.

*von Wulfing. 1948.

†Information supplied by K. F. Wiersum. See Research Contacts.

Acacia mearnsii, coast of New South Wales, Australia. (CSIRO Division of Forest Research)

• Rainfall. In its native range the tree occurs in areas with 500-700 mm rainfall. If introduced to areas with higher temperatures, more rainfall is necessary; e.g., in Indonesia minimum rainfall is 1,000 mm.

• Soil. Although it cannot tolerate calcareous soils, this tree is not very particular as to soil conditions and can grow on poor soils.

Establishment The species is easily established by direct seeding.

• Seed treatment. Dormancy is broken by immersing seeds in boiling water.

• Ability to compete with weeds. Unreported.

Pests and Diseases In general, disease and pest attacks are not serious and do not affect wood production except under wet conditions of more than 3,000 mm annual rainfall, which make the trees susceptible to insect attack and fungal diseases.

Limitations *Acacia mearnsii* coppices poorly. In Hawaii, it is a noxious weed and spreads prolifically between 600 and 1,200 m altitude in the 1,000-1,200 mm rainfall zones. In South Africa, it has been called "green cancer" because it is spreading out of control.

Related Species Two other species are so similar to *Acacia mearnsii* that all three have been endlessly confused in the literature. (Sometimes they have been reported as three varieties of *Acacia decurrens*; e.g., the black wattle being called *Acacia decurrens* var. *mollis*.) They are:

• *Acacia dealbata* (*A. decurrens* var. *dealbata*), silver wattle, a commonly cultivated tall tree (up to 33 m) whose leaves and small branches glisten with a "silver" fuzz. It is native to New South Wales, Tasmania, and Victoria.

• *Acacia decurrens* (*A. decurrens* var. *normalis*), king or green wattle, native to New South Wales.

Ailanthus altissima

Botanic Name *Ailanthus altissima* (Mill.) Swingle

Synonym *Ailanthus glandulosa* Desf.

Common Names Ailanthus, tree of heaven, China sumac

Family Simaroubaceae

Main Attributes Ailanthus is a "city tree." It is planted in urban areas over much of northern China. Peking, for example, uses it in urban forestry. The ailanthus (and other species) are cut down for lumber and firewood and new trees planted on an organized rotation. Ailanthus may help urban areas become more self-sufficient. It is more competitive than most weeds and will live and grow in environments almost devoid of other plant life. Hardy and tenacious, it flourishes in vacant lots, alleys, and trash heaps. The tree is very nearly indestructible and is apparently immune to pollution. Thus, it has potential for cultivation in heavily polluted areas where other species may grow poorly or not at all.

Description Ailanthus is a deciduous tree that grows with a crooked trunk and an open crown. It may reach 20 m but is more usually 6-10 m tall. Its leaves are long and pinnately compound. Individual trees are either "males" and produce pollen or "females" and produce seeds.

Distribution *Ailanthus altissima* is native to northern China and was introduced into cultivation in 1751. It is now cultivated widely in Europe and is also found in Morocco. It is the species most commonly planted for greenbelts around cities in the semiarid areas of Iran. It has become ubiquitous in the cities of northeastern United States and is naturalized over the eastern half of North America from southern Ontario to Florida and Texas.

Use as Firewood The yellowish ailanthus wood is fairly hard and heavy and is used for fuel and charcoal in several countries. The trees coppice very well.

Yield Although its yields are unreported, this is a fast-growing tree. In a 5-month growing season in the northeastern United States it may reach 3-4 m in height.

Other Uses

- Wood. *Ailanthus altissima* wood is difficult to split, but is easy to work and to polish. It is suitable for lumber (its wood has been likened to that of ash), furniture, and cellulose manufacture. It has reportedly been grown for lumber in New Zealand.
- Erosion control. Ailanthus is valued chiefly for shade, shelterbelts, and erosion control, particularly in cities where soils are poor and the atmosphere smoky. It has been used in Europe to cloak otherwise bare alpine slopes. It is reportedly used to check erosion around the Black Sea and in the mountains of Morocco.

Environmental Requirements

- Temperature. The tree of heaven is found in temperate to subtropical climates. It withstands frost and chilling winters such as those in New York.
- Altitude. 0-2,000 m or more.
- Rainfall. It grows in areas with low rainfall, 350-600 mm, and which may have up to 8 dry months.
- Soil. It grows on a variety of soils: hard, arid, wet, acid, alkaline, rocky, swamp, and depleted. Nothing seems to discourage it. A single square meter of soil beside a hot, dirty street provides a good habitat for ailanthus.

Establishment It regenerates abundantly from seed and easily from root cuttings, and reproduces profusely from root suckers.

- Seed treatment. Seed can be stored dry, sealed, and cold for 2 years without losing viability.
- Ability to compete with weeds. Very good.

Pests and Diseases It is not normally bothered by insects, but some small plantations in the United States have been defoliated and killed by tent caterpillars.

Limitations A tree of temperate zones, ailanthus thrives in subtropical areas such as Iran, but how well it will grow in truly tropical areas or in tropical highlands is unknown.

It is a prolific seeder and takes root anywhere a little earth has accumulated, so it could become a serious pest. (Planting root cuttings

Ailanthus altissima, Washington, D.C. (N.D. Vietmeyer)

only of male trees might avoid problems of uncontrollable self-seeding).

Ailanthus pokes up through cracks in concrete and worms its way out from under buildings and walls. It should not be planted where root suckers can invade nearby fields.

Related Species There are about seven species of *Ailanthus*. Three tropical species native to India are worth considering for use in developing-country firewood plantations.

- *Ailanthus excelsa* is cultivated in dry tracts of northern India. Its branches are lopped for fodder and its wood used as fuel. It is often planted in rows along the boundaries of farms and waterways.
- *Ailanthus grandis* is a lofty, fast-growing tree of northern India. It seems to hold promise as a forestry species. It has a straight, cylindrical bole, thin bark, a narrow crown, and is self-pruning. Its wood can be used for fuel, packing crates, matches, and newsprint pulp.
- *Ailanthus malabaricum* is a small tree of southern India, but is found elsewhere in tropical Asia as well. For example, it has been cultivated near Hanoi. If the bark is cut, a resin (known as mattipaul in India) exudes; when burned it provides a pleasant scent.

Alnus acuminata

Botanic Name *Alnus acuminata* O. Ktze.

Synonym *Alnus jorullensis* H.B.K.

Common Names Alder, aliso (Ecuador, Argentina), ramram, lambran, jaul (Costa Rica)

Family Betulaceae

Main Attributes *Alnus acuminata* is a quick-growing tree whose wood burns well. It grows well on steep mountain slopes and, because it is a nitrogen-fixing species, it is good for reforestation and soil reclamation. It reproduces itself freely on exposed bare-soil surfaces. This may prove to be a good species for producing firewood in pasture lands where livestock are grazed under trees.

Description This tree, which varies in height from 15 to 30 m under natural conditions and reaches 40 m in plantations, has a broad, spreading root system close to the soil surface. The trees have light-gray, sometimes silvery bark, lignified cones, and male and female flowers in catkins. The seeds are winged and spread easily in the wind.

Distribution *Alnus acuminata* is native to Central and South America. It is generally found at moderate-to-high altitudes growing along slopes, ravines, roadsides, and stream banks of the hill and mountain ranges from Mexico to Argentina. It is cultivated extensively on plantations along the central cordillera of Costa Rica, Colombia, Bolivia, and Peru. It has been successfully introduced into southern Chile (Valdivia Province) and into the South Island of New Zealand.

Use as Firewood *Alnus acuminata* wood has a specific gravity of 0.5-0.6. It has good, even-burning characteristics and has long been used for firewood in its native region. In Costa Rica the tree is considered too good for firewood, although pruned branches and the top-wood at the time of harvest are used for fuel. Its main uses are for timber, boxes, construction, and furniture. The tree coppices naturally, but it is not known if crops can be systematically reproduced by this method.

Yield *Alnus acuminata* is a fast-growing species; in plantations, it can grow to 25 m (with a diameter of 20 cm) in 10 years. Eleven-year-old trees in Costa Rica commonly average 38 cm in diameter and 16 m in height (30 m after 30 years). In rotations of about 20 years the annual yield of wood for fuel and industrial use is 10-15 m^3 per ha.*

Other Uses

- Wood. This species produces a straight-grained, fine-textured wood, which ranges from light to reddish brown and has a lustrous surface and no odor. It is often employed in bridges and pilings, furniture making, coffins, crates, and plywood manufacture. The wood has strong, impregnable fibers, a desirable attribute for making good pulp.
- Watershed protection. Because it grows so well on slopes and because its root system tends to be lateral and extended rather than deep and confined, it is very useful for controlling erosion in steep and unstable soils.
- Soil improvement. Although not legumes, *Alnus* species have nodules and are nitrogen fixing, and they enhance soil fertility and benefit crops grown in conjunction with them. Clusters of light-yellow nodules occur on the roots of plants as young as 2 months old and are found from the base of the root to the ends of the rootlets.

Wide-spaced *Alnus acuminata* in high elevation (2,000-3,000 m) pastures in the tropics have increased forage production severalfold in Latin America.

Environmental Requirements This species has given its best results on the humid lower mountain slopes in tropical latitudes.

- Temperature. *Alnus acuminata* occurs where mean annual temperatures range from 4°-27°C. It can withstand temperatures that dip briefly below 0°C.
- Altitude. This species is found on the moist flanks of mountains at elevations of 1,200-3,200 m. In the valleys, it has to be protected from cold dry winds or it will develop poorly.
- Rainfall. In its native habitat, it is found along roadways, ravine slopes, and small stream banks where there is light and adequate moisture. Total annual precipitation in the area is 1,000-3,000 mm or more.

*Information supplied by T. Lang. See Research Contacts.

Alnus acuminata, combined with pasture production. Jaul, Santa Cruz de Turrialba, Costa Rica. (P. Rosero)

• Soil. It is usually found on deep, well-drained loams or on loamy sands of alluvial origin. These usually make prime agricultural lands and *Alnus acuminata* (a good agroforestry species) is often grown in conjunction with pasture crops and dairy cattle, for example, in Costa Rica. However, it also grows adequately on many types of soils varying from gravel to sand to clay.

Establishment *Alnus acuminata* is easily grown from seed. It is usually raised in nurseries and planted to the field after 1 or 2 years. When sown in soils where *Alnus* does not already exist, the soil must be inoculated with the appropriate nitrogen-fixing bacteria. The plant is also easily propagated by root cuttings.

• Seed treatment. None required.

• Ability to compete with weeds. It does not compete well.

Pests and Diseases The plant is almost disease free, but its leaves are sometimes attacked by insects and it is susceptible to soil fungi, especially when it grows in soil with a high content of organic matter.

Limitations Seeds are available (in South America) in February, March, and August, but they must be planted quickly, or they lose their viability. For example, only 50 percent of 1-month-old seeds germinate.

Alnus nepalensis

Botanic Name *Alnus nepalensis* D. Don

Common Names Indian alder, Nepalese alder, maibau (Burma)

Family Betulaceae

Main Attributes *Alnus nepalensis* is a very fast-growing tree suitable for plantation cultivation in tropical highlands. It tolerates both shade and poorly drained soils and is probably a nitrogen-fixing species, even though it is not a legume.

Description A large tree, often reaching over 2 m in diameter, it has thick, silvery-gray bark. It is well known for the beautiful yellow flowers it develops in autumn.

Distribution *Alnus nepalensis* is native to the Burmese hills, the Himalayas, to China's Yunnan, Szechuan, and Kweichow provinces, and to Indochina.

It has been planted extensively in the hills of northern India (West Bengal, Kashmir, and Himachal Pradesh) and Hawaii. In the hilly areas of Burma it is effectively used in the afforestation work in abandoned *taungya* areas, where the seed is broadcast during the last years of cultivation.

Use as Firewood The wood of this species in Nepal is very light (specific gravity, 0.32–0.37), with strength properties comparable to those of red alder (page 80). Like that of *Alnus acuminata* (page 76), the wood of this alder dries rapidly and burns evenly, but rather quickly. The trees coppice, but successful regrowth seems to depend on the season. (In Hawaii, where there are no pronounced seasons, the trees coppice well year-round.)

Yield *Alnus nepalensis* grows large and quickly. In Kashmir, specimens reaching 33 m and 80 cm in diameter have been regularly recorded, though in other parts of the Himalayas diameters of 30–50 cm are more typical. Wood samples indicate that diameters increase about 2 cm per year. In Hawaii, 26-year-old trees were 50 cm in diameter when felled for wood testing. The logs were of low grade, with knotty wood close to the bark surface, because the tree is shade tolerant and retains its lower branches for a long period.

Other Uses

- Wood. The wood is a fair utility timber for use in unexposed conditions.

Environmental Requirements

- Temperature. The trees grow in cool, moist, tropical highland climates.
- Altitude. 1,000–3,000 m in its native habitat; 300–1,800 m in Hawaii.
- Rainfall. While it prefers stream beds, *Alnus nepalensis* also grows well in drier locales. In Hawaii it grows well in areas with rainfall in excess of 500 mm.
- Soil. It seems probable that, like other alders, this species has the capacity to fix atmospheric nitrogen so that its growth is not wholly dependent on the presence of soil nitrogen. It can withstand imperfect drainage and flooding, but not continuous waterlogging. Like *Alnus acuminata*, it grows best in deep, well-drained loams or loamy soils of alluvial origin, but also grows on a wide range of soils from gravel and sand to clay. The availability of soil moisture is more limiting than soil type.

Establishment The plant is raised from seeds in a nursery and transplanted to the field.

- Seed treatment. None required. Direct sowing is effective.
- Ability to compete with weeds. A fast starter, it can even stay ahead of sugar cane rattoon crops (although just marginally).

Pests and Diseases The leaves are sometimes damaged and stripped from the tree by beetle larvae. The trunk is sometimes attacked by borers.

Limitations Highly susceptible to wind damage.

Related Species Species worth considering for firewood plantations along with alders described in this report include:

- *Alnus cordata*, a Mediterranean species
- *Alnus firma*
- *Alnus glutinosa*, the European or black alder
- *Alnus hirsuta*
- *Alnus incana*, a native of the boreal forests of northern Europe and North America.

Alnus glutinosa and *Alnus incana* are native to climates with severe cold and may perform

Alnus nepalensis, lopped in alternate years for village fuel supply. Chirgaon, Himachal Pradesh, India. (Forest Institute, Dehra Dun)

Alnus nepalensis, Kohalia Forest Reserve, Hawaii, USA. Trees in this stand have reached 63 cm diameter and stems 18 m long (to a top diameter of 23 cm) in 23 years. The site is 900 m above sea level, and is extremely wet with an almost continuous fog-like cloud cover. (G.D. Pickford, Hawaii Division of Forestry, retired)

well in tropical highlands where unseasonal cold might destroy the red alder.

- Hybrid alders are being developed in northern Europe, eastern United States, Canada, and Japan, using *Alnus rubra* as a parent.

Alnus rubra

Botanic Name *Alnus rubra* Bong.

Synonym *Alnus oregana* Nutt.

Common Name Red alder

Family Betulaceae

Main Attributes The red alder is coming to be recognized as a premium firewood in the northwestern regions of the United States. It grows fast, and—although it is not a legume—it does have root nodules containing microorganisms (*Actinomycetes* species) that fix atmospheric nitrogen. This provides the trees a source of nitrogenous fertilizer and helps improve the site for subsequent agricultural or forestry uses.

Description Red alder is a tree up to 40 m tall, with smooth, gray bark and trunk diameter up to about 1 m thick. It has a narrow pyramidal head and pendulous branches; the leaves are dark and smooth on top, whitish underneath, and are edged with irregular teeth. The trees are readily distinguished by their oval cones resembling tiny pine cones.

Distribution Red alder is native to a low-elevation coastal zone of northwestern North America from Alaska to central California, where it is the most common broadleaf tree. It is little known elsewhere, but it is planted for windbreaks and as an ornamental in Europe and has been introduced in New Zealand.

Use as Firewood In western Canada and the United States, red alder is being considered as a potential energy crop. It is used now for domestic firewood, but the new idea is to use it for fuel to generate electricity. The wood is moderately dense (average specific gravity, 0.39), and the density varies little with growth rate, age of tree, or site. Its heat content is about 4,600 kcal per kg, and it makes good charcoal. Young red alder trees coppice quite well; three coppice crops have been harvested from the same rootstock in one experiment.

Yield Red alder is one of the most productive trees of North America. In unmanaged, natural stands on well-stocked good sites, mean annual increment (total stem) may approach 10-11 m^3 per ha for the 20- to 30-year rotations needed to produce pulpwood and small sawlogs. Yields in plantations are expected to be substantially higher. For short coppice and pulp-log rotations, estimated total stem yields are 17-21 m^3 per ha per year.

Other Uses

- Wood. Red alder wood is moderately dense, with a fine, even texture that makes it easy to work. One of the most easily stained and finished woods in the world, it is popular for furniture and paneling, although it is a soft wood that scars easily and wears poorly. Both heartwood and sapwood have the same color, which simplifies grading and sorting.
- Pulp. Red alder pulps well and is widely accepted by pulp mills in the western United States. Alder pulp is usually blended with conifer pulps to provide smoothness and softness to tissue, bond, envelope, and book papers.
- Land reclamation. The qualities that make red alder one of the first plants to invade cleared forests, landslides, and old roadbeds make it suitable for cultivation to heal eroding or derelict land. It produces usable wood and adds nitrogen to the soil at the same time. (Estimates of the amount range from 40 to more than 300 kg per ha annually.) In the State of Washington the plant is being used to rehabilitate coal mine spoils.

Environmental Requirements The red alder is a pioneer species. It is intolerant of shade, makes full use of the growing season, and has a self-sufficient robustness that enables it to prosper in a wide variety of environments.

- Temperature. Throughout its native range mean January minimum temperatures are about 0° and July maximum temperatures average less than 25°C. Temperature extremes range from -20° to 45°C.
- Altitude. It is native to altitudes below 750 m.
- Rainfall. Red alder occurs naturally in a humid to superhumid climate. Precipitation ranges from 600 to 3,000 mm annually, occurring primarily as rain in winter.
- Soil. It grows on soils varying from gravel or sand to clay, but is commonly found on moist sites: swamps, bottomlands, and along streams. Best growth is on deep, well-drained loams.

Alnus rubra, straight-boled, 10-year-old plantation averaging more than 10 cm diameter and 10–12 m in height, near the Oregon coast, USA. (D. Bergstrom)

Nodules on *Alnus rubra.* Although it is not a legume, red alder can fix as much as 300 kg nitrogen per ha per year when effectively nodulated. (H.J. Evans)

Establishment

- Seed treatment. None required.
- Ability to compete with weeds. Very good.

Pests and Diseases Red alder has few serious pests and diseases in its native habitat.

Limitations Foresters have had little experience with managing red alder. It has been largely ignored or considered a weed. Even today it is not considered a commercial species. However, that attitude is changing, spurred by increasing fertilizer energy costs and a growing appreciation for the red alder's potential. Because of its aggressiveness and vigor, the red alder can invade plantations of slower growing trees, overtop them, and become a serious pest.

Green logs and lumber are very susceptible to decay. If split and air dried soon after cutting, however, alder firewood may be stored for long periods under cover.

Eucalyptus globulus

Botanic Name *Eucalyptus globulus* Labill.

Common Names Southern blue gum, Tasmanian blue gum, fever tree

Family Myrtaceae

Main Attributes *Eucalyptus globulus* is the most extensively planted eucalypt species in the world. It is easy to establish, grows fast, closes canopy early, has straight stems, and is wind firm. Plantations do not need protection from livestock or wildlife because its foliage is unpalatable. The wood burns freely, leaves little ash, and carbonizes easily to produce good charcoal, which is already used in many countries.

Description Under favorable conditions *Eucalyptus globulus* attains about 60 m in height (specimens in Spain and Portugal have reached 70 m and are among the tallest trees in Europe) and 2.3 m in diameter, with a straight trunk as long as two-thirds of its total height. It has a smooth, whitish-blue bark and handsome, dark-green, glossy leaves. In open stands the trees tend to be heavily branched.

Distribution *Eucalyptus globulus*, a native of the eastern part of Tasmania, is now growing in plantations covering a total of 800,000 ha in dozens of countries, many of them having very different climates from its original habitat. About half the world's plantation area is in Portugal and Spain, but the plant can be found in Italy, southern France, Algeria, South Africa, East Africa, Central Africa, Ethiopia, Peru, Ecuador, Colombia, Uruguay, California, and India. The major successes have been in mild temperate climates and in cool tropical highlands. Elsewhere it fails.

Use as Firewood The wood of *Eucalyptus globulus* is heavy (specific gravity, 0.8-1.0). The calorific value of the air-dry wood is about 4,800 kcal per kg. The wood has good burning qualities; in India *Eucalyptus globulus* is widely cultivated for fuelwood and charcoal. The trees coppice vigorously at least twice, with yields usually falling off in the third coppice.

Yield *Eucalyptus globulus* shows remarkable early growth in height on favorable sites (for example, 20 m in 4.5 years in Tanzania, and 15 m in 3 years and almost 30 m in 10 years in India). In Australia, on reasonably suitable sites, it may grow 1-2 m a year for the first 5-10 years. Annual wood production of 10-30 m^3 per ha has been recorded on sites in Spain, Portugal, Italy, Peru, and elsewhere. The trees are usually grown on rotations of 5-15 years.

Other Uses

- Wood. When dry, the yellowish-brown timber is like oak both in strength and grain. It is suitable for light and heavy construction, poles, piles, and tool handles. It is only moderately durable. Its oil content makes it resistant to termites; in Ethiopia it is used for telegraph and utility poles.
- Land reclamation. Its dense and widespread root system is important for erosion control.
- Pulp. The plant is one of the best eucalypts for pulp and paper making.
- Oil and honey. The commercially important eucalyptus oil is extracted from the leaves of this tree. The flowers produce good honey; in Portugal honey farms thrive near stands of *Eucalyptus globulus.*

Environmental Requirements

- Temperature. *Eucalyptus globulus* is native to a temperate climate without extremes of heat or cold. Drought and cold limit its use; climate is the major limitation to its even more widespread use.
- Altitude. The trees have been successfully planted at altitudes up to 3,000 m (in East Africa).
- Rainfall. In its native range, rainfall is from 800 to over 1,500 mm, well-distributed year-round.
- Soil. This species adapts well to a variety of well-drained soils. However, it grows best in deep sandy-clay soils and good-quality loams. Shallowness, poor drainage, and salinity are usually the principal factors limiting its cultivation.

Establishment The tree is easily established and reproduces from self-sown seed in natural forest stands.

- Seed treatment. None required.
- Ability to compete with weeds. It will not successfully compete with grass.

Pests and Diseases Although growth has been good in most areas, the trees can become affected by a number of insects and fungi.

Eucalyptus globulus, South-eastern New South Wales, Australia. (CSIRO Division of Forest Research)

Limitations The dead leaves and bark that fall from the trees are very inflammable, but after a fire trees will sprout vigorously from buds protected by the bark of the trunk and branches.

Related Species Similar species with promise for firewood are:

- *Eucalyptus maidenii*
- *Eucalyptus bicostata*
- *Eucalyptus pseudoglobulus*
- *Eucalyptus nitens*
- *Eucalyptus viminalis*
- *Eucalyptus gunnii*
- *Eucalyptus dalrympleana.*

One of the most frost-tolerant members of the genus is *Eucalyptus pauciflora* var. *alpina* (*Eucalyptus niphophila*). It is not very productive, but it should be tested at high elevation in tropical highlands.* *E. fastigiata* and *E. macarthurii* are also worthy of trial. They can stand frost to about -5°C, provided it is not prolonged. They are more versatile and faster growing than *E. pauciflora* var. *alpina.*

*Information from Tasmanian Forestry Commission. See Research Contacts.

Eucalyptus grandis

Botanic Name *Eucalyptus grandis* Hill ex Maiden

Synonym *Eucalyptus saligna* var. *pallidivalvis* Baker et Smith

Common Names Flooded gum, rose gum

Family Myrtaceae

Main Attributes *Eucalyptus grandis* is a fast-growing tree, adapted to a wide range of soil types. It is relatively disease free but somewhat frost tender. The wood makes excellent fuel.

Description *E. grandis* is a straight, white-trunked eucalypt occurring in pure or almost pure stands, frequently with a rain-forest understory. In virgin natural stands, trees may reach 50-60 m tall, with trunk diameter up to 2 m. The tree grown in African and Brazilian plantations has been selected from several generations of cultivated crops and is markedly superior to wild types in yield and stem straightness. (It may be a hybrid of *E. grandis* and *E. saligna*.)

Distribution The species occurs in the coastal areas of eastern Australia from near Newcastle in New South Wales (30°S) into southeastern Queensland. Isolated populations occur in north Queensland near Mackay and on the Atherton tablelands (17°S).

The species is widely cultivated in South America, the East African highlands, and South Africa as well as in numerous small plantings in other countries. It is so important in Brazil that huge plantations are being established with an annual planting program on the order of 100,000 ha.

Use as Firewood While the wood is only moderately dense (specific gravity, 0.40-0.55), the fast growth and high-volume yield make the species particularly promising for firewood. It was introduced to East Africa as a railroad fuel earlier in the century; plantations near large towns are still used for domestic (and occasionally for industrial) fuel.

It is common practice to regenerate *E. grandis* forests by coppice from the stumps. Most of them will shoot within 3 months. They are then thinned to the best two or three shoots per stump. In some areas (Florida and Australia, for example), coppicing can only be done in spring and early summer.

Yield Height growth of 2 m per year and diameter increases of 2-3 cm per year are common. At Muguga in Kenya (a rather dry site), 6-year-old trees attained a height of 14 m, and 10-year-old trees reached 24 m. At Lushoto in Tanzania, 10-year-old trees (density, 500 per ha) have reached 37 m in height and 32 cm in diameter.

With intensive cultivation and fertilization, growth rates of up to 7 m in the first year after planting have been achieved in northern New South Wales.

Annual production is approximately 40 m^3 per ha under irrigation in Zimbabwe, between 17 and 45 m^3 per ha for good sites in Uganda, and up to 35 m^3 per ha in South Africa.

There is evidence, particularly in Kenya, that the second rotation from coppice will outyield the seedling rotation; the initial crop averaged 178 m^3 per ha at 6 years, while subsequent coppice crops averaged 277 m^3 per ha for the same period.

The usual rotations in Kenya are 6 years for domestic fuelwood, 10-12 years for industrial fuelwood, and 7-8 years for telephone poles.

Other Uses

- Wood. The pale-red timber of *Eucalyptus grandis* is softer and lighter than many eucalypts. It is easily worked and is extensively used for medium-quality joinery in offices and hotels. Small trees are sawn to make shipping crates for fruit. Because the trees are very straight, they are ideal for use in telephone poles and power transmission posts. The species has great potential for paper-pulp production and is occasionally peeled for veneer. Trees felled for timber need special treatment to avoid serious losses from splitting when they dry out.

Environmental Requirements *E. grandis* shows marked provenance variation within its natural distribution. Selecting the right provenance can significantly increase production. Volume differences of up to 100 percent occurred in trials in northern New South Wales. Most overseas collections came from the Coffs Harbour area, but Brazil is now showing interest in the north Queensland populations, partly because of their apparent resistance to disease.

- Temperature. In its native habitat, summer temperatures occasionally reach 40°C,

Rhabat Khalifa, Morocco. *Eucalyptus grandis*, 20 years old, 37 m tall. (E.R. Berglund)

while winter minimum may be -1° to -3°C. In plantations the species can survive frosts if the temperature falls gradually over several weeks. Sudden freezing is very damaging.

- Altitude. In Australia, natural populations range from sea level to approximately 800 m. Near the equator, plantations have been established at high altitude–2,000 m in India and up to 2,700 in the West Rift area of Kenya (but *E. globulus* is probably more productive at altitudes over 2,400 m).
- Rainfall. Mean annual rainfall in the native habitat varies between 1,000 and 1,800 mm with a summer/autumn predominance and a spring dry period. However, *E. grandis* has been successfully established in areas where the rainfall is in excess of 2,500 mm, or as low as 600 mm if irrigated.
- Soil. *E. grandis* prefers moist, well-drained soils derived from a variety of parent materials such as shales, slates, sandstones, some granite, and occasionally basalt. In its natural habitat these soils are generally rather infertile and have low phosphorus content. The plant is sensitive to boron deficiency–a problem on old tropical soils. This is easily and cheaply cured, but it may not be recognized because its symptoms are those of drought stress.

Establishment Seedlings are usually raised for 3-5 months and are about 20 cm high, but in some instances younger seedlings (6-8 weeks) have been successfully planted in southern Africa. Fertilizer applied at or about the time of planting can have spectacular results. The general trend is to plant at the start of the wet season. Spacing varies from about 2 m x 2 m to 5 m x 5 m. As with other eucalypts, measures must be taken against termites when planting on savanna sites.

- Seed treatment. Usually none required.
- Ability to compete with weeds. Grass and herbaceous weeds can severely limit growth. Mechanical or chemical site preparation is essential if rapid rates of growth are to be achieved and maintained.

Pests and Diseases Most exotic plantations of the species are as yet free of any serious pests and diseases. In Brazil the fungus *Diaportha cubensis* attacks the species, and termites will attack the young trees. Root rot has been a serious problem in Zambia.

Limitations The main limitations are vulnerability to frost and fire. *E. grandis* does not have lignotubers and is very sensitive to fire.

Related Species Similar eucalypts worth testing for firewood are:

- *Eucalyptus saligna*, which is closely related; the two are often confused but are usually separated on the basis of location. *E. grandis* grows on bottomland in Australia and *E. saligna* on slopes. As already noted, the two will hybridize.
- A natural hybrid between *E. grandis* and *E. robusta*, known as *E. grandis* var. *grandiflora* Maiden, is also a potential firewood.

Grevillea robusta

Botanic Name *Grevillea robusta* A. Cunn.

Common Names Silk oak, silver oak, roble de seda

Family Proteaceae

Main Attributes This tree is valuable in plantation cultivation for timber, but if cultivated at close spacings or in non-forest situations it could also be valuable for firewood. Trials for this purpose should be attempted. It grows quickly for such a large tree and is successful under a wide range of climatic and soil conditions.

Description A handsome Australian tree growing to 35 m high, *Grevillea robusta* has an attractive shape, long and beautiful, dense racemes of golden flowers, and finely cut foliage. It is briefly deciduous in late winter–sometimes the new leaves develop as the old ones fall. Its leaves are like big fern fronds, green on the upper side, silver below. They are covered with silky gray hairs.

Distribution *Grevillea robusta* is native to subtropical coastal areas of New South Wales and Queensland and has been successfully cultivated for shade or timber in semiarid, temperate, and subtropical climates in India, Sri Lanka, Kenya, Mauritius, Zambia, Malawi, Zimbabwe, Tanzania, Uganda, South Africa, Hawaii, and Jamaica.

Use as Firewood The wood is tough, elastic, and moderately dense (specific gravity, 0.57). It is used for fuel in Sri Lanka. The trees coppice poorly, but can be pollarded and do reseed themselves readily.

Yield Fast growing, it reaches a height of 20 m in 15-20 years in sites with favorable soil and climate. Early height growth averages 2 m a year on good sites. Yield data from Tanzania report a 14-year-old plantation (for timber) with a mean dominant height of 19 m and mean girth of 80 cm; total volume, including thinnings, was 217 m^3 per ha.

Other Uses

- Wood. The pale-pink or brown heartwood resembles oak. It is handsomely grained, strong, durable, and excellent for cabinetwork. The timber is also used in making railroad ties, plywood paneling, air-freight cases, and furniture, as well as in parquetry and turnery. The timber is considered to have economic and export potential for several countries.
- Beautification. Because of its height, good form, and attractive flowers, it is often planted as a street tree.
- Honey. The golden flowers are attractive to bees, making this an important honey plant.
- Shade. It has been used to provide light shade over coffee and tea plantations.

Environmental Requirements

- Temperature. The plant prefers warm, temperate to subtropical temperatures, generally with mean annual temperatures of about 20°C. Mature trees can withstand occasional light frost (-10°C), but young plants are frost sensitive.
- Altitude. It grows at a wide range of altitudes, from sea level to above 2,300 m.
- Rainfall. Annual rainfall in its natural habitat is from 700 to over 1,500 mm, most of which falls during the summer; however, the species has been introduced in many areas with annual rainfall of only 400-600 mm, with 6-8 dry months. In addition, it also grows in areas having as much as 2,500 mm of rainfall annually.
- Soil. The tree grows well in many types of soil, including sandy soils, loams of medium fertility, and acid soils. As it tends to be a deep-rooted species, deep soils are generally preferred. It does not tolerate waterlogging.

Establishment Silk oak is easily propagated from the great quantities of seeds it produces (although the seed is difficult to collect because of the size of the trees) from about 10 years of age. It regenerates naturally. Under normal storage conditions, seeds remain viable for only a few months after collection, but seeds dried and stored in a refrigerator are reported to keep well for 2 years. Normal plantation cultivation is by transplanting nursery-grown seedlings after they reach a height of about 0.6 m. Cuttings are also successful.

- Seed treatment. None required.
- Ability to compete with weeds. Good.

Pests and Diseases Several countries have reported dieback or growth reduction after about 20 years, particularly on dry sites. Trees in Puerto Rico have become infested with scale insects.

Grevillea robusta, 5-year-old plantation near Bangalore, India. (K.A. Kushalappa, Forest Department, Bangalore, Karnataka)

Grevillea robusta, naturalized stand, Huehue Ranch, Island of Hawaii, USA (elevation 500 m, annual rainfall 1,500 mm). (R.G. Skolmen)

Limitations The tree seeds itself so readily that it can get out of control. In Hawaii, where it was introduced in 1870, it is classed as a pest because it has run wild and shades out good forage grasses. It is being eradicated with herbicides.

The trees have brittle branches and can be damaged by high winds.

Inga vera

Botanic Name *Inga vera* Willd.

Synonym *Inga inga* (L.) Britton

Common Names Guaba, guama (Cuba); guaba del pais, guaba nativa (Puerto Rico); guama (Dominican Republic); pois doux à paille (Guadeloupe, Martinique); pan chock, river koko (Jamaica)

Family Leguminosae (Papilionoideae)

Main Attributes *Inga vera* grows rapidly and is a common shade tree for coffee and cacao plantations in Puerto Rico and throughout the West Indies. Its cultivation and characteristics are fairly well known and its wood is useful as fuel. Other *Inga* species are worth testing as firewood crops as well.

Description A medium-sized evergreen tree, *Inga vera* grows as tall as 20 m, with a diameter of about 30-45 cm. It has a wide, spreading crown of long branches and thin foliage.

Distribution The plant is found under extensive cultivation, particularly on coffee and cacao plantations in the Virgin Islands, Puerto Rico, Guadeloupe, Martinique, Haiti, Cuba, and the Dominican Republic. Related *Inga* species are used for shade and fuel throughout Mexico, Central America, and northern South America. It is common along riverbanks and in sheltered ravines.

Use as Firewood The wood of *Inga vera* is moderately heavy (specific gravity, 0.57). It makes excellent fuel and is utilized for charcoal throughout the West Indies. The trees coppice well.

Yield This is a very fast-growing species. Its trunk diameter sometimes grows in excess of 2.5 cm per year.

Other Uses

- Wood. The timber can be used for utility furniture, boxes, crates, light construction, and general carpentry. However, it is highly susceptible to drywood termites and decays readily when in contact with the ground.
- Shade. As already mentioned, the species is widely used as a shade tree for plantation crops.
- Honey. It makes a good honey plant, for the flowers are rich in nectar and attract bees.
- Food. The seeds are enclosed in a sugary, edible pulp.

Environmental Requirements

- Temperature. The tree is native to the humid tropics.
- Altitude. *Inga vera* is a lowland species commonly found in coastal regions and nearby foothills.
- Rainfall. This is a tree of moist areas, but in Puerto Rico it is also common on the dry southern coast. Thus, it seems to have some drought tolerance.
- Soil. In its native habitat this species grows well on many soil types, even limestone soils.

Establishment *Inga vera* is easily grown from seed.

- Seed treatment. The seed should be processed soon after collection (to avoid fermentation of the pulp). The pods should be macerated and the seeds separated, using copious amounts of water. The seeds germinate rapidly, but are short-lived, especially if dried out.

Pests and Diseases Unreported.

Limitations When planted as a shade tree in coffee plantations, the lower branches are always cut off and the canopy becomes so broad and dense that the trees must be spaced 9-12 m apart so that the coffee plants will receive adequate light.

Related Species Other *Inga* species worth testing as firewood crops include *Inga edulis*, whose pods are so sweet they are known as ice cream beans.

Inga vera, near San Juan, Puerto Rico. (C. Rivera)

IV Fuelwood Species for Arid and Semiarid Regions

Dry regions face more difficult fuelwood problems than either the humid tropics or the tropical highlands. Even though sunlight, temperatures, and soils would usually permit exceptional growth, aridity so limits plant life that the natural biomass productivity is less than in almost any other ecosystem. Yet some 450 million people inhabit the low-rainfall areas of developing countries, and a large portion of the earth's biomass–0.6 billion hectares–is considered to be semiarid to arid because it receives less than 500 mm of annual precipitation. In addition, there are seasonally dry tropical regions that may receive more than 500 mm annual rainfall but suffer 6 or more completely rainless months.

These hot, dry regions fall mainly into two great subtropical belts: a northern one that takes in the Sahara and Sahel, Middle East, South Asia, and parts of North America, and a southern one that includes southern and eastern Africa, Australia, and the South American dry zones of Peru, Chile, Argentina, and Brazil. However, there are pockets of aridity in unexpected areas such as the Hawaiian islands and the West Indies where aridity is created by a local rain shadow effect or porous (often sand or cinders) soils that do not hold water.

The fuelwood species described in this section are suitable for cultivation in hot, arid sites: scrub, open woodland, grassland, even sand dunes in some cases. They have shown a capacity to survive sites where annual rainfall is 500 mm or less (often much less) and where rainfall is extremely variable. Many avoid drought stress because their deep root systems penetrate to subsoil moisture; others use wide-spreading root systems to gather sparse moisture; some have both root types. (*Acacia, Prosopis, Haloxylon,* and *Tamarix* species are particularly notable for deep roots.) Many of the species described in this section are adapted to the high salinity often found in the soil moisture in arid areas. Many have small leaf blades (e.g., some *Acacia* species) or needlelike leaves (e.g., *Tamarix* species) to reduce transpiration during drought; others use different physiological mechanisms to conserve moisture by slowing evaporation through the leaves. Some of the plants described are unpalatable to animals or are so thorny that they discourage browsing animals.

Clearing forests for fuel, farming, logging, and grazing, together with in-

Excessive harvesting of firewood often breaks down the natural cycle of vegetation regeneration. This is particularly true in dry areas. Without trees to bind and shade the soil, grass growth thins and fails to protect against erosion and desertification sets in. Former woodlands become bare, shifting sand that is extremely difficult to reclaim. (A. Hutchinson)

vasions of peoples, armies, and livestock have destroyed virtually all original vegetation in these fragile dry habitats. Excessive exploitation causes desertification, the perhaps irreversible deterioration of the ecosystem's productivity. The increasing demand for wood fuel is speeding up this desertification. Pastoralists and major cities are casting further and further afield for fuel as all areas around them become stripped of trees and shrubs. As many arid environments are incapable of coping with this perturbation, man-created deserts result.

The oil-poor nations in these dry zones should seriously consider reforestation of arid and semiarid lands for fuel. The species outlined in this section are examples of successful adaptations to this kind of environment. Others that should also be examined include:

Acacia auriculiformis	page 34	*Guazuma ulmifolia*	page 48
Casuarina equisetifolia	page 38	*Leucaena leucocephala*	page 50
Eucalyptus grandis	page 84	*Terminalia catappa*	page 66

Acacia brachystachya

Botanic Name *Acacia brachystachya* Benth.

Common Names Umbrella mulga, turpentine mulga

Family Leguminosae (Mimosoideae)

Main Attributes Native to a vast area of arid and semiarid Australia, *Acacia brachystachya* is not well known elsewhere. It is considered a superior firewood species.

Description The umbrella mulga is a bushy shrub growing up to 7 m in height. It branches from just above ground level into a spreading crown.

Distribution *Acacia brachystachya* is native to almost the entire interior of the Australian continent (from southwest Queensland to central South Australia and across to the central coast of Western Australia).

Use as Firewood Like all members of the *Acacia* genus, this species also makes a good fuelwood. Its wood is hard and heavy.

Yield Unreported.

Other Uses

- Wood. The wood is durable and is used for the manufacture of small ornamental objects.
- Fodder. The foliage is sometimes eaten by stock and can serve as fodder in emergencies. It is not as palatable as the foliage of some other *Acacia* species.

Environmental Requirements

- Temperature. In its native habitat, temperatures range from 4°C to an average high of 58°C in the sun. But extremes as low as 0°C to as high as 69°C in the sun sometimes occur.
- Altitude. *Acacia brachystachya* is found near sea level to about 600 m in its native habitat.
- Rainfall. Where the species is native, annual rainfall is generally 200-300 mm. However, it is also found where rainfall is as much as 500 mm per year, though even here the threat of drought is ever present.
- Soil. *Acacia brachystachya* will grow on a variety of soils, including lateritic soils, friable clays, loams, and alluvium.

Establishment

- Seed treatment. Seed-coat dormancy can normally be broken by treatment with boiling water.
- Ability to compete with weeds. Unreported.

Pests and Diseases Unreported.

Limitations Seed production is low and irregular.

Related Species There are a number of tropical Australian *Acacia* species worth testing as firewood crops* including:

- *Acacia mangium*, a very fast-growing species found on the edge of rainforests. In cultivation it has reached 15 m in 3 years, with a diameter of 40 cm. It is also considered to be a good-quality timber tree.
- *Acacia lysiophloia*, more of a rangy shrub than a tree, grows in semiarid areas (rainfall 250-500 mm) on very poor soils. It has a sticky leaf, which makes it unpalatable to stock. It is sought out by stockmen as wood for their campfires.
- *Acacia holosericea*, a small tree growing in 500-1,000 mm rainfall, often in poor soils. It is very fast growing, often maturing in 2 years.

*Information supplied by R. Reid. See Research Contacts.

Acacia brachystachya, southern Queensland, Australia. (CSIRO Division of Forest Research)

Acacia cambagei

Botanic Name *Acacia cambagei* R. T. Bak.

Common Names Gidgee, gidyea, gidya, stinking wattle

Family Leguminosae (Mimosoideae)

Main Attributes *Acacia cambagei* is one of the most productive of the *Acacia* species that will grow in extremely harsh, arid environments.

Description A handsome, gray-foliaged tree up to 10 m tall, *Acacia cambagei* has a wide-spreading crown. However, under harsh conditions it grows with a sparse, open-canopied habit.

Distribution The tree is widespread in the arid and semiarid areas of inland, temperate, subtropical, and tropical Australia. It occurs in dense groves, particularly in areas of western Queensland and northern New South Wales. Small-scale planting trials have been established in India, North Africa, and the Persian Gulf.

Use as Firewood The wood is extremely heavy (specific gravity, 1.3). It makes excellent firewood and burns green or dry, with an intense heat, to a thick soft-white or pale-gray ash that may be up to 90 percent calcium oxide. The heat of burning is so intense that it is usual to mix other species with it; otherwise firebars will buckle.

Yield Unreported.

Other Uses

- Wood. The wood is one of the hardest and heaviest timbers in the world. Its grain is close and interlocked. It is noted for durability and termite resistance and is therefore widely used for fence posts. It also makes good picture frames and walking sticks.

Environmental Requirements

- Temperature. In its native habitat summer temperatures frequently top 34°C.
- Altitude. 75-500 m.
- Rainfall. In Australia, it grows in a region having only about 125-500 mm of annual precipitation, falling mainly in the summer, and with severe droughts, averaging 5 months long.
- Soil. It occurs on a wide range of soils. The most common types are clay loams with a thin sandy layer on top, but gidgee is also found on heavy-textured alluvials, heavy red and brown loams, stony ridges, and deep, fine sands. There is usually moderately good surface drainage.

Establishment There has been virtually no investigation of gidgee's reproductive requirements.

- Seed treatment. The seeds germinate rapidly without pretreatment and the use of boiling water to hasten germination can be harmful.
- Ability to compete with weeds. Unreported.

Pests and Diseases Unreported.

Limitations An unpleasant feature of this species is the offensive smell of the leaves (actually flat leafstalk phyllodes), especially during wet weather. It should not be planted near homes. Charcoal from gidgee wood has a high ash content (6-7 percent).

Acacia cambagei, north-western New South Wales, Australia. (CSIRO Division of Forest Research)

Acacia cyclops

Botanic Name *Acacia cyclops* A. Cunn. ex G. Don

Synonym *Acacia cyclopis*

Common Name Rooikrans (South Africa)

Family Leguminosae (Mimosoideae)

Main Attributes *Acacia cyclops* can grow in very dry areas with an annual precipitation of less than 300 mm. Because it tolerates salt spray, wind, sand-blast, or salinity, it is useful for stabilization of coastal dunes. It produces a dense, high-quality firewood.

Description It is a dense, evergreen, bushy shrub, often multistemmed, or grows as a small tree from 3 m up to 8 m, with a rounded crown. In windy coastal sites it forms a hedge less than 0.5 m high. The foliage comprises light green phyllodes, varnished when young, and growing in a downward vertical position. Pods mature in summer and are not shed, but remain on the tree, exposing their seeds to predators and dispersers.

Distribution Native to southwestern Australia, where it grows mostly on coastal sand dunes. Used for stabilization in South Africa, it is spreading on sand and sandstone into coastal bush and heathland.

Use as a Firewood A dense wood, the logs rarely exceeding 20 cm in diameter, it is a very popular firewood in South Africa and is sold regularly in Cape Town.

Yield Each tree yields about 12 kg dry mass at 10 cm basal diameter and about 60 kg dry mass at 15 cm basal diameter when growing in a sheltered site. A harvestable size may be reached in around 7-10 years. Near the coast, and in stressful conditions, *A. cyclops* remains stunted and hedgelike.

Other Uses

- Fodder. Goats and antelope browse the phyllodes. The seeds and their oily funicles are eaten by birds, primates, and rodents, and if crushed, might be suitable for cattle.

Environmental Requirements This species has a high light demand; it will not survive in deep shade.

- Temperature. Monthly means within the distribution range of this species vary from 5°C in winter to 31°C in summer. Slightly resistant to frost.
- Altitude. The species is generally found below 300 m in altitude.
- Rainfall. From 200 to 800 mm per annum.
- Soil. It grows on quartzitic or calcareous sand or limestone. It also is found in drier sites such as dune crests.

Establishment Direct sowing of pretreated seed. This species rarely coppices, and mature trees do no survive felling.

- Seed treatment. Abrasion, acid, and hot-water treatment are effective.
- Ability to compete with weeds. Unreported.

Pests and Diseases These are not an important factor in South Africa; in fact, the lack of seed destroyers is partly responsible for the weediness of the species.

Limitations This is an extremely weedy species spread by birds into indigenous vegetation. Once established over large areas, it is difficult to remove or replace. There is little vegetation cover beneath an *Acacia cyclops* thicket. The seeds remain viable in the soil for many years.

It is relatively slow growing. The pods are nondeciduous and are therefore not easily gathered. Unlike many *Acacia* species, it is not considered a valuable tannin or gum producer.

Acacia cyclops, Cape Province, South Africa. (P. Reyneke)

Acacia nilotica

Botanic Name *Acacia nilotica* (L.) Willd. ex Del.

Synonyms *Acacia arabica* auct. mult., non Willdenow; *Acacia arabica* var. *indica* Benth.; *Acacia nilotica* var. *indica* Hill.

Common Names Egyptian thorn; red-heat, kudupod, "sweet smell," babul (India); kiker, babar (Pakistan); lekkerruikpeul, ruikperul, sunt (Arabic)

Family Leguminosae (Mimosoideae)

Main Attributes In parts of tropical Africa as well as India and Pakistan, *Acacia nilotica* forests are extremely valuable sources of fuel, small timber, fodder, tannin, and honey. The plant is exceedingly drought tolerant and survives on many difficult sites. In India, it is one of the most important species in social and farm forestry.

Description *Acacia nilotica* is a moderate-sized tree that grows up to 20 m, but this is attenuated by site: it is usually no more than 10 m high and in very unfavorable locations is only a shrub. It has a flattish or umbrella-shaped crown and is easily identified by its bright-yellow, sweet-scented flower heads, its sweet-smelling gray or black pods, and its paired whitish spines at the base of each leaf. During the hot season the tree is in full leaf and its feathery foliage provides good shade.

Several subspecies of this plant are recognized. The two most widely grown for fuel are:

- *Acacia nilotica* subsp. *indica* (Benth.) Brenan; and
- *Acacia nilotica* subsp. *nilotica.*

Distribution Subspecies *indica* is native to the Sind, Punjab, and Deccan areas of Pakistan and India. It is now regenerated and cultivated throughout India and Pakistan, where there are some extensive, nearly pure forests (1,700 ha in East Khardesh, 5,000 ha in Poona, 6,000 ha in Sind, 5,000 ha in Punjab, and 36,000 ha in Hyderabad Division, for example).

Subspecies *nilotica* is native to Africa and is widely planted there—for example, along the Blue Nile in the Sudan, in the bushveld of Natal and Transvaal, and in Zambia and Botswana.

The species has been introduced to the West Indies as an ornamental and has become naturalized on many islands.

Use as Firewood The wood is a very popular fuel on the Indian subcontinent and large quantities are consumed as firewood and charcoal. It has also been used extensively to fuel locomotives and river steamers, and it powers the boilers of some small industries as well. The calorific value of sapwood is 4,800 kcal per kg, while that of heartwood is 4,950 kcal per kg. The word is heavy (specific gravity, 0.67–0.68) and the trees coppice occasionally.

Yield This is a fast-growing tree under favorable soil irrigation. In cultivation for industrial fuel in the Sudan, rotation varies from 20 to 30 years. In India and Pakistan, it is generally harvested on a 20-year rotation. The trees add about 2–3 cm in diameter each year.

Other Uses

- Wood. The hard, tough wood is resistant to termites, impervious to water, and is popular for railroad ties (sleepers), tool handles, carts, and oars. It is an attractive wood, good for carving and turnery, and is still used for boatbuilding, as it was in ancient Egypt. It is one of the best mining timbers in Pakistan.
- Fodder. The leaves and pods are widely used as fodder and, in arid regions of India, constitute the chief diet for goats and sheep. Pods contain as much as 15 percent crude protein.
- Tannin. The bark and pods are widely used in the leather industry; their tannin content varies from 12 to 20 percent.
- Gum. *Acacia nilotica* is probably the earliest commercial source of gum arabic, though this valuable commodity now comes mainly from *Acacia senegal.* The gum is still used in the manufacture of matches, inks, paints, and confectionery.

Environmental Requirements

- Temperature. *Acacia nilotica* trees withstand extremes in temperature, but are frost tender when young.
- Altitude. The tree will grow at elevations up to 500 m in the Himalayas.
- Rainfall. In general, the various subspecies can survive in very arid sites, but thrive under irrigation. In contrast, stands of subspecies *nilotica* in the Sudan and Pakistan are inundated with floodwaters for several

Acacia nilotica, Wad Medani, Sudan. The Blue Nile floods this area for several months each year. *A. nilotica* is a rugged species well adapted to periodic flooding followed by extended droughts. (P.J. Wood)

months each year and are restricted to such sites.

• Soil. It grows on a variety of soils, even poor ones. It prefers alluvium, but grows well on heavy, black cotton and clay soils as well.

Establishment *Acacia nilotica* is generally propagated by seed and rarely by seedlings. Direct seeding is common practice, but the resulting stands can be disappointing because of browsing animals, inadequate soil moisture, or weed competition.

• Seed treatment. Fresh seeds can be planted directly, but seeds that have been stored must be immersed in boiling water and soaked. In rural India and Pakistan, the pods are fed to goats and the scarified seeds are either recovered from the dung, or the goats are confined in the area that is to be reseeded.

• Ability to compete with weeds. Young seedlings require full sun and frequent weeding.

Pests and Diseases The trees are sometimes affected by wood borers in Africa, India, and Pakistan. Bruchid beetles can seriously attack seed in the pods. Pathogenic fungi are also known.

Limitations The trees are extremely thorny and could become a major problem. Thus, this species should be introduced only to those arid areas where the need for firewood is absolutely critical.

Acacia saligna

Botanic Name *Acacia saligna* (Labill.) H. Wendl.

Synonym *Acacia cyanophylla* Lindl.

Common Names Golden wreath wattle, orange wattle, blue-leafed wattle

Family Leguminosae (Mimosoideae)

Main Attributes *Acacia saligna* is an extremely rugged tree and has proved widely adaptable to barren slopes, derelict land, and exceptionally arid conditions in Australia and North Africa. It grows rapidly and tolerates drought. It is used for reclaiming eroded hillsides and wastelands and for stabilizing drift sands as well as for fuel.

Description The plant is a dense, bushy shrub, usually between 2 and 5 m tall, which can also grow treelike up to 8 m tall with a single main stem (diameter up to 30 cm). In spring its usually drooping branches are clad in beautiful and abundant yellow flowers.

Distribution *Acacia saligna* is native to the southwestern corner of western Australia. It was introduced to South Africa in the 1840s in an attempt to stabilize the shifting sand dunes. It has also been planted in Uruguay, Mexico, Israel, Iran, Iraq, Jordan, Syria, Greece, Cyprus, and North African countries.

Use as Firewood Plantations for fuel have been established in some Mediterranean countries. But, according to one report from South Africa, the wood is "sappy, light, and not a popular fuelwood." The plant can withstand some shade and can be grown as an understory beneath pines or eucalypts in energy plantations or village fuel and fodder areas.

Yield *Acacia saligna* grows quickly, often reaching up to 8 m tall with a spread as great as its height in just 4 or 5 years. In very dry situations, growth rate is slower. Annual yields vary from 1.5 to 10 m^3 per ha, depending on site. In the Mediterranean countries, the fuelwood from this species is harvested on a coppice rotation system of 5-10 years.

Other Uses

- Sand-dune fixation. This is one of the best woody species for binding moving sand.
- Windbreaks. It is useful for windbreaks, amenity plantings, beautification projects, and roadside stabilization in semiarid regions.
- Fodder. The leaves (actually phyllodes) are palatable to livestock when fresh or dried into hay. They are especially used as a supplementary feed for sheep and goats. Crushed seeds have been fed to sheep without ill effects. Regrowth of established bushes is so good that *A. saligna* can be completely grazed off without harming the plants.
- Gum. The damaged bark exudes copious amounts of a very acidic gum that seems to show exceptional promise for use in pickles and other acidic foodstuffs.

Environmental Requirements *Acacia saligna* can grow throughout the tropical and the warm temperate regions of the world.

- Temperature. In its native habitat, the summer temperature ranges from about 23° to 36°C. Winter temperatures are 4°-9°C. The plant does not withstand frost and grows best where the winter and summer means are between 13° and 30°C respectively.
- Altitude. From near sea level to about 300 m, with isolated occurrences at higher elevations.
- Rainfall. *Acacia saligna* is particularly drought hardy. It grows where annual rainfall is as low as 250 mm, though it probably prefers the more semiarid regions (annual rainfall 350-600 mm). It is also known to grow well where annual rainfall is as high as 1,000 mm.
- Soil. This acacia grows mainly on sandy, coastal plains, but it is found in a wide variety of environments from swampy sites and riverbanks to small, rocky hills (often granitic) and the slopes of the coastal ranges. It occurs on many soil types, especially poor acid or calcareous sands. It will grow under the most dry and adverse soil conditions and in moderately heavy clays and a range of podzols. The plant tolerates salt spray, soil salinity, and alkalinity.

Establishment Seeds germinate readily; young plants can often be found under mature trees in the hundreds. Seedlings are easily raised in a nursery and established in the field. This species develops root suckers and coppices freely.

- Seed treatment. The seeds are normally

Tripoli, Libya. *Acacia saligna*, only 5 months after its planting, in a grid system, on sand dunes. (FAO photo)

Sand dunes stabilized by *Acacia saligna* plantation at Salamis Forest, Cyprus. (Ministry of Agriculture and Natural Resources, Nicosia)

treated with boiling water, but nicking the seed coat, soaking in sulfuric acid, and exposing the seeds to dry heat are also effective.

- Ability to compete with weeds. Remarkable.

Pests and Diseases *Acacia saligna* supports a diverse and abundant range of herbivores that cause damage to the plant.

Limitations Because of its hardiness and profuse reproductive abilities, *Acacia saligna* has become a serious menace in parts of South Africa by invading and displacing indigenous vegetation. It infests water courses (which may decrease the water available for purposes such as irrigation) and has proved very difficult to eradicate.

Acacia senegal

Botanic Name *Acacia senegal* (L.) Willd.

Synonym *Acacia verek* Guill. & Perr.

Common Names Gum acacia, hashab (Arabic), gum arabic tree

Family Leguminosae (Mimosoideae)

Main Attributes Although not fast growing, *Acacia senegal* produces excellent fuelwood, and it is so robust that it is often the only woody species to survive in dry areas. *Acacia senegal* is the source of gum arabic, the bark exudate that has been used commercially for at least 4,000 years. The tree can survive the most adverse conditions–hot, dry wind and sandstorms on the poorest soils of rock and sand. This species is ideal for reclamation of refractory sites and shifting sand dunes.

Description *Acacia senegal* is a small thorny tree that grows as high as 13 m, although it is commonly little more than a shrub. The species has many geographical races. The pattern of variation between them is not yet well understood, but all are likely to be excellent fuelwood sources. Some of them, however, don't produce gum. The natural life of the tree is usually 25-30 years. The prickles typically grow in threes at each node–the central one recurved, the two laterals (sometimes absent) upward. Flowers are in whitish spikes.

Distribution The tree is characteristic of the drier parts of the southern Sahara and is found throughout the Sahelian zone from Senegal to Somalia. It is an important forest resource of the Republic of Sudan, which supplies about 80 percent of the world's gum arabic. The principal plantations are in the Sudan and Senegal, where the local farmers harvest the crop for supplementary income. In addition, it is now cultivated in India, Pakistan, and Nigeria.

Use as Firewood In Senegal and Mauritania the hard and heavy wood of *Acacia senegal* is considered the best firewood and is the main species for charcoal. Villagers are willing to pay up to 5 percent more per donkey load for *Acacia senegal* wood than for that of other species such as *Acacia tortilis* or *Leptadenia pyrotechnica*.* One "head load" of wood (about 22.5 kg) is found to equal 72,000 kcal. Firewood cutters often girdle trees and let the gum run and the branches dry for a month before the wood is carried to the kitchens. The trees coppice well. In Upper Volta fuelwood plantations have been established around the provincial capital of Dori, and more extensive plantations are planned for cultivation around small settlements and wells.

Yield Average production of wood from natural forests is about 5 m^3 per ha. It has been reported that the yield of wood from gum plantations where the trees are widely spaced is fairly low, with annual increments of about 0.5-1.0 m^3 per ha. But if the trees are grown at a density suitable for firewood production, the yield should be much greater.

Other Uses

- Wood. *Acacia senegal* wood is used for poles and agricultural implements. The root fibers are utilized for rope and fishnets and for lining wells.
- Gum. Gum arabic is used in foods and beverages, in pharmaceutical preparations, in confectionery, and in a wide range of industrial applications.†
- Fodder. *Acacia senegal* foliage and pods are rich in protein and are an important feed during the rainy season and early dry season for camels, sheep, and goats throughout the tree's native range.
- Seeds. Seeds are dried and preserved for human consumption as vegetables.
- Erosion control and soil rehabilitation. *Acacia senegal* is a nitrogen-fixing species that probably increases soil fertility. It is highly suitable for agroforestry systems and is already widely grown in combination with watermelon, millets, forage grasses, and other crops. It can be used in desertification control to reestablish a vegetation cover in degraded areas as well as for sand-dune fixation and wind-erosion control.

Environmental Requirements

- Temperature. Where the tree grows in the Sudan, average temperatures are between 14° and 43°C; in India, temperatures are between -4° and 48°C.

*Hammer. 1977.

†See National Academy of Sciences, 1979, for details.

Acacia senegal, Sudan. (G. Sarlin, Centre Technique Forestier Tropical)

• Altitude. It grows from 100 to 1,700 m in East Africa.

• Rainfall. *Acacia senegal* is very drought resistant. It can grow under subdesert conditions where annual rainfall is as low as 200 mm, with 8-11 dry months in the year, but prefers 300-450 mm. It will also survive with annual rainfall as high as 800 mm.

• Soil. The tree will grow naturally in sand; it also grows in clay, except where the rainfall is high enough to cause waterlogging (800 mm or more).

Establishment *Acacia senegal* is easily raised from seed. Natural regeneration in existing stands is often nonexistent; the species does, however, regenerate easily in fallow lands and some degraded soils, often from stump coppicing.

• Seed treatment. Overnight soaking is effective.

• Ability to compete with weeds. The seedlings need weeding for the first 2 years.

Pests and Diseases Pods are affected by insects, which can severely affect the viability of seed, and the roots are susceptible to attack by termites during extreme droughts. Mature trees are remarkably resistant to insect attack, but seedlings are susceptible.

Limitations Young plants may be severely damaged by goats, sheep, and other animals.

This species may form thorny thickets and could become a serious pest. Both Australia and South Africa have policies to eradicate and prevent further introduction of "this noxious thorny weed."

Acacia seyal

Botanic Name *Acacia seyal* Del.

Synonym *Acacia fistula* Schweinf.

Common Name Talh. The wood is known as shittim wood.

Family Leguminosae (Mimosoideae)

Main Attributes *Acacia seyal* is a resilient, drought-tolerant tree of African drylands that resists forest and grass fires. Its wood is dense and highly prized for firewood in parts of the world where few other plants survive. Its foliage provides good animal feed.

Description A typical acacia of the African savannas, *Acacia seyal* is a shrub or small tree, growing up to 12 m high, often flat-topped and sometimes branching from near the base. It has sharp thorns, which are usually straight and paired, yellow fragrant flowers, feathery leaves, and curved pods with constrictions between the seeds. Its rough bark is cream to greenish-yellow (var. *fistula*) or dark gray, reddish brown, or black, and it flakes off to expose a powdery undersurface.

Distribution *Acacia seyal* is native to the Sahelian zone from Senegal to Sudan. It is also found in Egypt and in eastern and southern Africa from Somalia to Mozambique and Namibia.

Distinct varieties occur in different regions:

- *Acacia seyal* var. *seyal*. Northern tropical Africa and Egypt.
- *Acacia seyal* var. *fistula*. Eastern Africa from Sudan to Mozambique.

Use as Firewood The tree is considered to produce the best firewood in Chad and provides practically all the fuelwood brought into the capital, N'Djaména (Fort-Lamy). It is used in the Sudan to make a fragrant fire over which women perfume themselves.

Yield Unreported.

Other Uses

- Wood. The timber is hard, shock resistant, and seasons reasonably well. It has been used for centuries; ancient Egyptians made coffins from it, some of which still exist.
- Forage. Over much of Africa the leaves, pods, and flowers of *Acacia seyal* are a major source of early dry-season fodder for sheep and goats. It is considered the best fodder plant in northern Nigeria and in Sahelian savannas, where it grows in quantity. In the dry season in the western Sudan, Fulani drive their cattle to the districts where it grows.
- Gum. Although darker in color and generally inferior to that of *Acacia senegal*, the gum exuded from the wounded bark of *Acacia seyal* has some economic value and is exported to India and Europe. It is edible when fresh, though it has a slightly acid taste. With systematic tapping, the yield and color improve greatly, but the trees have not yet been deliberately cultivated for gum.

Environmental Requirements

- Temperature. Hot.
- Altitude. It is found in elevations up to 2,100 m in tropical areas, but it is generally a lowland species found at lower elevations than *Acacia senegal*.
- Rainfall. *Acacia seyal* usually is found in the drier woodland or grassland savannas (350 mm annual rainfall and above). It can, however, withstand inundation better than other acacias and is sometimes found along riverbanks.
- Soil. The tree is often found on stony ground. It seems to thrive on most soil types, even heavy clay where it forms pure stands (var. *fistula* withstands waterlogging better than var. *seyal*). The plant probably fixes atmospheric nitrogen through bacteria-filled nodules on its roots.

Establishment Planting is by seed, but truncheons (large cuttings) are reputed to strike roots readily in moist soil.

- Seed treatment. Like most acacia seeds, those of *Acacia seyal* require scarification for good germination. For a small number of seeds, the simplest method is to nick the seedcoat; for larger quantities, the seeds can be dipped briefly in boiling water or soaked in concentrated sulfuric acid.
- Ability to compete with weeds. Unreported.

Pests and Diseases *Acacia seyal* is reported resistant to attack by insects, but felled logs may be severely damaged by wood borers.

Limitations The trees are thorny and rather slow growing.

Acacia seyal, Upper Volta. This natural stand is a source of firewood, poles, and forage. (H. J. von Maydell)

Felleta cattle grazing in a pure stand of *Acacia seyal* on clay soils in Darfur Province, Republic of the Sudan. (G.E. Wickens)

Acacia tortilis

Botanic Name *Acacia tortilis* (Forsk.) Hayne

Synonyms *Acacia raddiana* Savi, *Acacia spirocarpa* Hochst. ex A. Rich., *Acacia heteracantha* Burch.

Common Names Umbrella thorn (Africa), haaken-steekdoring (South Africa), Israeli babool (India), sayal or samor (Egypt and Sudan), seyal (Arabic)

Family Leguminosae (Mimosoideae)

Main Attributes The umbrella thorn is a distinctive acacia that supplies fuel to much of arid and semiarid Africa and the Middle East. It is a drought-resistant species that can be established in a wide array of habitats under extremely arid conditions.

Description *Acacia tortilis* is a medium-sized tree (4–15 m tall), sometimes with several trunks that spray upwards and outwards, fountain-like, that support a flat-topped umbrella of feathery foliage. Under extreme aridity it becomes a small shrub, often barely 1 m tall. Under heavy grazing it is frequently reduced to a number of trailing, seemingly unconnected branches radiating from a low sand mound. Its thorns are a distinguishing feature; there are two kinds–long, straight, and white, and small, brownish, and hooked. The fragrant white flowers are borne singly or in clusters. Pods are contorted or spiraled like a coil spring.

Distribution Four distinct subspecies are known in different ecological zones:

- Subspecies *tortilis* — Sahel, Middle East
- Subspecies *raddiana* — Sudan, Middle East, Sahel
- Subspecies *spirocarpa* — Eastern Africa, Sudan
- Subspecies *heteracantha* — Southern Africa

Because *Acacia tortilis* (subspecies *raddiana*) has grown so well when introduced to Jodhpur, India (from Israel), fuelwood plantations are now being established elsewhere in Rajasthan, as well as in other parts of India (for instance, Haryana, Gujarat, Tamil Nadu, and Andhra Pradesh). The different subspecies seem to have different ecological tolerances, which is important to consider when choosing a subspecies for plantations.

Use as Firewood The dense, red heartwood of this species has high calorific value (4,400 kcal per kg) and makes superior firewood and charcoal. It is one of the main firewoods (and charcoals) used in Khartoum, Sudan, for example. The plant coppices well, so there is no need to replant trees after every harvest.

Yield *Acacia tortilis* is fast growing for an arid-zone plant. In comparison-plant trials at the Central Arid Zone Research Institute, Jodhpur, India, it was noted that *Acacia tortilis* grew twice as fast as indigenous acacias and that the plant withstood arid conditions better than *Acacia nilotica*, *Acacia senegal*, and *Prosopis cineraria*. A 12-year-old plantation (3 m x 3 m spacing) yielded 54 t of fuel per ha.

Other Uses

- Wood. Wood from *Acacia tortilis* is used for fence posts and for manufacturing small implements and articles.
- Fodder. Pods are produced prolifically. They fall to the ground and are devoured by wild herbivores and goats, sheep, and other domestic livestock. They provide sustenance for wildlife in East Africa's national parks and have a 19-percent protein content. The foliage is also palatable. It is, for example, the major dry-season (9 months) fodder for sheep and goats in the whole Sahara-Sahelian belt in the Sudan. The thorny branches are used to pen cattle, goats, and sheep.
- Sand stabilization. *Acacia tortilis* is excellent for sand stabilization. In Rajasthan, it has stabilized over 800 ha of shifting sand dunes. It has also been used on a limited scale for sand dune stabilization in the Sudan.

Environmental Requirements

- Temperature. This species grows well in hot, arid climates with maximum temperatures as high as 50°C; the subspecies *raddiana* grows where minimum temperatures are close to 0°C. Plants less than 2 years old are easily damaged by frost and require protection.
- Altitude. *Acacia tortilis* is best adapted to the lowlands.
- Rainfall. *Acacia tortilis* thrives where rainfall is up to 1,000 mm. However, depending on the subspecies, it is also extremely drought resistant and can survive in climates with less

Acacia tortilis subspecies *raddiana*, Negev Desert, Israel. (M. Evenari)

than 100 mm annual rainfall and long, erratic dry seasons.

• Soil. The tree favors alkaline soils. It grows fairly well in shallow soil, less than 0.25 m deep, though it develops long lateral roots that can become a nuisance in nearby fields, paths, and roadways. In shallow soil, the plants remain shrubby and must be widely spaced to allow for their lateral root growth.

Establishment The umbrella thorn is easily raised from seed and the seedlings can be established in plantations with little loss.

• Seed treatment. The seeds are dipped in hot water and soaked overnight to ensure quick and uniform germination. Seeds can also be treated with concentrated sulfuric acid.

• Ability to compete with weeds. Seedlings require initial weeding to facilitate faster growth.

Pests and Diseases Seed production is often severely reduced by insects (bruchids). Trees are susceptible to attack by caterpillars, beetles, and blight diseases that infest other Mimosoideae in an area. Wild herbivores graze new shoots and young seedlings.

Limitations Thorniness.

Related Species Other African acacias that can compete in drought tolerance on sandy soils are *Acacia nubica* and *Acacia ehrenbergiana.*

Adhatoda vasica

Botanic Name *Adhatoda vasica* Nees

Synonym *Justicia adhatoda* L.

Common names Vasaka, adhatoda, adusa, Malabar nut tree, and many others

Family Acanthaceae

Main Attributes *Adhatoda vasica* thrives where other vegetation fails, because it is not browsed by goats or other animals. It can be cultivated in gardens and as a hedge plant.

Description It is a lush, evergreen, many-branched shrub that usually grows to a height of 2.5 m, but can reach 6 m. The large leaves, bright green on the upper surface and pale below, have an unpleasant odor when crushed. The flowers are white or purple and are arranged in dense, leafy, clustered spikes. The fruit is a four-seeded capsule that explodes to disseminate the seeds.

Distribution The species is found in waste places throughout the plains and submontane tracts of India. It also occurs in Sri Lanka, Malaysia, and Burma. It is grown experimentally in southern Florida. Introduced into Curaçao in 1969, it flourishes on that desert island with only early watering until well established. It has been occasionally grown as an ornamental in Cuba.

Use as Firewood *Adhatoda* is a particularly desirable wood for quick, intense, long-lasting, clean cooking fire. It burns with little or no smoke, odor, or sparks. The wood is moderately hard and is used in the manufacture of gunpowder charcoal.

Yield Unreported, but it is a fast-growing plant and coppices well.

Other Uses

- Drugs. The leaves and roots are well-known drugs in the Ayurvedic and Unani systems of medicine and are recommended for a variety of ailments such as colds, bronchitis, asthma, fever, and jaundice. European physicians in former times have successfully employed the dried leaves in treating typhoid and diphtheria.
- Green manure. The leaves are used as green manure in paddy fields because of their potassium nitrate content.
- Insecticides and fungicides. An infusion of leaves is used for destroying termites, flies, mosquitoes, and other noxious insects. They contain the alkaloid vasicine, which is toxic to cold-blooded creatures (including fish) but not to mammals. The leaves are used in packing or storing immature fruits to speed ripening and development of natural color, as well as to inhibit fungi and repel insects.
- Dye. When boiled, the leaves give a durable yellow color used for dyeing coarse cloth and skins. They are also employed in a preparation applied to pottery before firing to impart a special black hue.
- Herbicide. The leaves "prevent the growth of lower aquatics and check the development of parasitic vegetation."*

Environmental Requirements

- Temperature. The shrub requires a tropical or subtropical climate; it is killed to the ground by brief periods of frost but recovers rapidly.
- Altitude. It grows well up to 1,300 m in the Himalayas.
- Rainfall. The plant tolerates a wide range of precipitation. Curaçao averages 500 mm of rain annually; southeast Florida averages 1,650 mm.
- Soil. The shrub is found on riverbanks, dry slopes, forest margins and disturbed areas, and in dry and moist deciduous forests. The soil in which the shrub grows well in Curaçao is weathered diabase; in southern Florida, oolitic limestone.

Establishment *Adhatoda vasica* can be propagated by seeds or cuttings.

- Seed treatment. None required.
- Ability to compete with weeds. No data. Florida seedlings were greenhouse grown or grown in the field under clean cultivation. Having herbicidal action, the plant should be a good weed-competitor.

Pests and Diseases None observed in Florida or Curaçao and none reported in other areas.

Limitations Since it is unpalatable to livestock and has the ability to colonize waste places and disturbed areas, this plant may become a pest when introduced to new areas.

*Information supplied by C. K. Atal. See Research Contacts.

Adhatoda vasica, southern Florida. (P.K. Soderholm)

Albizia lebbek

Botanic Name *Albizia lebbek* (L.) Benth.

Synonyms *Mimosa lebbeck* L., *Mimosa sirissa* Roxb. (the genus name is sometimes spelled *Albizzia*)

Common Names Lebbek, karana, East Indian walnut, siris-tree, kokko, frywood, woman's tongue tree, acacia amarilla

Family Leguminosae (Mimosoideae)

Main Attributes This is a robust, adaptable tree that produces small timber and useful fuel, fixes nitrogen through nodules on its roots, and is good for reforesting dry, alkaline soils.

Description *Albizia lebbek* is a handsome, deciduous tree with a spreading umbrella of feathery foliage. It may reach 30 m tall and a diameter of 1 m. It produces white flower heads with striking green stamens. Its long, straw-colored pods rustle in the breeze. It is completely leafless for part of the year.

Distribution *Albizia lebbek* is one of the best-known trees of India. It is native also to Bangladesh, Burma, and Pakistan and has been cultivated in tropical and subtropical regions in North Africa, the West Indies, South America, and Southeast Asia. There are extensive plantations in Nepal and in Central and South India.

Use as Firewood The wood is dense (specific gravity, 0.55-0.6) and makes good fuel. The calorific value of moisture-free heartwood is 5,200 kcal per kg. The trees coppice fairly well.

Yield Although some other *Albizia* species (e.g., *A. falcataria*) are among the fastest-growing trees measured, *Albizia lebbek* grows only moderately fast. In India, if it is managed in rotations of 10-15 years or more, it produces about 5 m^3 per ha annually.

Other Uses

- Wood. The heartwood is finding increasing use in furniture and as construction timber for houses. While not easy to work by hand, it carves and polishes well and resembles rosewood.
- Beautification. In India, this attractive tree is often planted for shade in gardens and along roadsides and in tea, cardamom, and coffee plantations. It is highly regarded by beekeepers for the light-colored honey its nectar provides.
- Fodder. The young foliage contains about 20 percent protein and is fed to livestock. One tree, it is alleged, may provide 20 percent of a water buffalo's annual feed, or 27 percent of a cow's. The leaves also make useful green manure.
- Erosion control. The tree is a good soil binder and therefore is often planted along embankments.

Environmental Requirements

- Temperature. After the first year the trees can tolerate light frost and drought.
- Altitude. From sea level to 1,600 m in India.
- Rainfall. In its native tropical and subtropical grasslands, annual rainfall varies from 500 to 2,000 mm and the summers are wet.
- Soil. The tree thrives on a variety of soils, though it prefers moist conditions and grows best on well-drained loam. The trees tolerate salt spray and grow well near the seashore.

Establishment *Albizia lebbek* produces seed prolifically and is easily propagated by seed—even by direct sowing. (Seeds remain viable for 4 or 5 years at ambient temperatures.) It can also be propagated by stem cuttings or root-shoot cuttings. It sprouts root suckers vigorously when its roots are injured.

- Seed treatment. Seeds are immersed in boiling water and allowed to cool and soak for 24 hours. Germination is 50-90 percent.
- Ability to compete with weeds. Seedlings need to be regularly weeded during their first 2 years of growth.

Pests and Diseases Wildlife and livestock relish the foliage and must be kept away. In India, some fungus diseases attack the leaves and pods.

Limitations Seedlings can be destroyed or severely damaged by browsing animals. The tree is not wind-firm, as its roots are close to the surface. The lightweight pods create much litter.

Albizia lebbek, south Florida, USA. (J. Morton)

Anogeissus latifolia

Botanic Name *Anogeissus latifolia* Wall.

Synonym *Conocarpus latifolia* Roxb.

Common Names Axle-wood tree, bakli, dhau, dhawra, dhausa (Hindi)

Family Combretaceae

Main Attributes This rugged tree produces fine fuelwood and provides other valuable products. It is therefore very important to the economics of semiarid areas of India. It can be produced on dry, rocky hills, ravines, and denuded wastelands that are too dry for normal agriculture.

Description On deep soil the tree is large—sometimes attaining a bole 15 m long and 1 m in diameter—but on rocky slopes it remains stunted and rounded. It is conspicuous by its hard, shiny, gray bark and its leaves that turn red in the autumn.

Distribution The tree is common to most of the dry, deciduous forests of India and the open grasslands of Sri Lanka. It is not formally cultivated; natural regeneration supplies the current demands for its wood. The trees have not been tested outside their native habitat.

Use as Firewood Throughout India, the purplish-brown wood of *Anogeissus latifolia* is harvested from natural stands near urban areas and sold as firewood. It is very dense (specific gravity, 0.9) and has a calorific value of about 4,900 kcal per kg (ash content, 4 percent). It also makes excellent charcoal. The trees coppice well and are fire hardy.

Yield Cut stems have about four rings per cm of radius.

Other Uses

- Wood. The wood is hard and polishes well and is used for poles, rafters, farm implements, and especially for the shafts and axles of carts.
- Gum. The trees produce a light-colored and valuable gum (known as gum gatty or Indian gum) that is twice as viscous as gum arabic. It is used in India and is exported to Europe and North America for use in pharmaceuticals and calico printing.
- Tannin. The bark and leaves are rich in tannin (up to 19 percent), which is used in the Indian leather industry.
- Dye. The leaves yield a black dye that is used commercially.
- Fodder. The foliage is fed to animals.
- Silk. Increasingly, raw silk comes from tasar silkworms and *Anogeissus latifolia* is one of several tropical plants on whose foliage they feed.
- Pulp. The wood's potential for paper pulp has recently been recognized.*

Environmental Requirements

- Temperature. Tropical and subtropical.
- Altitude. *Anogeissus latifolia* is found up to 1,300 m elevation throughout much of mountainous India.
- Rainfall. This species grows in the dry, tropical forests. Some Rajasthan forests where it grows receive 600 mm, but they have long periods of water deficit because of irregular rainfall and high evapotranspiration (generally exceeding the precipitation). In this area the monsoon failed each year from 1965-1969, causing the trees to die back, but they have since resprouted with healthy vigor.†
- Soil. These trees are usually found in dry sandy or rocky soils.

Establishment The trees are easily propagated by seed. In India seedlings are planted on staggered contour ridges and also in pits filled with weathered soil.

- Seed treatment. None required.
- Ability to compete with weeds. Unreported.

Pests and Diseases These are usually not a problem except during extreme drought when the trees are stressed, lose vitality, and can become infected with leaf spot and stump-rot fungi, as well as insects.

Limitations Although it will produce fuel in uncommonly dry and difficult sites, this species grows more slowly than many of the other species in this book.

Related Species Other *Anogeissus* species worth testing for firewood are *A. leiocarpus* and *A pendula*, which often grow in association with *A. latifolia* on the subcontinent.

*Guha et al. 1974.

†Verma. 1972.

Anogeissus latifolia, Kerala, India (P.M. Ganapathy)

Anogeissus leiocarpus, Upper Volta. (H.J. von Maydell)

Azadirachta indica

Botanic Name *Azadirachta indica* A. Juss.

Synonyms *Melia indica* Brand., *Melia azadirachta* L.

Common Names Neem, nim

Family Meliaceae

Main Attributes Neem is potentially one of the most valuable of all arid-zone trees. It can grow in arid and nutrient-deficient soil and is a fast-growing source of fuelwood. Moreover, it has many commercially exploitable by-products and environmentally beneficial attributes. Although in Asia neem leaves are often used as fodder, the ecotype in West Africa is ignored by cattle, sheep, and even goats, which makes it easy to establish.

Description Neem is a deep-rooted, medium-sized tree, broadleafed and usually evergreen, except in periods of extreme drought. It has a short bole with wide spreading branches forming a rounded or oval crown. It has moderately thick, gray bark and its reddish heartwood is hard and durable.

Distribution Native to the dry forest areas of India, Pakistan, Sri Lanka, Malaya, Indonesia, Thailand, and Burma, neem has been widely cultivated in the arid regions of India and Africa. It thrives in the dry areas of the tropics and subtropics. It has grown well in plantations in the Sudan and Sahelian zones of Africa as well as in Sierra Leone, Malawi, Zimbabwe, Tanzania, Zanzibar, and the non-Sahelian areas of Guinea, Nigeria, and Ghana.

Use as Firewood Neem has long been used for fuel in India and Africa. It has become the most important plantation species in northern Nigeria and is planted for fuelwood and poles around the large towns. The calorific value of its wood is reputed to be only slightly less than that of Enugu coal in Nigeria. The wood is relatively heavy with specific gravity varying from 0.56 to 0.85 (average, 0.68). The tree coppices freely and early growth from coppice is faster than growth from seedlings.

Yield The rate of development of young neem plants after the first season is fairly rapid. As a rule, the trees put on an average annual girth increment of 2.3–3.0 cm, though more rapid growth is easily attained. In 4 different test plots, tree height varied from 4 to 7 m after the first 3 years and from 5 to 11 m in 8-year-old stands.

In West Africa, cropping is usually done on an 8-year rotation, with the original spacing between the plantation trees most commonly 2.4 m x 2.4 m. In Ghana, first rotation yield was 30–38 cords (108–137 m^3) of fuelwood per ha, and in Samaru (northern Nigeria), the yield of 8-year-old neem ranged from 19 to 169 m^3 per ha.*

Other Uses

- Wood. Neem is related to mahogany and neem wood resembles Cuban mahogany. It resists decay and insects and is tougher than teak. It is excellent for construction and furniture making. Neem poles are straight and strong, and seldom attacked by termites.
- Energy. Neem seeds contain up to 40 percent oil, which is used as fuel for lamps and as a lubricant for machinery. The pulp that surrounds the seeds is reputedly a promising substrate for generating methane gas.
- Windbreak, shade. Neem has been used successfully as a windbreak and as a source of shade for cattle. It is a splendid street tree for the arid tropics. In the Sudan it is a common avenue tree.
- Soil improvement. Leaves and twigs have been successfully used as a mulch and fertilizer in Sri Lanka, India, and Burma. Neem cake (the residue left after extracting oil from the seeds) is reportedly an excellent fertilizer, several times richer in plant nutrients than manure. Neem has successfully reclaimed arid wastelands, particularly in India.
- Industrial chemicals. Neem bark contains 12–14 percent tannins. In recent trials in India, these compared favorably with conventional tanning chemicals. Neem oil is a useful ingredient in soaps and disinfectants and in pharmaceuticals and cosmetics.
- Insect repellent. The seeds and leaves yield azadirachtin, a compound that appears to be a promising new insect repellent. It is a systemic pesticide that is absorbed into a plant and works from within. Japanese beetles and many other insect pests (even the desert locust) will starve before they will eat plants treated

*Gravsholt. 1967.

Azadirachta indica, native specimen. Northeast Thailand. (Sa-ard Boonkird)

Fifty-year-old neem shade tree in the main street of Maiduguri, northeastern Nigeria. (R. W. Fishwick)

Coppice shoots, 2 years old, 10 m tall. (R. W. Fishwick)

Kano, Nigeria. Ten-year-old neem plantation. (R.W. Fishwick)

Neem and *Acacia albida* trees surround the village of Kongoussi, Upper Volta. (E. Ernst)

with extracts of neem seed. (Azadirachtin also appears to be a nematode repellent.)

Environmental Requirements

- Temperature. Neem survives great heat—even shade temperatures up to 44°C. Trees in Dade County, Florida, have occasionally withstood temperatures near 0°C or lower.
- Altitude. It will grow at altitudes from 50 to 1,500 m.
- Rainfall. The neem tree is most successful in arid tropical and subtropical zones having a mean average annual rainfall of 450-1,150 mm. However, it will tolerate as little rainfall as 130 mm per year. It is useful even where the rainfall is over 500 mm, for it can tolerate long dry seasons.
- Soil. The tree is undemanding and grows well on most soils, including dry, stony, clay, and shallow soils. It will not grow on seasonally waterlogged soils or in deep dry sands where the dry-season water table lies below 18 m. The roots seem to have an unusual ability to extract nutrients and moisture from even highly leached, sandy soils. The optimum pH is 6.2 or above, although neem will grow well at pH 5, bringing surface soil to neutral pH by its leaf litter. It does not grow well on saline soils.

Establishment Although neem can be easily raised in a nursery and transplanted (as pot plants, or seedlings) early in the rains of the second season, direct sowings of fresh seed in the shelter of existing vegetation have also proved successful, though the initial growth is usually slower. In the field, the plants establish an extensive root system before aerial growth becomes rapid.

Neem Firewood. Above: Accra Plains, Ghana. (E.S. Ayensu) Right: Kano, Nigeria, 4-year-old coppice growth in background. (R.W. Fishwick)

In northern Nigeria, neem interplanted among groundnuts, beans, and millets in farmlands showed markedly superior growth. When the crop was harvested, a healthy stand of neem seedlings was left behind.

Seed bearing begins at about 5 years. Seed production and natural regeneration are profuse.

- Seed treatment. None required.
- Ability to compete with weeds. Neem is intolerant of grass competition and needs thorough weeding, especially in dry areas.

Pests and Diseases Neem has few serious pests. Occasional infestations by *Microtermes* and *Lorantus* species of insects have been recorded in Nigeria, but the attacked trees almost invariably recover. Plantations of neem in West Africa are strikingly insect free, evidently due to the tree's insect-repellent constituents. However, in imperfectly drained soils the taproot tends to rot and the trees gradually die off.

Limitations Neem seeds do not retain their viability very long and have to be sown within 2 or 3 weeks after harvest. Although neem needs light, young seedlings are liable to suffer from strong direct insolation and a light shade is desirable during the first season of growth. The seedlings are killed by frost and fire.

If neem is grown among other crops, it needs careful control, for it may aggressively invade neighboring plants.

Growth and utilization of neem are limited by lack of knowledge about the tree and its specific climatic requirements. A number of strains (provenances) will have to be tested to achieve the forms best adapted to local conditions.

Cajanus cajan

Botanic Name *Cajanus cajan* (L.) Millsp.

Synonym *Cajanus indicus* Spreng.

Common Names Pigeon pea, congo pea, red gram, gandul

Family Leguminosae (Papilionoideae)

Main Attributes Pigeon pea, a food crop with tall woody stalks, has seldom been considered a producer of firewood. Nevertheless, it offers the promise of a crop that within 3-9 months produces both food and fuel for family use. Its cultivation is already well known. In India, about 2.3 million ha are devoted to growing this crop. The stalks are an important by-product for the rural home because they make excellent firewood for the family's daily needs. Information gathered from the farmers indicates that the value of the stalks is roughly equivalent to that of the grain.

The plant is adapted to lands normally unsuited for other crops because of infertility, aridity, or topography. It is one of the best nitrogen-fixing legumes and costs little to produce. In 1978 it was selected as one of the most promising "new" crops for the United States.*

Description The pigeon pea is a woody shrub that can grow as tall as 3.6 m. There are many diverse types that vary in shape (for example, tall, open, and upright, or dwarf, compact, and bushy), growth period, and in the color, shape, and size of pods and seeds. Many cultivars have been selected or bred for high seed yield, but there has been no consideration of their relative qualities for use as firewood. The plant's nitrogen-fixing capacity is reportedly excellent.

Distribution The pigeon pea's origin is not well known, but the plant is probably native to northeastern Africa. It was cultivated in ancient Egypt and has been used widely in Africa and Southeast Asia since prehistoric times. It reached the Americas and Pacific in colonial times. Today, India is responsible for over 90 percent of the world's production, but the crop is also popular in the West Indies and throughout the tropics at a longitude between 30°N and 30°S.

*Theisen, Knox, and Mann. 1978.

Use as Firewood The spindly stalks are extensively used as fuel for cooking in Indian villages. In the past the stalks were employed for making charcoal used in the production of gunpowder. Generally, the thick main stem is used for firewood and the thin straight branches are used for thatch and basket making. The plant is also used for firewood in Chirazulu District, Malawi, an area with a particularly severe shortage of fuelwood.†

Yield On average, 2 t of woody stalks are obtained per ha per growing season.‡ The plants mature and produce seeds in 100-300 days, depending on cultivar, location, and time of sowing, but they are perennials and can be cultivated as such. When cut off at ground level, the plants do not resprout, but regrowth is satisfactory if the plants are cut at heights above 0.15 m. The crop is planted very densely with at least 30,000 plants per ha.

Other Uses

- Food. The pigeon pea is, of course, normally grown as a pulse crop. The dry seeds contain about 22 percent protein and are an important protein food in many tropical areas. Also, the green seeds and the immature pods are often eaten as fresh vegetables.
- Forage. The pods, husks, and foliage can be used for feeding animals. The plant has also been cultivated for feeding silkworms and the lac insects from which shellac is obtained.
- Amenity planting. The pigeon pea very rapidly produces dense ground cover that protects soil from erosion. It is sometimes planted in double rows as a windbreak and makes a hedge that also provides food and fuel.

Environmental Requirements

A range of cultivars is available that adapt the pigeon pea to many different environments. Only tall-growing types should be planted for fuelwood. The following details are based on experience growing the pigeon pea as a food crop.

- Temperature. The crop is cultivated in areas with average temperatures as high as 35°C, but the most favorable growing temperatures

†Information supplied by J. E. M. Arnold. See Research Contacts.

‡Information supplied by D. Sharma. See Research Contacts.

Tall pigeon pea varieties in experiments at the International Crops Research Institute for the Semi-Arid Tropics, Hyderabad, India. (D. Sharma) See also pictures pages 8 and 13.

seem to be between 18° and 29°C. It is killed by frost.

- Altitude. Various cultivars of pigeon pea are grown from sea level to high altitude (up to 3,000 m in Venezuela, for example). The plant does not thrive in seashore areas subject to salt spray.
- Rainfall. Average annual rainfall between 600 and 1,000 mm is most suitable. However, the pigeon pea can be grown in humid areas (even over 2,500 mm annual rainfall) and is renowned for its drought tolerance. Indeed, it has been selected as one of the most promising food crops for the semiarid tropics. It gives economic yields of seeds in areas where rainfall averages about 400 mm annually.
- Soil. Although it cannot withstand waterlogging, the pigeon pea can be grown in a wide range of soils. Its deep taproot and extensive lateral root system allow it to tolerate low fertility and low moisture. (The upright types, which are probably best for firewood, have the deepest roots.) It thrives in light sandy soils, but grows best in neutral deep loams. Some cultivars tolerate problem soils with excess salt, soluble aluminum, or manganese. Inoculation is not needed to get good nodulation in most sites.

Establishment The crop is established from seed sown directly in the field. It can be intermixed among other crops.

- Seed treatment. None needed. Fresh seed germinates well (85-95 percent). In humid regions the seed may lose viability after 4 months.
- Ability to compete with weeds. Requires weeding during the first 4-8 weeks.

Pests and Diseases Many troublesome insect pests and diseases (wilt and rust, particularly) are known to attack the succulent foliage. Local extension services should have details.

Limitations Pigeon pea seeds can bear fungi and must be treated with fungicide before shipment.

The plants initially grow slowly, although a small amount of nitrogen fertilizer boosts early growth. The crop cannot be produced in shaded sites.

When cultivated commercially as a pulse crop, the pigeon pea is grown as an annual or biennial because productivity declines after the first year. When grown for forage or green manure it is usually maintained no more than 5 years. The plant will die in about 10-12 years.

Cassia siamea

Botanic Name *Cassia siamea* Lam.

Common Names Yellow cassia, minjri, muong, angkanh, kassof-tree, Bombay blackwood, cassia (French)

Family Leguminosae (Caesalpinioideae)

Main Attributes *Cassia siamea* has long been cultivated for firewood on plantations in tropical areas. It is inexpensive to establish if seeded directly into the plantation site. The trees are fairly resistant to termites, grow rapidly in full sunlight, and may be harvested for fuel within a few years.

Description This medium to large ornamental is an evergreen with a dense crown of foliage and smooth, gray bark. It bears large, attractive bunches of yellow flowers. Its long pods hang in clusters.

Distribution *Cassia siamea* is native to Southeast Asia from Indonesia to Sri Lanka. It has been introduced to the West Indies, Central America, Florida, East and West Africa, and southern Africa. Formerly, *Cassia siamea* was the most widely planted plantation species in Africa, particularly from 1910 to 1924. It was much planted in Ghana, western Nigeria, Zambia, Tanzania, and Uganda for both fuel and poles.

Use as Firewood Its dense, dark-colored wood (specific gravity, 0.6–0.8) makes excellent fuel, although it is "smoky." The trees coppice readily and continue yielding well for four or five rotations. The plant has become naturalized on the Accra plains of Ghana and now provides employment for many woodsmen and a continuous supply of firewood for the area.

Yield Although the tree is not large, it grows fast and produces much small-sized wood suitable for fuel. It can attain heights of 5 m in 3 years and 15 m in 10 years (diameter, 15 cm). Annual production can be as much as 15 m^3 per ha. The plantations are usually harvested every 7 years, though the rotation may be as short as 5 years on favorable sites or as long as 10 years in dry climates.

Other Uses

- Wood. The heartwood makes an attractive timber that is used for cabinetmaking.
- Revegetation. The plant serves as dense windbreaks with virtually no undergrowth. It is also useful for reforesting denuded hills and cut-over areas, and in northern Nigeria it has been used to reclaim abandoned tin-mining sites. In India it is used as the host for sandalwood, *Santalum album*, a parasitic tree.

Environmental Requirements

- Temperature. The species cannot withstand cold but thrives in tropical heat.
- Altitude. It is generally a lowland species.
- Rainfall. *Cassia siamea* grows in a wide range of climates: humid, subhumid, dry, and arid. However, it is most prevalent in monsoonal areas where annual rainfall is 1,000 mm or more and the dry season lasts 4 or 5 months. In drier areas (those with 500–700 mm annual rainfall), the tree will grow after its second or third year only if its roots have access to deep soil moisture. In the Sudan, for example, it is grown along riverbanks and canals and in irrigated plantations.
- Soil. The plant grows best in deep, well-drained, relatively rich soils. It can tolerate soils containing laterite and limestone, provided drainage is not impaired.

Establishment *Cassia siamea* grows easily, and plantations can be established by direct seeding.

- Seed treatment. Seeds require no treatment if they are fresh. Older seed must be scarified—hot water or concentrated sulfuric acid is usually employed.
- Ability to compete with weeds. Weeding is necessary for the first year or two. In the Sudan, the tree does not grow well in areas with *Imperata cylindrica*, although in Malaysia it is reported to grow up through *Imperata* grass and kill it.

Pests and Diseases The trees are susceptible to attack by scale insects.

Limitations The young trees are browsed by livestock and wildlife and must be protected.

The yield of wood from *Cassia siamea* is usually not as good as from *Gmelina arborea* or some *Eucalyptus* species, so that while it is still widely used for firewood, it is being replaced in some forestry programs.

The seeds, pods, and foliage of *Cassia siamea*

Cassia siamea, Malanga, Angola (R.J. Poynton)

are highly toxic to pigs, which must be kept out of the plantations because they relish them (cattle and sheep are apparently not affected).

Wood may contain a yellow powder that is highly irritant to the skin.

Related Species

- *Cassia marginata.* A vigorously coppicing, fast-growing species that is proving popular as a firewood source in Haiti.
- *Cassia spectabilis.* A fast-growing, coppicing species native to Central America and northern South America.

Colophospermum mopane

Botanic Name *Colophospermum mopane* (Kirk ex Benth.) Kirk ex J. Leonard

Synonym *Copaifera mopane* Kirk ex Benth.

Common Names Mopane, balsam tree, butterfly tree, turpentine tree, Rhodesian ironwood; mopani (India)

Family Leguminosae (Caesalpinioideae)

Main Attributes Claimed to produce "the best firewood in Africa," mopane grows in areas with such poor soil quality and with such bad drainage that they are unable to support much else, including useful grasses.

Description Mopane can be a moderately tall tree (up to 23 m high) with a deeply fissured dark-gray trunk up to 1 m in diameter. But where it is heavily browsed it becomes a rounded shrub. Mopane's kidney-shaped leaves come in pairs and look almost like butterflies. They tend to hang folded and straight down, and thus produce little shade.

Distribution Mopane is a tree of lowland areas, especially river valleys, and of central and southern Africa. In Mozambique, Zimbabwe, Zambia, Botswana, Angola, and Namibia, it forms pure stands called "mopane woodland." In 1965 it was introduced to the Rajasthan area of India, and has proved to be one of the few exotic species to regenerate from self-sown seeds under arid Indian conditions.

Use as Firewood The dark-red to blackish wood is very hard and heavy (specific gravity, more than 1.0). The heartwood burns slowly and quietly and gives off great heat. A log left on the fire in the evening may still be burning the following morning. After bush fires, fallen mopane trees are often found smoldering for weeks. The residue of fine white ash is negligible. The trees coppice vigorously after fire, frost, or felling.

Yield In the arid climate of Rajasthan when grown in sandy soils, mopane grew slowly, reaching a height of about 5 m in 10 years.

Other Uses

• Wood. The wood is durable. Mopane poles are used in hut construction and for mining props, fence posts, piles for bridge construction, and for railroad ties. It is also used to make carved ornamental articles and parquet.

• Fodder. Mopane leaves are scented and are readily consumed by livestock and wildlife. The flavor does not taint milk or meat, even if the animals have fed on them exclusively. The leaves retain high food value even after they have fallen. Mopane could be a prime protein source for game farming in regions where cattle cannot survive because of tsetse fly or long droughts.

Environmental Requirements

• Temperature. Mopane is native to hot, tropical climates. It withstands high summer temperatures that frequently surpass 36°C. However, it withstands only mild frosts; heavy frosts may cause the branches to die back.

• Altitude. Grows best below 900 m.

• Rainfall. Grows well where annual rainfall is 200–450 mm; sometimes found where annual rainfall is above 800 mm. Mopane survives in the Kaokoro area of Namibia where rainfall is a meager 125 mm.

• Soil. Mopane generally grows in shallow, compacted, clay, alkaline, and badly drained soils. But it grows best in rich, deep, alluvial soils. The plant has remarkable ability to tolerate dry, saline sites. Although it is leguminous, the mopane does not seem to nodulate in the wild. It is important for stabilizing dry, alkaline soils.

Establishment The plant reproduces well by seedlings. Young plants produce root suckers as well.

• Seed treatment. None required.

• Ability to compete with weeds. Except under very dry conditions, mopane competes poorly with other species and its seedlings may be suppressed by grasses.

Pests and Diseases In Africa, mopane trees are often defoliated by the "mopane worm," a hairless, black caterpillar. (Dried or roasted, the caterpillar has a mild turpentine-like flavor and is a protein-rich delicacy for some rural Africans.)

Also closely associated with the mopane tree in Africa is the "mopane bee," a small and irritating insect that occurs in large numbers and

Colophospermum mopane, Jodhpur, India. (H.S. Mann)

always makes straight for the moisture around one's eyes and nose.

Limitations Mopane is a very difficult tree to fell because of the wood's extreme hardness. Felling by axe is a slow process and even a lightweight chainsaw is defeated. It becomes even harder and more difficult to cut when the wood dries out.

The species is not cultivated in commercial plantings, so there is little information on its propagation, protection, or yield.

Emblica officinalis

Botanic Name *Emblica officinalis* Gaertn.

Synonym *Phyllanthus emblica* L.

Common Names Emblic, Indian gooseberry, aonla, mirobalano, nelli, amla, Malacca tree

Family Euphorbiaceae

Main Attributes This is a food and medicinal crop much appreciated in tropical Asia. When virgin land is cleared, this is the one wild tree always left standing because its fruit is so esteemed for relieving thirst. In Thailand, buses stop in the countryside to allow passengers to dismount and pick the fruit. The tree has potential as an important firewood crop.

Description *Emblica officinalis* is a medium-sized, deciduous tree growing to a height of 25–33 m. It has slender branchlets, feathery foliage, clusters of small, almost inconspicuous, greenish-yellow flowers, and round, hard, pale-green or red acid fruits.

Distribution This tree is found both wild and cultivated throughout much of tropical Asia: central and southern India, Sri Lanka, Burma, Malaysia, and China. It is often cultivated in gardens and home yards for its fruit. Individual trees have been cultivated in Florida, Hawaii, the West Indies, Cuba, Puerto Rico, Trinidad, and Panama. In Rajasthan, India, the species is raised for afforestation purposes.

Use as Firewood The red, close-grained wood serves as a good fuel and makes excellent charcoal. It is dense (specific gravity, 0.7-0.8), burns with a calorific value of about 5,200 kcal per kg, and has an ash content of about 2 percent. The stems coppice easily when the trees are cut back.

Yield It is a fast-growing tree, producing 2.7 rings per 5 cm of diameter. Seedlings that went through an unusually cold winter grew to nearly 3 m in just 2 years in Florida.

Other Uses

• Wood. The wood from *Emblica officinalis* is durable even under water, and is used for agricultural implements and for lining wells.

• Food. The acid-tasting fruit of this species is one of the richest known natural sources of vitamin C. Two of its fruits 2.5-4.5 cm wide, weighing a total of 100 g, contain nearly 12-15 times the amount found in an orange. The fruits are also rich in pectin. They are used fresh, dried, or pickled in cooking, preserves, relishes, and candies. Healthy trees 15 years old in India produce 200 kg of fruits.

• Fodder. The foliage and fruits are relished by livestock.

• Green manure. In India, the branches are lopped for green manure, particularly for neutralizing excessive soil alkalinity in betel nut and cardamom plantations.

Environmental Requirements

• Temperature. The trees are best adapted to a climate with a distinct winter and summer. They are frost sensitive when young (up to 3 years) but withstand freezing temperatures when mature. At the other extreme, mature trees tolerate temperatures as high as 46°C, while young plants must be protected from such heat.

• Altitude. The tree grows wild from sea level to 1,800 m elevation in India.

• Rainfall. The emblic is native to humid climates, but has grown well on dry sites in the West Indies.

• Soil. The plant can be cultivated in alkaline and poor soils where most fruit crops fail. However, it prefers a deep, moist loam soil, and flourishes on alluvial soil. Hard pan at a depth of 1 m may halt growth after 10-12 years.

Establishment *Emblica officinalis* is propagated by both budding, cuttings, and inarching. Seedlings do not come true to type. The fruits are small and of inferior quality. Budded trees begin to fruit after 6 years.

• Seed treatment. Treatment with hot water (80°C) for 5 minutes ensures 80 percent germination in about 10 days. The seeds are not viable for long.

• Ability to compete with weeds. It competes fairly well.

Pests and Diseases The trees are attacked by a leaf rust, ring rust, and fruit rot in India, but not in Florida. In India, the pest *Bestonsa stylophora* sometimes attacks the trees. The most serious enemy is the bark-eating caterpillar (*Inderbella* sp.), which can be controlled by twice-a-year spraying with insecticide. Seedlings can be destroyed by insects, rats, or squirrels.

Emblica officinalis, Lucknow, India. (G.S. Srivastava)

Emblica officinalis, Mirzapur District, Uttar Pradesh, India. (J. and P. Hubley)

Limitations The tree requires honey-bee pollination to avoid heavy flower and fruit drop. Poor pollination may result in 70 percent flower loss. The branches are brittle and often break with the weight of a heavy crop. Young plants may need irrigation every 10-15 days during especially dry seasons.

Eucalyptus camaldulensis

Botanic Name *Eucalyptus camaldulensis* Dehnh.

Synonym *Eucalyptus rostrata* Schlecht.

Common Names Red river gum, red gum, Murray red gum, river gum

Family Myrtaceae

Main Attributes As firewood, the timber from *Eucalyptus camaldulensis* has few equals. It is also a good charcoal wood, and the steel industry in Argentina, for example, relies on its charcoal for steelmaking. The tree will grow in many climates from tropical to subtropical; it has the ability to thrive on relatively poor soils and in areas where dry seasons are prolonged. *Eucalyptus camaldulensis* and *Eucalyptus globulus* (page 82) are the most widely planted eucalypts in the world.

Description *Eucalyptus camaldulensis* is found over almost all the mainland of Australia. In the tropical north it is usually a slender-stemmed tree up to 25 m tall, with erect or spreading branches and a smooth bark that may be pink, cream, or white. In the subtropical south it is a thick-stemmed tree (sometimes with a diameter over 2 m) up to 40 m tall, with wide-spreading branches and a flaky bark. The trees are often crooked (more so than *E. grandis* and *E. globulus*).

Distribution About 500,000 ha of *Eucalyptus camaldulensis* have been planted throughout the world. It is the dominant eucalypt around the Mediterranean (for example, Spain has 114,000 ha and Morocco has 87,000 ha). Plantations are also found in Pakistan, Uruguay, Argentina, Kenya, Nigeria, and Tanzania. In Australia, it is the most widely distributed of all eucalypts and occurs along or near almost all of the seasonal watercourses in the arid and semiarid areas.

Use as Firewood When fully dry, the wood is an outstanding fuel, highly valued in Australia. It is moderately dense (specific gravity, 0.6), with a fuel value of 4,800 kcal per kg. The species is being planted in fuelwood projects in Upper Volta and Senegal.

Some of its provenances coppice well for six or more rotations.

Yield With the right provenance on a favorable site, *Eucalyptus camaldulensis* grows very fast. Mean annual growth increments of 2 m in height and 2 cm in diameter can be maintained for the first 10 years. Annual wood yields of 20-25 m^3 per ha have been reported from Argentina, 30 m^3 per ha from Israel, 17-20 m^3 per ha from Turkey for the first rotation (from seedlings) and 25-30 m^3 per ha in subsequent coppice rotations. On good sites, plantations are managed on coppice rotations of 7-10 years. But on poor dry sites, annual yields drop off to between 2 and 11 m^3 per ha and rotations of 14 or 15 years may be needed.

Other Uses

- Wood. The reddish heartwood is moderately strong, durable, and resistant to termites. It is useful in general construction and is the most important inland hardwood in Australia.
- Shelterbelts. In dry areas, the species is commonly planted along roadsides, in shelterbelts, and in farm woodlots. In the Sudan, it is planted to protect crops from blowing sands. With its heavy and widespreading crown it is a good shade tree.
- Honey. Honey produced from the nectar is clear or pale in color, with a mild, pleasant flavor.
- Pulp. The wood is sometimes used for paper pulp, although it is harder, heavier, and more deeply colored than the wood from *E. grandis* or *E. globulus*.

Environmental Requirements Because the species occurs over such an extensive natural range, seed collected from different localities may produce trees with very different appearance, growth rate, health, environmental tolerance, and wood quality. It is critically important to get seed from a climatic zone similar to that of the planting area. Four outstanding provenances* are those from: Katherine (Northern Territory) and Petford (Queensland) for tropical climates; Lake Albacutya (Victoria) for Mediterranean climates; and Broken Hill (New South Wales) for arid climates.

- Temperature. The trees withstand high summer temperatures and are hardy down to

*For more information see Jacobs, 1977.

3°C, though some provenances can withstand -5°C and as many as 20 frosts a year.

- Altitude. Although mainly a tree of river plains, some provenances can be grown in highland areas (above 1,200 m in Zimbabwe, for example).
- Rainfall. In its native habitat the species is found both in areas of low and high rainfall (200-1,250 mm). A lower limit for commercial plantations is 400 mm annual rainfall, although the trees will grow well in drier areas if there is seasonal flooding or a high water table. Provenances from the tropical north of Australia are from a summer rainfall area, those from the temperate south are from a winter rainfall area.
- Soil. The tree adapts well to a wide variety of soils, although, as noted, the provenance choice is very important. For example, seed from Wiluna (Western Australia) and Port Lincoln (South Australia) tolerate calcium soils better than most. Salt tolerance also varies with seed origin. Most provenances, however, will tolerate periodic waterlogging.

Establishment *Eucalyptus camaldulensis* produces good seed crops every year or two. Seed germination is high and seeds are long-lived when sealed and in dry, cold storage. Seedlings are usually started by nursery planting in containers. When transplanted to the field, spacings as close as 2 m x 2 m are used for fuelwood.

- Seed treatment. None required.
- Ability to compete with weeds. Poor; extensive weeding is mandatory.

Pests and Diseases The leaves are not favored by livestock or wildlife—an advantage. Mature trees are also fairly fire resistant, although young ones are susceptible. Young trees and those weakened by drought can be badly infected by moth larvae, eucalyptus snout beetle, termites, and eucalypt borer.

Limitations Selecting seed of the right origin is the single most important factor for successful reforestation. Seeds must be selected with particular thought to latitude, altitude, temperature, rainfall, soil type, and pest resistance.

Villagers in Upper Volta have selected neem (page 114) and *Acacia senegal* (page 102) over *Eucalyptus camaldulensis* because the wood burns fast and smokes heavily; also the tree kills other plants around it.

Related Species Similar species with promise for firewood plantations include *Eucalyptus tereticornis,* for tropical latitudes.

Eucalyptus camaldulensis, 7-year-old plantation, Central Coastal Plain, Israel. (R. Karschon)

Eucalyptus citriodora

Botanic Name *Eucalyptus citriodora* Hook.

Common Names Spotted gum, lemon-scented gum

Family Myrtaceae

Main Attributes Because of its fast growth, excellent bole form, and good timber quality, this adaptable species is being increasingly cultivated. Several hybrids between *Eucalyptus citriodora* and other *Eucalyptus* species have proved successful and are worth considering for fuelwood plantations. The hybrid with *E. torelliana* has shown considerable promise in Nigeria, for example.

Description *Eucalyptus citriodora* is an attractive tree with a white, red, or faintly bluish bark. It grows to about 45 m in height, with a straight white trunk about 1.3 m in diameter and an open, graceful crown of narrow, pendulous foliage.

Distribution *Eucalyptus citriodora* occurs naturally in only two locations: the central and northern coasts of Queensland, Australia. However, it has adapted to cultivation in a number of countries with widely differing climates and soil types. Good results have been obtained in Portugal and in much of Africa. It has also done well in Brazil, India, and Hawaii.

Use as Firewood *Eucalyptus citriodora* has long been used as a fuel in Australia. The hard, heavy wood (specific gravity, 0.75-1.1) burns steadily. Its charcoal has an ash content of 1-2 percent. This is the principal species employed in Brazil for charcoal used in steel production.

Yield The species is fast growing, typically increasing in height by 3 m per year for the first several years and growing even faster in the best localities. Plantations in Tanzania, harvested on an 8-year coppice, produced 15 m^3 per ha per year.

Other Uses

- Wood. The wood is very heavy, strong, and tough, with a relatively low shrinkage in drying for its density. It is moderately resistant to decay and termites. It makes a first-class saw timber and is used for general construction, poles, tool handles, railroad ties, and other purposes.
- Perfume. The leaves of *Eucalyptus citriodora* yield a lemon-scented oil, rich in citronellal, used in the perfume industry.
- Honey. In Kenya, *Eucalyptus citriodora* is a favorite of beekeepers because of the quality of the honey and the amount produced.
- Beautification. The species has been planted in Portugal and North Africa as an ornamental.

Environmental Requirements

- Temperature. In Queensland, natural habitat of the species, the climate ranges from tropical to subtropical. The trees withstand both high temperatures (29°-35°C mean monthly maximum) and light frosts. However, as a seedling, the plant is delicate and frost sensitive.
- Altitude. It occurs at altitudes from sea level up to 900 m in Queensland, but in Sri Lanka it has been grown at elevations as high as 2,000 m. In Hawaii, it grows from sea level to about 500 m; other eucalypts do better at high elevations.
- Rainfall. In its native habitat it tolerates dry seasons of 5-7 months. The minimum required annual rainfall is 600 mm, but for best growth, over 900 mm is desirable.
- Soil. In its native habitat, this tree occurs on rolling countryside where the soils are generally poor and gravelly, including podzols, residual podzols of lateritic origin, and infertile clays. It seems to prefer well-drained soils.

Establishment Most planters raise seedlings in a nursery for later transplanting, but in Zimbabwe the seed has been successfully sown in ash on ground that had just been burned.

- Seed treatment. None required.
- Ability to compete with weeds. Although the tree needs protection when young, seedlings 0.3-0.5 m tall planted on wet sites in Hawaii have beaten out most competition.

Pests and Diseases None reported.

Limitations *Eucalyptus citriodora* produces good seed crops only irregularly in Australia, so planting material is relatively hard to get at times.

Because it has large, brittle branches, it is not recommended for urban or household planting where the trees would be allowed to grow large.

Eucalyptus citriodora, Queensland, Australia. (Queensland Department of Forestry, Brisbane)

Eucalyptus gomphocephala

Botanic Name *Eucalyptus gomphocephala* A. DC.

Common Name Tuart

Family Myrtaceae

Main Attributes In many countries this Australian eucalypt is a popular tree crop for sands or sandy soils, especially those high in lime or limestone, which is toxic to many other species. It has proved outstanding for planting in winter rainfall areas on soils where, for example, even *Eucalyptus camaldulensis* fails.

Description Commonly, tuart is a tall tree, reaching up to 42 m in height and having a diameter of over 2.3 m. The trunk is short, often no more than half the tree's height, and is frequently crooked or forked. The tree is characterized by fibrous, dull, pale-gray bark, with thick, shiny leaves. It has both a deep taproot (to 6 m if the water table is that deep) and extensive surface roots. It therefore exploits both deep groundwater and surface water.

Distribution The only natural tuart forest in the world is on a narrow strip of sand plain (overlaying limestone), often barely 1.5 km wide, about 200 km long, extending behind the coastal dunes just near Perth in Western Australia. The largest cultivated areas are in North Africa. Morocco has over 66,000 ha of plantations; others are in Tunisia and Libya. The plant has also shown promise for Italy, Greece, Turkey, Israel, Cyprus, Ethiopia, and Uruguay.

Use as Firewood Tuart's grayish-yellow wood is hard, tough, durable, and one of the densest known woods. It burns well and makes good firewood.

Yield Annual yields of 21-44 m^3 per ha have been obtained on irrigated fertile soils in Morocco. On more difficult sites, annual yields of 6 or 7 m^3 per ha are more typical. For firewood production, rotations of 7-10 years are used. The trees regenerate readily from coppice.

Other Uses

- Wood. Tuart wood seasons and works well and is excellent for construction, fence posts, and stakes.
- Environmental protection. Tuart has been widely planted to stabilize sand dunes and for windbreaks (it is wind firm), soil protection, and as an avenue and shade tree.

Environmental Requirements

- Temperature. Absolute minimum is -4°C. The tree is native to a temperate climate with winters that are usually frost free and summers that have cooling afternoon sea breezes that temper the heat.
- Altitude. It has shown promise at altitudes up to 2,000 m in Ethiopia.
- Rainfall. Absolute minimum, 300 mm annually. Tuart is showing good results in North Africa and Turkey in both semiarid (350-600 mm) and subhumid (600-1,000 mm) zones. In its native habitat, precipitation is generally about 700-1,000 mm annually, with 6 dry summer months.
- Soil. Native tuart grows on neutral yellow or brown sands overlaying limestone. It grows well on free-draining sands and will tolerate up to 25 percent of active calcium. It tolerates slightly saline soils, but not waterlogging. It will tolerate lower fertility than *E. camaldulensis.*

Establishment The plant is easy to raise from seed.

- Seed treatment. None required.
- Ability to compete with weeds. To suppress weed growth, cultivation by dry-farming techniques is required in the first 1-2 years after planting.

Pests and Diseases A borer attacks drought-stressed specimens. Young plantations are susceptible to fire.

Limitations The tree is unsuitable for dry areas where it is very prone to eucalypt borer (*Phoracantha semipunctata*). Also, it has poor frost tolerance and is unsuitable for high lime soils.

Ettouazite, Morocco. *Eucalyptus gomphocephala* 30 years old, 27 m tall, and 37 cm diameter, growing in arid, calcareous, stony soil. (E.R. Berglund)

Eucalyptus microtheca

Botanic Name *Eucalyptus microtheca* F. Muell.

Synonym *Eucalyptus coolabah* Blakely & Jacobs

Common Names Flooded box, coolibah, coolabah

Family Myrtaceae

Main Attributes Coolibah is the tree of arid areas of inland Australia that is featured in "Waltzing Matilda," the famous song of the jolly swagman. It produces one of the strongest and hardest timbers in the world, and makes a very good fuel. The tree flourishes in dry lands because it resists drought, high temperatures, and alkaline soils.

Description This variable species may, depending on site, be a stunted bush barely 3 m tall or a tree over 20 m tall, with a bole 1 m in diameter. With its spreading branches, it is highly regarded in central Australia for the shade it provides, often where no other trees can survive.

Distribution It is found over a wide geographical range in all the Australian states except Victoria and Tasmania. Typically, it is found in open woodlands, seasonally inundated land around the edges of swamps or lagoons, or on the extensive floodplains of inland rivers.

The coolibah has been grown with success in the Sudan, Iran, Iraq, Pakistan, Tanzania, Nigeria, and Egypt.

Use as Firewood Coolibah is cultivated for fuelwood in the Sudan and other countries. There are many woodlots of it in the Gezira Agricultural Scheme, and they are being continually expanded to meet Sudanese firewood needs. Coolibah charcoal is good, but has a relatively high ash content (between 2 and 6 percent). The tree coppices well.

Yield On suitable sites with irrigation, growth may reach 2.5 or 3 m per year. In the Gezira, it is harvested after 8 years from a seedling crop and after 6 years from subsequent coppice crops.

Other Uses

- Wood. Because of interlocking grain, coolibah wood is difficult to work and is unsuitable for construction, but it is strong, resists decay and termites, and makes durable poles and fence posts.
- Shelterbelts. Coolibah is a good tree for shelterbelt planting; it is wind firm and almost free from insect and fungal pests. It is valuable for soil conservation, erosion control, and shade in torrid regions that desperately need all three.

Environmental Requirements

- Temperature. Coolibah's native habitat is hot most of the year, with mean maximum temperatures in the hottest months from 35°-38°C; mean minimum temperature in the coldest month is about 5°C. It can withstand a few frosts a year.
- Altitude. It grows mainly from 80 to 340 m, and up to 700 m.
- Rainfall. The main area of coolibah's native habitat is arid or semiarid, with annual rainfall as low as 200 mm and with a dry season of up to 7 months. But the plant tolerates higher rainfall, and in southeastern Queensland, coolibah occurs in areas with annual summer rainfall of 500 mm. Rainfall in the northern part of its habitat is monsoonal, with up to 1,000 mm of rain annually.
- Soil. The soils of the sites where it grows are usually clays or silty clay loams, often alkaline. Generally, they are heavy soils, characteristic of seasonally inundated watercourses.

Establishment The trees are easily propagated from seed. Seedlings are ready for transplanting to the field after about 6 months, when they reach about 40 cm high. Direct sowing is not satisfactory.

- Seed treatment. No special treatment is required, but seeds must be exposed to light during germination.
- Ability to compete with weeds. The tree requires complete weeding until it is well established, because the extremely small seeds give rise to puny seedlings.

Limitations There seems to be considerable genetic variation in this species—a study has been started by CSIRO Division of Forest Research in Canberra—and provenance trials would be advisable before any large planting program is undertaken.

The trees are fire-tender.

Eucalyptus microtheca, Rockhampton, Queensland, Australia. (CSIRO Division of Forest Research)

Eucalyptus occidentalis

Botanic Name *Eucalyptus occidentalis* Endl.

Common Names Swamp yate, flat-topped yate

Family Myrtaceae

Main Attributes This is a eucalypt that can grow under very saline and alkaline soil conditions. It is a possible candidate for unusually difficult sites.

Description In Australia *Eucalyptus occidentalis* attains heights of 28 m and diameters of up to 0.8 m. It has a flat-topped or umbrella-like crown.

Distribution The species is native to the semi-arid wheatbelt zone of western Australia and to parts of the adjacent subhumid zone. Good results have been obtained on small-scale plantations in Iran, Morocco, Algeria, Sri Lanka, California, and Hawaii. It has grown well in the dry Negev Desert of Israel, an area where even *E. camaldulensis* barely survives.

Use as Firewood Swamp yate produces a hard, heavy wood that burns steadily with a hot fire.

Yield Its growth is generally slow, though in San Jose, California, under horticultural conditions, it grew 5 m in 3 years.

Other Uses

- Wood. The timber is of high quality and can be used in construction and for products requiring heavy and strong wood.
- Shade. The tree is recommended by the Forests Department of Western Australia for shade planting; it is suitable for park use because its light shade permits grass to grow to the base of the tree.

Environmental Requirements

- Temperature. In its native habitat, summers are mild to hot, with brief periods over 38°C; the winters may have up to 20 frosts a year and a mean minimum of about 2°C for the coldest month.
- Altitude. The species is found at 50–300 m in altitude.
- Rainfall. This is a drought-tolerant species. Rainfall in its natural habitat ranges from 300 to 760 mm, with moderate variability. However, the tree is often found on alluvial flats subject to seasonal flooding and adjacent to salt lakes.
- Soil. Soils are generally clays, but occasionally they have a sandy surface layer with poor drainage. This is a good tree for salty soils and has done well on soils with up to 8 percent chlorides.

Establishment

- Seed treatment. Unreported.
- Ability to compete with weeds. Unreported.

Pests and Diseases Unreported.

Limitation The tree is not particularly fast growing.

Eucalyptus occidentalis, Western Australia. (D. Watkins)

Eucalyptus occidentalis, Cyprus. (M. Ch. Iacovides)

Haloxylon aphyllum

Botanic Name *Haloxylon aphyllum* (Minkw.) Iljin

Synonym *Haloxylon ammondendron* (C. A. Mey.) Bunge

Common Names Black saksaul (the Russian name could also be spelled saksaoul or saxaul; it is correctly pronounced säk-sạ-o͞o͞l.); odzhar (Turkmeni)

Family Chenopodiaceae

Main Attributes This robust plant combines the attributes of a xerophyte, halophyte, and mesophyte because it is drought tolerant, salt tolerant, and is also adapted for growth under medium moisture conditions. It can regulate its life processes to fit harsh and changing environmental conditions. Because of its adaptability and hard wood, it seems well worth testing for firewood in appropriate cold or hot deserts. (See also *Haloxylon persicum*, the white saksaul, page 138.)

Description This small tree can attain heights of 5-8 m and trunk diameters of 20-40 cm. The trunk generally has a large, irregular base and the limbs are also of irregular form, ribbed and bent, with very thin gray or grayish-brown bark. The large branches can attain the same diameter as the main trunk. Instead of leaves, the plant uses thickened leafstalks (cladodes).

Distribution Black saksaul is found in temperate deserts of Central Asia from western China and Mongolia to the Caspian Sea as well as in the hot deserts of the Middle East, Asia Minor, and North Africa. It often occurs naturally on heavier soils than white saxaul.

Use as Firewood In Soviet Central Asia, saksaul wood was widely used as a primary material for charcoal and locomotives and also was the basic source of fuel for the local population (black saksaul wood is considered comparable in effectiveness to brown coal). With the general electrification and gasification of populated areas its use is now quite limited.

The wood is very hard and brittle, with a specific gravity of 1.02.

Yield In locations with favorable growing conditions the yield of combustible wood is 15-40 t per ha. As growing conditions deteriorate, the yield drops, however. The trees take 5-7 years to form their open, irregular canopy of foliage and flower and set seed abundantly from age 7 onward.

In tests in Iran these trees have been coppiced at 50-cm height on a 5-year rotation. Under these experimental conditions, the yield was equivalent to a growth of 1.4 t per ha per year.

Other Uses

- Stabilizing deserts. In the Soviet Union, black saksaul is cultivated on large tracts to combat wind erosion and to halt desert creep. Plantings are carried out in primary forests, especially where soil protection and water conservation are important.
- Forage. The foliage is grazed by livestock and the tree is planted in grassland to raise the forage yield. It is also being widely planted for shelterbelts to protect grazing lands in the North Caspian area.

Environmental Requirements

- Temperature. In Central Asia, black saksaul withstands burning hot summers (50°C) and subfreezing winters (-35°C).
- Altitude. It grows mostly at low altitudes in Central Asia.
- Rainfall. The tree survives in some areas with less than 100 mm annual rainfall. Its taproot grows vigorously during the first years of life and penetrates soil strata as deep as 7 m to find moisture. In the Soviet Union its seeds germinate as the snow starts melting; the roots elongate so rapidly that they keep pace with the moisture as it sinks into the soil.
- Soil. The species is native to desert soils poor in humus and nutrients and rich in salts (chlorides and sulfates).

Establishment In establishing protective stands, planting stock consists of year-old seedlings with an open and closed root system of the taproot type, with a height above ground of not less than 50 cm. The planting norm is 1,000 per ha, with a survival rate of 60-80 percent.

- Seed treatment. Under Central Asian conditions the seeds of the black saksaul are harvested in November after the first frost. Storage of seed until the second silvicultural season is

Haloxylon aphyllum near Chardzhou Region, Turkmenistan, USSR.
(M. I. Ishankuliyev)

possible in hermetically sealed polyethylene packets with a seed moisture of 4 percent.

- Ability to compete with weeds. Excellent.

Pests and Diseases Although many insects affect black saksaul and great damage can be done by fungi, the wood production is usually relatively unaffected.

Limitations Since it is not a pioneer sand stabilizer, sowing and planting on moving dunes result in low survival.

Haloxylon persicum

Botanic Name *Haloxylon persicum* Bunge

Common Names White saksaul; ak-sazak (Turkmeni)

Family Chenopodiaceae

Main Attributes This is a salt-tolerant and extremely drought-resistant tree that is easy to establish. It lives in sand dunes and is a basic fuel of desert peoples in Central Asia, particularly in the Soviet Union.

Description A tall shrub or small, gnarled tree with jointed, brittle stems, it can grow up to 7 m high. It has a stout, rugged stem and light-gray bark. When covered by drifting sand the lower part of the trunk sends out horizontal auxillary roots. They can reach several meters long before turning downward to reach depths of 1–2 m. The tree often branches close to the ground, especially when the plant is stressed. In sandy areas it often becomes bushy and grows to only 1.5-2.5 m tall. Instead of leaves, the plant has leathery cladodes (flattened leaf stalks).

Distribution White saksaul is native to the Sinai, Israel, Arabia, Iraq, Iran, Afghanistan, and Central Asia. In Soviet Central Asia and Mongolia, it grows primarily in the same geographical zones as black saksaul (page 136). It forms stands on sand hills and sand ridges.

Use as Firewood In desert areas of Central Asia, white saksaul is the major supplier of fuelwood. The Soviet population in the area uses about 160,000 t per annum and the railroads about 18,000 t. The wood is excellent fuel and in thermal efficiency compares favorably with bituminous coal. It burns down to long-smoldering cinders. It is also used extensively for making charcoal.

Yield In Turkestan, the standing stock of native white saksaul is reported to average 1-6 m^3 per ha. Yields in plantations have not been reported.

Other Uses

- Forage. The foliage of *Haloxylon persicum* is used as a forage for camels and sheep, particularly in winter. In Soviet Central Asia, the tree is used to increase the forage production of grazing lands. The shoots have a good nutrient content.
- Sand fixation. The ability of the lower trunk to establish new roots when covered by sand makes this an excellent plant for stabilizing sandy areas.

Environmental Requirements

- Temperature. This desert tree is less resistant to cold than black saksaul.
- Altitude. Unreported.
- Rainfall. The tree is sometimes found in areas with only 100 mm of annual rainfall. It is more drought resistant than black saksaul.
- Soil. It grows in heavy, loose, wind-borne sands and will not grow well where the sand is powdered and compacted. It has a highly developed root system and is able to obtain moisture from a large area and from depths of 5 or 6 m. It is known to grow on dunes as high as 40 m. It is less salt tolerant than black saksaul. Young plants and new shoots cannot tolerate soil salinity higher than 1 percent, but mature trees can endure 5-6 percent salt in the groundwater.

Establishment Seeds are extremely small and light and do not keep well in storage; they must be sown while fresh.

- Seed treatment. Same as with black saksaul.
- Ability to compete with weeds. Young plants can be smothered by weeds, but litterfall salinizes the soil beneath the canopy of mature trees and kills all competition.

Pests and Diseases Same as those of black saksaul.

Limitations Moving dunes must be stabilized, for example with grasses, before white saksaul can be effectively established.

Haloxylon persicum, near Ashkhabad, Turkmenistan, USSR. (A. G. Babayev)

Saxaul woods (*Haloxylon persicum*) in Turkmenistan, USSR. (Novosti Press Agency, London)

Parkinsonia aculeata

Botanic Name *Parkinsonia aculeata* L.

Common Names Jerusalem-thorn, palo de rayo, Mexican or blue palo verde, horse bean tree, retama, sessaban (Arabic), Hanson sessabani (N. Nigeria), Barbados flower fence

Family Leguminosae (Caesalpinioideae)

Main Attributes This fast-growing tree is easy to plant and cultivate; it adapts to a variety of environments and soil types and is extremely drought resistant. Its dense, hard wood burns well.

Description The Jerusalem-thorn is a small, often crooked tree, growing to 10 m tall, with a green trunk up to 40 cm in diameter. It has thorny, gracefully drooping branches. Its foliage is "transparent" because the tiny leaflets are borne on long, flat leaves that look like blades of grass. Numerous bright yellow flowers occur in loose racemes.

Distribution The plant is native to a vast New World area extending from southwestern United States (Texas, New Mexico, and Arizona) to Argentina. It has become naturalized in Hawaii, South Africa, and most of India. It is cultivated as an ornamental in Florida, Cyprus, India (in dry Deccan areas and the Punjab), Jamaica, Israel, Uganda, and South Africa.

Use as Firewood The wood is used for firewood and charcoal in countries such as Mexico and Puerto Rico. It is close-grained, hard, heavy (specific gravity, 0.6), and brittle. The trees regrow vigorously even after drastic pruning.

Yield The tree is noted for fast growth. Young plants, fertilized, will grow up to 1 m annually.

Other Uses

- Beautification. Because of its unusual foliage and vivid flowers, it makes an attractive ornamental that withstands pruning and can even be trained as a hedge.
- Erosion control. Since it grows in arid climates and in sandy soils, the Jerusalem-thorn can be used to afforest eroding and sandy soils.
- Fodder. Its seeds have been used in the past in Mexico for food, and the seed pods are greedily eaten by animals. Young branches are lopped to feed goats and sheep.

Environmental Requirements

- Temperature. It grows in tropical and subtropical climates. It withstands high temperatures (for example, up to 36°C) and light frosts. (In the United States it survives the cold as far north as Georgia.) For best growth it requires full sun and will lean toward the sun; in shaded sites its growth is retarded.
- Altitude. It is generally found at altitudes below 1,300 m.
- Rainfall. The tree grows well where annual rainfall is as high as 1,000 mm, but its greatest potential is in dry areas receiving as little as 200 mm annually and having dry seasons as long as 9 months.
- Soil. Jerusalem-thorn occurs naturally in poor, gravelly, or sandy alluvial soil and in desert grasslands and canyons. It flourishes in oolitic limestone of southern Florida, even when given no care. It tolerates saline sites, but grows poorly in soils subject to waterlogging. Although this species is a legume, it is not now known whether it fixes nitrogen; young plants respond to fertilizer, however.

Establishment The plant produces seed prolifically and it grows easily from seed, root or shoot cuttings, or air-layers.

- Seed treatment. The seeds are either soaked in water for 3 or 4 days or scarified and then soaked in warm water for 1 day.
- Ability to compete with weeds. Unreported.

Pests and Diseases The tree is generally free of disease and insects but can be attacked by twig borers and by snow scale; new growth is subject to fungal die-back in winter months in humid climates. Young plants are damaged by termites.

Limitations The tree is thorny and reproduces so easily that it can escape from controlled cultivation and become a weed.

Related Species

- *Cercidium microphyllum*
- *Cercidium floridum*

Cape Verde Islands. *Parkinsonia aculeata* is planted in arid, upland area for forage and firewood. (P. Freeman)

Parkinsonia aculeata, Sobra Arboretum, Sudan. (H.A. Musnad)

Pinus halepensis

Botanic Name *Pinus halepensis* Miller

Common Names Aleppo pine, pino carrasco, sanaoubar halabi

Family Pinaceae

Main Attributes *Pinus halepensis* is a drought-tolerant pine that withstands aridity and poor soils better than most timber species that grow in a Mediterranean climate. Under favorable conditions, it grows rapidly and with good form.

The closely related *Pinus brutia* and *Pinus eldarica*, which are probably subspecies of *Pinus halepensis*, also seem to be exceptionally promising.

Description *Pinus halepensis* is a round-headed pine that may reach 27 m tall, with a diffuse crown, spreading branches, and clusters of cones.

Distribution *Pinus halepensis* is the most widely distributed pine of the Mediterranean basin, particularly in the western part. It is found in all countries around the Mediterranean, except Libya and Egypt. It is also being planted in warm temperate, semiarid areas of Argentina, Chile, Uruguay, Mexico, the Soviet Union, South Africa, and Australia.

Use as Firewood The wood is coarse and resinous and makes a useful fuel.

Yield In Israel and Jordan, the annual production is between 3-5 m^3 per ha per year, according to the site fertility, but mean annual increments of 8-12 m^3 per ha per year have been reported in plantations elsewhere. In Italy, good strains produce 5-6 m^3 per ha per year. The average height growth is 0.3-0.5 m per year up to 50 years.

Other Uses

- Timber. The wood is used in general construction and for packing cases and posts.
- Resin. The trees may be tapped for resin. In Algeria and Greece this is commercially important.
- Ornamental. The tree is a common ornamental species in all Mediterranean countries.
- Soil conservation. Because of its robustness, it is good for afforestation and soil conservation where soils are poor and droughts long. It is used widely in Australia for shelterbelts.

Environmental Requirements

- Temperature. The tree withstands high temperatures, but it will also withstand brief and occasional cold spells of -18°C to -20°C. (*Pinus eldarica* is particularly adapted to frost.)
- Altitude. In its native habitat, *Pinus halepensis* grows at altitudes below 1,000 m, but it sometimes can be found up to 2,000 m.
- Rainfall. This pine is suited to areas of Mediterranean climate having 250-800 mm of rainfall and 7-8 rainless months.
- Soil. Aleppo pine is able to grow in fairly shallow, poor, eroded soil. It will tolerate a high carbonate content; in fact, it is often found on shallow limestone and similar soils. It is not suitable for saline and swampy soils, but thrives better than any other pine on heavy clay soils. It may suffer phosphorus and nitrogen deficiency on deep podzolized sands.

Establishment Good seed crops occur most years. Germination is good (about 90 percent) and seeds retain viability up to 10 years if kept dry, sealed, and cool. Sowing is generally done in spring, the plants raised in pots or beds and planted at the onset of the winter rains.

The pine regenerates freely after fires.

- Seed treatment. None required. Chilling (for 4 weeks) improves the rate of germination.
- Ability to compete with weeds. The trees are sensitive to weed competition when young, and during the first 2 years weeding is often necessary. When planted on small terraces, strip or patch weeding may be adequate.

Pests and Diseases No serious diseases are known, though "damping off" can be troublesome if nursery work is careless. Several fungi may attack the young tree. The tree has only limited resistance to browsing.

Limitations Young seedlings suffer from drought; mortality can be severe during the first summer after germination. Careful management to attain high survival the first year is the key to successful establishment.

It is important to choose seed most suited for the planting site, and international trials

Pinus halepensis, 20-year-old plantation, Menashe Hills, Israel. (Jewish National Fund, Jerusalem)

have been set up to establish the most desirable provenances.

Related Species Two closely similar species also have promise for firewood plantations.

- *Pinus brutia.* A native of the hills of southern Turkey, northern Syria, and northern Iraq. Grows on a variety of soils from those with high lime content to acid podzols. Found where rainfall is as low as 250 mm. Withstands 35°C heat. The trunk remains straight, even on unfavorable sites.

- *Pinus eldarica*. The natural stands of this species occur in the USSR south of the Caucasus mountains and in Afghanistan. Plantations exist in Iran and Pakistan. In plantations, and on suitable sites, it is said to grow faster than either *P. halepensis* or *P. brutia*, and trials to quantify this are underway in different parts of the world. However, its ability to tolerate frosts, high temperature (40°-45°C), and low rainfall (200-250 mm) make it of great interest for the winter rainfall dry zones of the world.

Pithecellobium dulce

Botanic Name *Pithecellobium dulce* (Roxb.) Benth. (Genus sometimes spelled *Pithecolobium dulce*)

Synonyms *Mimosa dulcis* Roxb.

Common Names Manila tamarind, Madras thorn, quamachil, guamuchil, kamachile, black-bead, bread-and-cheese tree, opiuma (Hawaii)

Family Leguminosae (Mimosoideae)

Main Attributes The Manila tamarind is fast growing, endures drought, and withstands heavy cutting. It survives both heat and shade and is able to grow on poor soils and denuded lands in dry climates and on seacoasts even with its roots in brackish or salt water. It is a widely appreciated, easily established, multipurpose plant that produces useful fuelwood.

Description A large, nearly evergreen tree that grows up to 20 m or more in height, the Manila tamarind has a broad crown (to 30 m across) and a short bole (to 1 m thick). At the base of each leaf is normally found a pair of short, sharp spines, though some specimens are spineless.

Distribution It is native to a vast region that extends from the Pacific slopes of Mexico and southern California through all of Central America to Colombia and Venezuela. It has been widely planted and naturalized in many tropical regions, particularly in the warmer and drier regions of the Philippines and India. It has been introduced to the Sudan, Tanzania, and other dry areas of tropical Africa, largely the coastal regions. Further, it has been commonly planted and runs wild in southern Florida, Cuba, Jamaica, Hawaii, Puerto Rico, and St. Croix.

Use as Firewood The reddish-brown wood is usually hard, heavy, and strong, though it is also brittle and rather difficult to cut. It is used in India, Africa, and Central and South America as a fuel, but it smokes considerably and is not the best quality firewood. Calorific value, 5,200-5,600 kcal per kg. In parts of India it is planted and harvested as fuel for brick kilns. The tree coppices vigorously.

Yield In favorable soils and climates, the Manila tamarind may reach a height of 10 m in 5 or 6 years.

Other Uses

- Wood. The wood is durable, finishes smoothly, and is used in several countries for general construction purposes and for posts.
- Shade, hedges, and ornamental use. This attractive species makes a good highway tree. With regular trimming it makes dense, almost impenetrable, thorny hedges that keep out livestock and form useful shelterbelts.
- Food. The pods are harvested in Mexico, Cuba, and Thailand and are customarily sold on roadside stands. They contain a thick, sweetish, but also acidic pulp that is usually white, but sometimes red. It is eaten raw or made into a drink similar to lemonade.
- Forage. The pods are devoured by livestock of all kinds; the leaves are browsed by horses, cattle, goats, and sheep; and hedge clippings are often gathered for animal feed. The plants withstand heavy browsing.
- Seeds. The seeds contain a greenish oil (20 percent), which, after refining and bleaching, can be used for food or in making soap. The presscake residue is rich in protein (30 percent) and may be used as stockfeed.
- Miscellaneous. An extract from the bark is used in the Philippines for tanning; it produces a light-colored leather. The flowers are visited by bees and yield good-quality honey. The wounded bark exudes a mucilaginous gum somewhat like gum arabic.

Environmental Requirements

- Temperature. Warm subtropical and tropical, though it can withstand both shade and heat.
- Altitude. Up to 1,800 m in Mexico and 1,500 m in Burundi.
- Rainfall. This species is suitable for most dry regions. It is drought resistant and in low rainfall areas develops an extensive root system. In Burundi it grows well at 800 m elevation where annual rainfall fluctuates between 450 and 600 mm, spread evenly year-round. In southern Florida rainfall averages 1,650 mm or more.
- Soil. Manila tamarind has great adaptability and grows on most soil types, including clay, oolitic limestone, and rather barren sands. It can also be found in wet sands that have a brackish water table.

Pithecellobium dulce planted as roadside trees in Tainan, Taiwan. (Taiwan Forestry Research Institute)

Establishment Manila tamarind is readily propagated by cuttings or seed. It fruits at an early age. Seed may be stored for about 6 months (though it must then be protected from insects).

- Seed treatment. None necessary. Germination takes only 1 or 2 days.
- Ability to compete with weeds. Readily outgrows competition.

Pests and Diseases Normally pest damage is insignificant, but the tree can become affected by leaf spot diseases and a number of defoliating and boring insect pests. It is a favorite host of the thornbug.

Limitations Once introduced, it may hold an area firmly. In Tamil Nadu, India, repeated attempts to replace this plant with other species have failed. It is classed as a pest in Hawaii because it infests pastures and shades out more-desirable forage plants.

Because the trees are often top heavy and shallow rooted, in heavy windstorms branches may break off or the whole tree may topple.

Both the thorniness and an irritant sap that causes severe eye irritation in tanners and long-lasting welts on skin are limitations that have caused the tree to be abandoned as a street tree in southern Florida. Any injury to roots gives rise to suckers that are exceedingly thorny.

Prosopis alba

Botanic Name *Prosopis alba* Griseb.

Common Names Algarrobo blanco, ibopé or ibope-pará, tacu

Family Leguminosae (Mimosoideae)

Main Attributes A valuable fuel, *Prosopis alba* is also used for windbreaks, roadside planting, fodder, and timber.

Description This is a round-crowned tree, 5-15 m tall, sometimes with a trunk as large as 1 m in diameter and a short bole with many branches. The sapwood is yellowish and the heartwood dark brown. It is a thorny tree but thornless varieties are available. Like all *Prosopis* species it can fix nitrogen. Seeds need to be inoculated with mesquite rhizobia.*

Distribution *Prosopis alba* grows in the arid zones of northern Argentina, Paraguay, and Bolivia.

Use as Firewood The tree is mainly used for firewood.

Yield Ten-year-old plantations of *Prosopis alba* in Argentina, spaced 2 m x 2 m on a fair site, produced 7 m^3 per ha annually.

Other Uses

- Timber. The wood is difficult to work. It is, however, used for flooring, wine casks, shoe lasts, and paving blocks.
- Fodder. The pods are eaten by cattle. They contain about 25 percent glucose sugar and 10 percent protein.
- Food. The fruit is milled into flour that is used in baked goods for human consumption.
- Amenity planting. *Prosopis alba* is a valuable tree for windbreaks and roadside planting.

Environmental Requirements

- Temperature. *Prosopis alba* grows in areas with an average winter temperature of 15°C; it is not frost hardy. (Although the tree can withstand a few hours of mild frost, prolonged cold weather (-6°C) kills most young seedlings.) As is true with all *Prosopis* species, prolonged high temperatures are required to initiate and sustain flowering, nectar secretion, and pod production.
- Altitude. The plant is native to flatlands and low sierras (up to 1,000 m at the latitude of 30°S).
- Rainfall. *Prosopis alba* is very drought resistant, growing in dry climatic areas with 100-500 mm rainfall.
- Soil. It grows best on sands with a high clay content and tolerates some salt in the soil.

Establishment The tree is seeded directly in the field with remarkable success. However, it is better to sow it in a nursery and transplant it to the field when 2-3 months old. As is true with most *Prosopis* species, quick germination (3-4 days) requires high temperatures (night, 26°C; day, 32°C). Seeds are planted in long beds (approximately 2 x 2 x 40 cm). They are ready for planting in 2 months and should be planted in spring immediately after the last frost-free date or at the onset of the rainy season.

- Seed treatment. As is the case with many *Prosopis* species, the seeds are hard to extract from the gummy pulp in the pods. A heavy-duty sausage grinder with coarse holes has recently been found to work very well. This also scarifies the seeds.
- Ability to compete with weeds. Good, with weeds such as grasses and shrubs.

Pests and Diseases Not severe, but bruchid beetles attack the seeds in the pods.

Limitations All *Prosopis* species are cross pollinated. Research on clonal propagation methods (e.g., cuttings) are needed so that plantations with known properties can be obtained.

* Available from J. Burton, Nitragin, 3101 W. Custer Avenue, P.O. Box 09186, Milwaukee, Wisconsin 93209, USA.

Thornless *Prosopis alba*, Imperial Valley, California, USA, only 9 months after transplanting from seedlings (25 cm tall, 2 mm diameter). This specimen is well over 2 m tall. (P. Felker)

Prosopis alba, Imperial Valley, California, USA. This 6-year-old irrigated ornamental is 8 m tall and 23 cm in diameter. (P. Felker)

Prosopis chilensis

Botanic Name *Prosopis chilensis* (Mol.) Stuntz

Common Names Algarroba, kiawe (Hawaii), algarrobo blanco, algarrobo de Chile, mesquite

Family Leguminosae (Mimosoideae)

Main Attributes *Prosopis chilensis* is a South American tree that is a useful shade and fodder plant for hot dry zones. It is very drought resistant, well adapted to light soils, and is probably nitrogen fixing. It has proved suitable for planting in subdesert regions of Africa. It is the most drought resistent and consistently high yielder of biomass of all the *Prosopis* species tested in North America.*

Description *Prosopis chilensis* grows 8 to 15 m high. It has a shallow, spreading root system and branches freely. Its flowers are greenish yellow and it has a slender seed pod.

Distribution *Prosopis chilensis* is native to the Pacific coast of Peru, central Chile, and eastern Argentina. It is naturalized in Hawaii and is common in thickets along the beaches.

Use as Firewood *Prosopis chilensis* wood is hard, heavy, and strong (specific gravity, 0.80-0.92).

Yield This tree is not normally considered fast growing like *Eucalyptus* or leucaena, but it survives and grows where most other species fail and it has shown high biomass productivity (over 14 t per ha per year; see illustration in recent tests in California).* Wood yields have not been recorded.

Other Uses

- Feed. The pods, or beans, are an excellent feed containing much sugar. They are ground into a meal for use in concentrated rations. The foliage also provides animal feed in arid regions.
- Wood. *Prosopis chilensis* heartwood is rich dark brown, often with a purplish hue. It is rather coarse textured and irregularly grained, but is easy to work, finishing smoothly and taking a natural polish. The wood is very resistant to decay.
- Ornamental. Thornless varieties are becoming popular for landscaping in the southwestern United States.

Environmental Requirements

- Temperature. The tree withstands extremely high desert temperatures. It is poorly adapted to cold and requires temperatures of about 27°C for good growth. Seedlings may tolerate an odd mild freeze of -5°C but cannot withstand lower temperatures or even many (10 or more) mild frosts.
- Altitude. In southern Peru, the tree is found at elevations of up to 2,900 m. In India, it is recommended for planting as a fodder tree at altitudes ranging from 340 to 1,230 m.
- Rainfall. In Africa, it is recommended for unirrigated subdesert areas receiving 200-400 mm annual rainfall† and suffering 8-11 dry months a year.
- Soil. Unreported.

Establishment *Prosopis chilensis* is propagated by seed. For one establishment procedure, see *Prosopis alba*.

- Seed treatment. Seeds must be scarified in hot water or concentrated sulfuric acid before planting.
- Ability to compete with weeds. Unreported.

Pests and Diseases The trees are largely disease free, although bruchid beetles usually destroy much of the seed crop.

Limitation *Prosopis chilensis* trees are aggressive and may easily become pests, especially when pastures are overstocked and the grass cover depleted.

*Information supplied by P. Felker.

†Laurie, M. V. 1974. *Tree Planting in African Savannas*, Food and Agriculture Organization of the United Nations, Rome: p. 72.

Prosopis chilensis, Riverside, California, USA. Thornless variety grown as ornamental 2½ years after transplanting as a seedling 25 cm tall. (P. Felker)

Thornless *Prosopis chilensis*, Imperial Valley, California, USA, 9 months after transplanting. First-year productivity of this stand (1.6 m spacing and partially irrigated) was 12.3 t per ha (dry weight). (P. Felker)

Prosopis cineraria

Botanic Name *Prosopis cineraria* (L.) Druce

Synonym *Prosopis spicigera* L.

Common Names Jand, khejri (India); ghaf (Arabia); jandi (Pakistan)

Family Leguminosae (Mimosoideae)

Main Attributes *Prosopis cineraria* is a useful tree with great vitality. Owing to its long taproot (over 3 m), it is able to find moisture and survive where other plants succumb and it withstands hot winds and dry seasons. The plant deserves detailed study, genetic selection, and more widespread use.

Description A thorny tree, *Prosopis cineraria* grows 5-9 m high and has an open crown. It is leafless for a short period before flowing. Because of its deep taproot, it does not compete for moisture with crop plants, which may be grown close to its trunk.

Distribution *Prosopis cineraria* occurs in the dry and arid regions of northwestern India in Punjab, West Rajasthan, Gujarat, Uttar Pradesh, in dry parts of central and southern India, and extends into Pakistan, Afghanistan, Iran, and Arabia. In Abu Dhabi, plantings totalling 2,000 ha have been made on flat, silty, gravelly plains and in shifting sand dunes.

Use as Firewood *Prosopis cineraria*'s rather scanty purplish-brown heartwood is preferred in its native regions over other kinds of wood for cooking and heating. It is an excellent fuel and gives high-quality charcoal (5,000 kcal per kg). The tree coppices readily.

Yield In standing crops, the amount of fuel per ha varies from 7 to 70 m^3 and averages 21 m^3 stacked. Annual yields of stacked firewood up to 2.9 m^3 per ha have been reported.*

- Rainfall. It grows in zones having a long dry season and from 75 to 850 mm rainfall.

Other Uses

- Fodder. *Prosopis cineraria* is considered to be the best browse plant for cattle, sheep, and camels in Jodhpur, India, on the basis of availability, palatability, and nutritive value.
- Afforestation. It is important for afforestation in arid and semiarid zones, where it can survive even under extremely dry conditions. It holds promise for easy establishment in shifting sand.
- Timber. It is used for house building, posts, tool handles, and boat frames, although the poor form of the unimproved trees currently limits this use.
- Agroforestry. This species is revered by farmers in India and Pakistan because of the increased fertility beneath its canopy.†

Environmental Requirements

- Temperature. The plant withstands both slight frost (-6°C minimum) and high temperatures (40°-50°C maximum shade temperature).
- Altitude. *Prosopis cineraria* is a low-altitude tree.
- Soil. The tree grows on alluvial and coarse sandy often alkaline soils. It is common on moderately saline soils. It is also found on black cotton soil in open forest and on dry stony lands. It reportedly tolerates high alkalinity (pH 9.8).

Establishment *Prosopis cineraria* reproduces freely by root suckers and establishes well from seed. Seeds remain viable for decades.

- Seed treatment. The seeds require soaking in water 24 hours prior to sowing. For establishment procedure, see *Prosopis alba*.
- Ability to compete with weeds. One or two weedings appear necessary the first year. The seedlings need early pruning to force them to grow straight.

Pests and Diseases One fungus and five insect species are known to attack the tree.

Limitations The trees are thorny, and if planted in subhumid areas they may become pestilential weeds.

Related Species Other *Prosopis* species with promise as firewood include:

- *Prosopis farcta* a species native to North Africa, the eastern Mediterranean, Iraq, Iran, Afghanistan, Pakistan and the U.S.S.R. (Transcaucasia and Turkestan).

*Khan. 1955.

†Shankar et al., 1976; Singh and Lal, 1969.

Prosopis cineraria on slowly moving sand dune, Al Ain, Abu Dhabi. (P.J. Wood)

Prosopis cineraria, northwestern India. (Forest Research Institute and Colleges, Dehra Dun, India)

Prosopis juliflora

Botanic Name *Prosopis juliflora* (Swartz) DC

Synonym *Mimosa juliflora* Swartz

Common Names Mesquite, algarroba

Family Leguminosae (Mimosoideae)

Main Attributes *Prosopis juliflora* is a highly esteemed fuelwood source in several tropical countries and it is also valued for the shade, timber, and forage it can provide. It is planted where other, more valuable forest species cannot be successfully grown.

Description A thorny, deciduous, large-crowned, and deep-rooted bush or tree, *Prosopis juliflora* may grow to 10 m or more, depending on the variety and site. The leaves are dark green. The long fleshy pods are straw colored when ripe.

Distribution The species is native to Central America and northern South America. The tree has been planted in many arid zones of the world. It is widely propagated in Africa and Asia, particularly in India.* The species was introduced into India over 100 years ago, planted mainly for stabilizing dunes, but also for fuel.

Use as Firewood *Prosopis juliflora* wood is hard and heavy (specific gravity, 0.70 or higher). It is excellent for firewood and makes superior charcoal. Because of its high heat value, the wood has been termed "wooden anthracite." It burns slowly and evenly and holds heat well.

Yield *Prosopis juliflora* grows fast. On a 15-year rotation, the expected yield is 75-100 t per ha; on a 10-year rotation, it may be 50-60 t per ha. The trees coppice readily.

Other Uses

- Wood. The wood is very durable and is used for fence posts, door and window frames, and other light carpentry.
- Honey. The flowers are a valuable source of nectar for high-quality honey.
- Food and fodder. The pods are eaten by livestock and may also be ground into flour for human consumption.

Environmental Requirements

- Temperature. *Prosopis juliflora* grows in very warm, dry climates. Some varieties are not frost hardy.
- Altitude. The tree is found growing from sea level to 1,500 m.
- Rainfall. It grows in areas with an annual rainfall of 150-750 mm. The plant's roots penetrate to great depths in search of soil moisture.
- Soil. The tree grows on a variety of soils. It does well on sandy soils and will grow on rocky terrain provided that root growth is not impeded.

Establishment *Prosopis juliflora* reproduces easily by root suckers and seeds.

- Seed treatment. To overcome the seed-coat dormancy, the seeds must be mechanically scarified and either treated with 20 percent sulfuric acid for 1 hour, or soaked in concentrated sulfuric acid for about 20 minutes, or covered with boiling water and allowed to cool and soak for 24 hours.

Germination of the treated seeds is 80-90 percent.

- Ability to compete with weeds. Good.

Pests and Diseases Bruchid beetles often damage much of the *Prosopis juliflora* seed crop.

Limitations *Prosopis juliflora* is an aggressive invader and is sometimes a nuisance. This is a species to be tried only in very arid problem sites. Elsewhere, the problems it causes may be immense.

Related Species Other species of *Prosopis* that are useful for firewood include:

- *Prosopis algarrobilla*
- *Prosopis caldenia*
- *Prosopis tamarugo* (page 156)
- *Prosopis pallida* (page 154)
- *Prosopis chilensis* (page 148).

*Taxonomy of the genus *Prosopis* is confused. The name *P. juliflora* has been used in the past to describe species native to Texas and nearby states. These are now known as *P. glandulosa* and *P. velutina*. Plants that were distributed around the world under the name *P. juliflora* are probably mislabeled in many cases.

Prosopis juliflora, Dubai, United Arab Emirates. (H.A. Salman)

Prosopis juliflora firewood for roadside sale, Haiti. (M.D. Benge)

Prosopis pallida

Botanic Name *Prosopis pallida* (Humboldt & Bonpland ex Willdenow) H.B.K.

Synonym *Prosopis limensis* Bentham

Common Names Algarrobo or algarroba, algaroba, huarango (Peru), algarrobo americano (Puerto Rico), kiawe (Hawaii)

Family Leguminosae (Mimosoideae)

Main Attributes A valuable tree for forage, beautification, and for shelter under arid conditions, *Prosopis pallida* is also a useful fuelwood source. It has remarkable tolerance to salt and can be irrigated with water up to half as salty as seawater.* It deserves research on genetic selection and improved silvicultural methods.

Description *Prosopis pallida* is a tree (or shrub on sterile soils) 8-20 m high, with a trunk to 60 cm in diameter. The leaves are medium to small in size and a pallid greyish-green when dry. It is usually spiny, although in Hawaii many of the trees are thornless.

Distribution *Prosopis pallida* is native to drier parts of Peru, Colombia, and Ecuador along the Pacific coast. It has been naturalized in Puerto Rico and the Hawaiian Islands and has also been introduced for cultivation in India and Australia.

Use as Firewood Like that of other *Prosopis* species, the wood has high calorific value. It is used largely for charcoal.

Yield Unreported.

Other Uses

- Fodder. Both the leaves and pods of *Prosopis pallida* are fed to cattle, donkeys, and other livestock and are eaten by wildlife.
- Food. Pods of *Prosopis pallida* are sweeter than those of most other *Prosopis* species. In its native habitat they are made into a sweet syrup used to prepare various drinks.
- Afforestation. This is a potentially important species for plantations in hot, dry regions, especially where salinity makes the cultivation of other species difficult.

Environmental Requirements

- Temperature. *Prosopis pallida* is more frost sensitive than *Prosopis chilensis* (page 148), *Prosopis alba* (page 146), or *Prosopis cineraria* (page 150). It will tolerate -2°C, but year-old seedlings have been killed by a -6°C frost in California.
- Altitude. It thrives from sea level to 300 m.
- Rainfall. The tree requires from 250 mm to 1,250 mm per year. It requires more moisture than *Prosopis chilensis* or most *Prosopis alba* provenances.
- Soil. It is not particular as to soil, growing in old lava flows as well as on coastal sand.

Establishment The tree is grown from seed. When planted in new locations, the seeds should be inoculated with mesquite rhizobia.

- Seed treatment. Unreported.
- Ability to compete with weeds. Unreported.

Pests and Diseases Termites and the wood-boring beetle *Clytus cornis* are attracted to it. *Psyllid* insects, which feed on terminal shoots and leaf tips, infect *Prosopis pallida* more than other *Prosopis* species.

Limitations *Prosopis pallida* may become an invader and form thickets, which can be a costly nuisance. Because it has shallow roots, wind storms easily blow many of the trees down.

*Information supplied by P. Felker.

Prosopis pallida as a shade and ornamental tree, Oahu, Hawaii, USA. (J. L. Brewbaker)

Prosopis pallida, Piura, Peru. In the dry coastal region of Peru, *Prosopis* pods are commonly fed to cattle, donkeys, and other livestock. They are also added to soups and are made into "algarrobina," a sweet syrup used to prepare drinks. (C. López-Ocaña)

Prosopis tamarugo

Botanic Name *Prosopis tamarugo* F. Phil.

Common Name Tamarugo

Family Leguminosae (Mimosoideae)

Main Attributes *Prosopis tamarugo* is the only tree able to survive on the arid salt flats of northern Chile. It produces the only available forage, timber, and firewood in the area. It has promise for reforesting deserts, especially those with thick surface-salt incrustations.

Description The tamarugo is a deciduous, open-crowned tree, reaching 8-15 m in height, with a trunk up to 80 cm thick. The flowers are golden yellow and occur in long cylindrical spikes. The pod is curved and contains about seven seeds embedded in a brownish pulp. The tree produces a dense mat of lateral roots and a quick-growing taproot that may be 6 m deep on a plant that is only 15 m tall.

Distribution In the part of the Atacama Desert in northern Chile known as Pampa del Tamarugal, the tamarugo stands were so overexploited for firewood that they were reduced to scattered woodland. This area is an inland salt desert about 40 km wide and 300 km long at about 1,000-1,500 m in altitude. In 1965 the Chilean government initiated a project to plant tamarugo, and there are now about 25,000 ha covered with tamarugo forests.

Use as Firewood The wood burns well.

Yield Tamarugo is relatively slow growing under desert conditions, but it nevertheless holds remarkable potential. After 15 years it reaches about 10 m in height in the harsh tamarugal area.

Other Uses

- Wood. The wood is used for furniture, although it is very heavy and difficult to work because it is very hard.
- Forage. Sheep or goats feed on the fallen leaves and pods and occasionally browse the young shoots. Care must be taken that they do not destroy the lowest branches, which are important in the plant's water economy because they shade the dense lateral root zone near the soil surface.

Environmental Requirements

- Temperature. It tolerates a wide temperature range, from -12° to 36°C. Low night temperatures help create a very high relative humidity and conditions to condense humidity as dew, enabling it to absorb atmospheric water.
- Altitude. The tree's natural habitat has a range of 1,000-1,500 m, but elsewhere it could probably grow at other altitudes.
- Rainfall. The climate of its native habitat is very dry. For years there may be no rain at all; the normal annual average is less than 10 mm. Fog is very rare, although it occurs occasionally from December to February. Under salt-desert conditions, rain may in fact be injurious, as it would leach the salt from the surface into the root zone, increasing the difficulty of water extraction. Heat radiation on cold nights causes condensation and water absorption by the leaves from atmospheric moisture, important to the plant's survival.
- Soil. The tree is found on salty-sandy or clay loam soils, sometimes with a 40-cm salt incrustation on the surface. There is some evidence that a subsurface water reserve is created by the lateral roots, which exude moisture absorbed by the leaves during cold nights from the atmosphere.*

Establishment The tree coppices easily and regenerates from seed. Fresh seed is produced from October to January and one may expect 30-45 percent germination. Where there is a surface salt incrustation, a pit is dug in the salt and the plant is placed with a mixture of manure and soil in a hole at the bottom. Great care is taken not to damage the roots, which normally are quite long, or the lower branches that shade the soil surface when the plant grows.

- Seed treatment. The seeds germinate well after soaking in water 48 hours, but some researchers scarify them with sulfuric acid to improve germinability.
- Ability to compete with weeds. Its roots exude a natural herbicide that makes growing other plants near to tamarugo trees impossible.

Pests and Diseases The seed is frequently attacked by bruchid beetles, but the wood is unaffected.

*Information supplied by F. Sudzuki.

Before: The Pampa de Tamarugal in Chile's Atacama Desert is fearsome. Average annual rainfall is often a mere 70 mm, the water table is at 20 m depth or below and a crust of salt almost half a meter thick covers the area . . .

After: *Prosopis tamarugo* is planted through the salt, and now thousands of hectares are converted into "forests" useful for forage, firewood, and lumber. (Pictures courtesy of M.A. Habit and M. Sarquis, respectively)

Limitations Seedlings must be kept watered during the first year of their establishment until the first shoot appears, after which they do not need watering.

If planted in more equable climates than those of the harsh Tamarugal area, this plant might spread uncontrollably and become a serious weed.

Tamarix aphylla

Botanic Name *Tamarix aphylla* (L.) Karst. and related species

Synonyms *Tamarix articulata* Vahl, *Tamarix orientalis* Forssk.

Common Names Tamarisk, athel tree, salt cedar, eshel

Family Tamaricaceae

Main Attributes Tamarisks are aggressive, hardy shrubs or trees of desert, steppe, and seashore. There are over 50 species, and most tolerate salty soils, poor-quality water, drought, and high temperatures. Several types can be used to afforest sand dunes and saline wastelands and for shelterbelts in hot, dry desert areas. They coppice after heavy cutting.

Description Tamarisk taxonomy is extremely complex and confusing because the plants vary widely in growth habits and phenology; individual specimens of some species can usually be identified only by a specialist. *Tamarix aphylla* is the only easily identifiable species and it is fairly representative of the other tree species of the genus. It is a heavily branched tree, 8-12 m tall. Its slender branches and fine bluish-gray foliage produce a light, feathery effect. The leaves are reduced to tiny "scales" that ensheath the wiry twigs and are well equipped to withstand desiccation. The plant has a deep and extensive root system. Like other *Tamarix* species it excretes salt. Salty "tears" drip from glands in the leaves at night, so that the soil beneath is covered with salt.

Distribution Species of *Tamarix* are found throughout the Old World from the Canary Islands to China. There are 23 species in Iran alone. *Tamarix aphylla* originates in the central Sahara. It has been spread (vegetatively) to Pakistan, India, Afghanistan, the Middle East, and North Africa, as well as to Eritrea, Somalia, Kenya, and Ethiopia. It is also widely planted in Australia. The saltcedar of Mexico and southwestern United States is probably *Tamarix chinensis* Lour., a native of China. The attractive shrub *Tamarix gallica* (French tamarisk) is native to Mediterranean seacoasts where it grows in sand dunes and dry saline soils. *Tamarix nilotica* is a small shrub of the eastern Mediterranean that has successfully revegetated moving sand dunes, for example in the Negev Desert of Israel.

Use as Firewood Most tamarisk wood is hard and durable. It is slow to catch fire, but burns reasonably well. The ash content can be high (4-6.6 percent in some Israeli specimens). The wood of *Tamarix aphylla*, for example, is used for firewood and charcoal in Asia Minor and India. Leaf litter and small branches will not burn, because of their high salt content. The plants coppice satisfactorily.

Yield Unreported.

Other Uses

- Wood. The light-colored wood is useful for carpentry, turnery, plows, furniture, and the manufacture of fruit boxes.
- Afforestation. Tamarisk is valuable for erosion control, especially for stabilizing sand dunes, because it grows fast and resists being buried by shifting sand. It has an extensive root system and sheds twigs abundantly, forming a compact litter that improves the sand's water-holding capacity.
- Windbreaks. Tamarisk species are outstanding for use as shelterbelts in arid zones. Being evergreens, they are effective year-round. Many miles of highway and railroad in the deserts of Southern California and Arizona are protected from wind and blowing sand by lines of tamarisk.
- Firebreaks. Salt drip from tamarisk kills all ground vegetation beneath the leaves, and litter from the trees is too salinized to burn. Thus, strips of tamarisk can be grown to stop wildfires, a major hazard during hot, dry summers in arid areas. They will also halt the spread of fires along highways or railroads caused by sparks or cigarettes.

Environmental Requirements

- Temperature. Most tamarisks grow well where temperatures are high. Some species will tolerate frost in the winter. In the Mojave Desert of California, where tamarisk is used for shelterbelts, summer temperatures may reach 50°C and winter temperatures -10°C.
- Altitude. Unreported.
- Rainfall. *Tamarix chinensis* in California, *Tamarix nilotica* in Israel, and *Tamarix aphylla* all have survived where average annual rainfall is

Tamarix aphylla, 4 years old and 6–7 m tall, used as a shelterbelt, Darling Downs, Queensland, Australia. (Queensland Department of Forestry, Brisbane)

a mere 100 mm. But their optimum growth is probably at 350–500 mm. *T. aphylla* is recommended in Australia for cultivation in areas as dry as 175 mm.

• Soil. Some tamarisks are among the most salt-tolerant trees, although this varies widely with the species. They will regenerate in very adverse sites, even those with salt appearing on the surface, and some of them withstand salt spray. They grow well not only in sand or sandy soils, but in Queensland, Australia, have performed satisfactorily in heavy clays that swell when wet and shrink and crack when dry.

Establishment Tamarisk seeds lose viability within a few days, so for all practical purposes trees are established exclusively by cuttings taken in spring from the previous year's growth. The process is easy and inexpensive. Long, well-rooted cuttings can be planted directly into the sand, even on moving dunes.

• Ability to compete with weeds. Removal of competition is necessary to facilitate establishment.

Pests and Diseases Many insects feed on tamarisk or cause gall formation.

Limitations Most tamarisks cause salinization of the upper soil layers (even on nonsaline soil). Their extensive roots extract all soluble salts from the soil, excrete the salts on the leaves, and return them to the topsoil with litter fall. They commonly reduce the growth of nearby plants because of this and also because their roots collect moisture and plant nutrients over a large distance. The effect is significant when crops are grown close to shelterbelts or shade trees. They may, for example, reduce crop yields for up to 50 m.

Tamarisks transpire abundantly, and when their roots penetrate deeply they may cause great losses of otherwise useful groundwater. For this reason, they are held in low esteem in the western United States. In areas with shallow water tables, their deeply penetrating roots use up more groundwater than other phreatophytes tested in Arizona.

In one year a single tamarisk tree can produce more than 500,000 seeds. Most germinate, producing jungle-like growth, in less than 5 years.

Zizyphus mauritiana

Botanic Name *Zizyphus mauritiana* Lam. Also spelled *Ziziphus*

Synonyms *Zizyphus jujuba* (L.) Lam., non Mill.

Common Names Indian jujube, Indian plum, Indian cherry, Chinese date, ber, jujubier (France), beri, bor, nabbak el fil (Arabic), and many others

Family Rhamnaceae

Main Attributes The jujube tree can withstand severe heat, frost, and drought; it is planted in dry areas and on sites unfit for other crops. Its wood makes good fuel. Although seldom recognized as a fuelwood source, the tree is already widely cultivated for its fruit in the tropics.

Description This normally spiny species grows to about 12 m or more in height, with a spreading crown and a trunk diameter of about 0.3 m. In severe sites it is more commonly seen as a compact shrub only 3 or 4 m tall. Its fruits look like glossy crabapples, light reddish-brown in color.

Distribution The plant is native to South Asia and grows wild, particularly in sub-Himalayan hill country. It is now found in many parts of Asia, Australia, the West Indies, tropical America, and Africa (for example, Senegal, Gambia, the Ivory Coast, Ghana, northern Nigeria, and western Sudan). It is usually found growing in dry areas.

Use as Firewood The wood is hard and heavy, having a specific gravity of 0.93. The tree's drooping branches are easily accessible for harvesting. It is an excellent fuel tree and makes good charcoal, with a heat content of almost 4,900 kcal per kg. The trees coppice well and grow vigorously from stumps and root-suckers.

Yield Noted generally for its fast growth, the tree is considered one of the most rapidly growing species in the Sahel.

Other Uses

- Wood. The timber is hard, strong, fine grained, and reddish in color. It works well and takes a good polish. It is most often used for making agricultural implements, sandals, tent pegs, golf clubs, and other products that need a durable, close-grained wood.
- Fruit. The tree bears heavily, sometimes producing thousands of nutritious fruits per plant year after year. They are eaten either fresh, dried, or pickled and can be made into a floury meal, a butter, or a cheese-like paste, which is used as a condiment. The juice can be made into a refreshing drink.
- Living fence. The tree is useful as a living fence; its spiny stems and branches deter livestock.
- Tannin. The bark is sometimes used for tanning purposes.
- Silk. Jujube leaves can be used to feed the tasar silkworm. Tasar silk, highly prized, is the only one commercially exploited in the tropics.
- Lac insects. This is one of the few trees that can be used to host lac insects. The resinous encrustation from these insects is used to produce shellac.
- Fodder. Its foliage makes a good fodder for cattle, camels, and goats. Livestock avidly devour the fruit.

Environmental Requirements

- Temperature. Although it is an extremely robust plant, able to withstand severe heat as well as frost and drought, it does require tropical or near-tropical climatic conditions for best growth.
- Altitude. The jujube is a low-altitude species for cultivation below about 600 m in elevation.
- Rainfall. In Puerto Rico, it grows where rainfall is up to 2,000 mm, but it is widespread in areas with annual rainfall of 300-500 mm.
- Soil. The species has no particular soil requirements and thrives in a wide variety of soils, even in oolitic limestone.

Establishment Although the tree responds to cultural treatment, traditionally it receives little care. Propagation is usually by direct seeding, but it is easily propagated vegetatively by grafting, inarching, or top-working (these are essential for maintaining quality cultivars).

- Seed treatment. To quicken germination, the seeds are sometimes cracked.
- Ability to compete with weeds. Unreported.

Pests and Diseases It is susceptible to insect pests such as fruit flies and leaf-eating caterpil-

Zizyphus mauritiana, Lucknow, Uttar Pradesh, India. (T. N. Khoshoo)

lars, but these probably do not greatly affect the yield of firewood.

Limitations The plant is disseminated by birds and animals feeding on the fruits, which can lead to a profusion of dense, prickly clumps. Some thornless varieties are known in Assam.

As a fruit tree the jujube has found little favor outside of India because the hard stone clings to the flesh. Also there is great variability in fruit quality; some are extremely astringent. Superior types could be of real value in the tropics if they were available.

Related Species

- *Zizyphus jujuba* Mill. The common or Chinese jujube is a similar fruit tree adapted to more temperate areas.
- *Zizyphus nummularia* DC. A shrub of desert areas of northwest India and Pakistan. A good browse plant. Heartwood has calorific value of 4,400 kcal per kg.

Zizyphus spina-christi

Botanic Name *Zizyphus spina-christi* (L.) Desf.

Synonym *Rhamnus spina-christi* L.

Common Names Christ thorn, kurna, nabbag or sidr (Arabic)

Family Rhamnaceae

Main Attributes *Zizyphus spina-christi* is a spiny bush or tree that strongly resists both heat and drought. It develops an extremely deep taproot and has extraordinary regenerative power. Wherever it grows, it is used for fuel.

Description This is an evergreen, medium-sized (3-10 m tall), long-lived tree. Its slender branches have many short, curved spines, and are said to have been used for Christ's crown of thorns.

Distribution The Christ thorn is native to a vast African area stretching from Mauritania through the Saharan and Sahelian zones of West Africa to the Red Sea. It is also native to the eastern Mediterranean, Iran, eastern Turkey, and the Arabian Peninsula.

Use as Firewood The red or dark-brown wood is much used for fuel. It is hard and dense and burns with an intense heat. The bushes coppice well.

Yield Unreported.

Other Uses

- Wood. Christ thorn wood is used for spear shafts, posts, roofing beams, and household utensils, and is a good cabinet wood. It is said to be termite proof.
- Fodder. The fruits are greedily eaten by sheep and goats, and the foliage by camels. This feed keeps stock healthy in the Sudan when often no other feed is to be found.
- Erosion control. Because it develops an extremely deep taproot and spreading laterals, it is useful for stabilizing sand dunes and other unstable soils. It also makes useful windbreaks and shelterbelts.
- Living fence. It can be grown to form a stock-proof living fence.

Environmental Requirements

- Temperature. It is very resistant to heat.
- Altitude. Up to 1,500 m.
- Rainfall. It grows in desert areas with about 100 mm rainfall annually, but grows best in wadis where groundwater is available. It also can be found in less-arid areas, especially on alluvial plains.
- Soil. It extends into dry, desert areas, but prefers alluvial plains with deep soils.

Establishment Propagation is invariably by seeds, but it can also be propagated by cuttings.

- Seed treatment. The hard, woody shells should be cracked with a hammer and the shelled seeds soaked overnight in lukewarm water.
- Ability to compete with weeds. Unreported.

Pests and Diseases Unreported.

Limitations This gregarious plant can form spiny, impenetrable thickets. It should be tested only in very dry areas where few others species can survive.

Zizyphus spina-christi, Umm Dam, Kordofan Province, Sudan. (G.E. Wickens)

APPENDIX 1*

Using Fuelwood Efficiently

The fuelwood situation in developing countries can be improved by the planting of more trees and by better management of existing forest resources. However, another approach to be considered is reducing the demand for firewood through introduction of more efficient burning equipment. The possibilities of this approach seem particularly attractive because firewood is most scarce in those regions of the world where it is also burned most inefficiently.

Roughly 80 percent of the fuelwood consumed in developing countries is used for domestic purposes: cooking, space heating, and hot water.† Many traditional cooking stoves and open fireplaces waste wood mainly because they focus the flames poorly on the cooking surface.

Improving wood-burning devices seems to be one of the best ways to alleviate the twin curses of firewood scarcity and forest depletion. Often, all that is is required is very minor redesigning of existing stoves.

Improved stove models are insulated to prevent heat loss through the walls. The hearth can be closed to regulate air intake and improve combustion conditions. The flow of air and hot gases through the stove is directed to concentrate heat on the cooking surface. A chimney is usually incorporated into the design to provide draft, the motive power that provides air for combustion. (If poorly designed, however, chimneys can decrease efficiency by creating excessive draft.) Cooking pots are arranged to fit properly, preventing leaks and heat loss.

Improved stoves can probably achieve an overall efficiency of between 20 and 30 percent and they have the potential for reducing wood requirements. In addition to reducing the devastation of the world's trees, the wide dissemination of such stoves could reduce the time, energy, and cash that Third World women now spend acquiring fuel and cooking. It will also help eliminate smoke-filled homes, sooty hands, and the pain and eye defects caused by smoke, fumes, and sparks.

However, the claims about efficient stove designs have seldom been substantiated in unequivocal tests. Therefore, it is important that any new stove be field tested on site before it is widely promoted for local use. Perhaps the best measure of performance is the weight of fuel needed to cook a number of typical meals. This result should be compared with those of other stoves, including traditional ones.

Many factors, other than efficiency, complicate the acceptability of a

*This appendix was compiled in collaboration with W. Magrath, Volunteers in Technical Assistance (VITA).
†Arnold. 1978.

cooking stove: cost, availability of materials, size and type of wood available, family size, cooking practices, and types of dishes to be prepared. These vary greatly from region to region, which means that any given stove design may not be accepted or used efficiently outside the area where it was designed.

The other approximately 20 percent of fuelwood used in developing countries is burned at similarly low efficiencies in small- and medium-scale industries such as food processing and manufacturing.* There are numerous ways in which modern engineering principles can be applied to improve the fuel efficiencies of these technologies. These applications, such as tobacco processing, sugar refining, lime production, and brick manufacturing, are particularly important because of the contribution they make to the development of infrastructure, the growth of rural incomes, and foreign exchange earnings.

In many cases fuelwood savings can be realized even without changes in equipment. Perhaps the best way to do this is simply to dry or season wood before burning. When burned, moist wood produces less-useful heat and more smoke. To ensure that wood is dry it should be split (in the case of large logs) and stored for 6 months. Since economic pressures often prevent proper seasoning of wood, carefully designed credit schemes to ease cash flow problems might enable the populations in developing countries to save millions of tons of firewood annually.

In charcoal production, efficiency can be improved by improving the skills of individual entrepreneurs† and/or by introducing new equipment, such as kilns and retorts.‡ Research is underway in many parts of the world on small-scale pyrolytic converters that burn agricultural and forestry wastes and produce charcoal as well as burnable gas and oils.

The rest of this appendix depicts some wood-burning devices claimed to be fuel efficient.

*Arnold, 1978.
†International Labor Organization, 1975.
‡Little, 1972 and 1978.

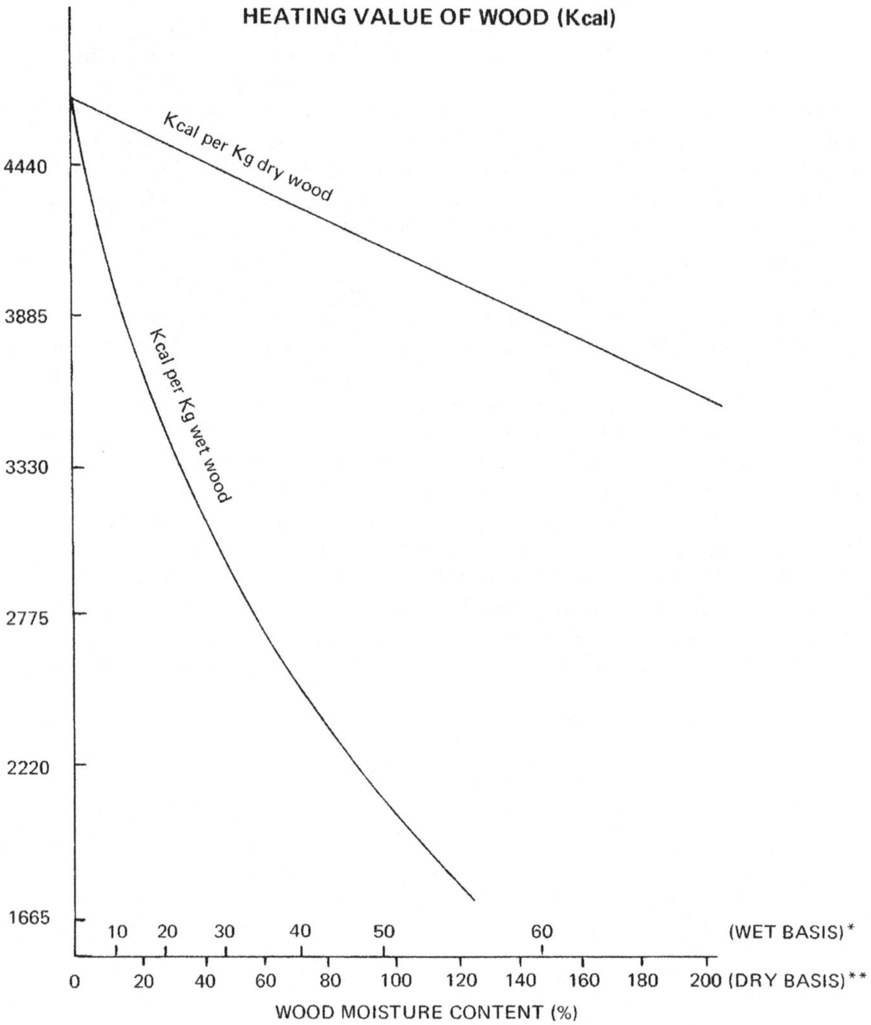

EFFECT OF WOOD MOISTURE CONTENT ON ITS HEATING VALUE

$$\text{*WET BASIS} = \frac{\text{Wet Weight-Dry Weight}}{\text{Wet Weight}}$$

$$\text{**DRY BASIS} = \frac{\text{Wet Weight-Dry Weight}}{\text{Dry Weight}}$$

Improved wood drying. Whatever the stove design, proper drying of wood fuel (to a moisture content of around 20 to 25 percent) will reduce the quantity of wood needed for a given heating requirement by 20 percent or more. This is so because wet wood consumes some of its energy to evaporate the moisture in the wood. The potential gain from improved wood drying could be significant, and provision for firewood drying should certainly be incorporated in any firewood program. (Diagram courtesy of J. Bene)

"Fireless" cooking. If a heavy pot of food is brought to a boil and then placed inside a heavily insulated box, it continues to cook in its own heat. No additional fuel is needed, thus saving much firewood. The system is particularly useful for beans, sauces, stews, and other foods that are cooked slowly over several hours. Hot stones can be added for further cooking. (Low Energy Systems, 3 Larkfield Gardens, Dublin 6, Ireland)

Indian Chulah. Named from the Hindi word for "cooking fireplace," the chulah has been the target of much research in India over the past 40 years. The chulah can be built of sand and clay to accommodate different sizes and numbers of cooking pots. Experiments with chulah have shown efficiencies as high as 30 percent. However, flue losses can be quite high and some tests have shown these stoves to use more fuel than their traditional counterparts that have no flues. (VITA)

The Roti stove. A seemingly efficient cooking system has been in use in the remote Indonesian islands of Roti and Sumba. The stoves are horizontal. Pots are hung in the flames so that the heated surface is exceptionally large: no heat is wasted up a flue–there is none.

Only one description of this cooking system has ever appeared in English. It is by Captain Cook (or perhaps by Sir Joseph Banks, as it appears in the 1773 Hawkesworth edition of *Cook's Journals*). This is how he described it:

"They dig a hollow under ground, in a horizontal direction, like a rabbit burrow, about two yards long, and opening into a hole at each end, one of which is large and the other small: by the large hole the fire is put in, and the small one serves for a draught. The earth over this burrow is perforated by circular holes, which communicate with the cavity below; and in these holes are set earthen pots, generally about three to each fire, which are large in the middle, and taper towards the bottom, so that the fire acts upon a large part of their surface. Each of these pots generally contains about eight or ten gallons, and it is surprising to see with how small a quantity of fire they may be kept boiling; a palm leaf or a dry stalk, thrust in now and then, is sufficient: in this manner they boil all their victuals, and make all their syrup and sugar."

Because of the quantity of heat transferred in this cooking system, meals can be cooked using palm fronds–a very poor fuel. The system certainly seems to warrant Captain Cook's enthusiasm and it deserves testing for possible use in villages everywhere. (J.J. Fox)

The Lorena stove, developed at the Estación Experimental Choqui, Quezaltenango, Guatemala, has attracted considerable attention around the world. Constructed of a mixture of clay and sand, the stove can be built for very low cost. The stove was designed by actively involving potential users. The stove greatly reduces the amount of fuel required, is designed to accommodate traditional cooking pots, and is built almost entirely of local materials.

A cottage industry has developed around the stove. Trained stovemakers have been able to make a living by building stoves in people's homes. In this way, and a variety of other promotion mechanisms, an estimated 2,000 stoves have been built in about 2 years. The Lorena design has been replicated or adapted in West Africa, Java, Nepal, the United States, and throughout Latin America. (Aprovecho Institute)

A

B

C

This model shows the main features of the Lorena stove. Picture A: With the top removed, the tortuous path the flame must travel is visible. It flows around the walls of each cooking pot as it passes. Small bumps in the firepath floor force the heat up onto the cooking pot surfaces, which hang in the firepath. The thick insulating adobe is shown in pictures B and C: The metal dampers are raised and lowered to control the fire. (W. M. Todd)

Southeast Asian charcoal stove. These neat, simple, and efficient cookstoves are found typically in Thailand, Laos, and elsewhere in Indochina. Produced as a cottage industry, they use mainly local materials. An outer bucket made from scrap tin sheets gives the stove durability and provides attachment for a handle. A middle layer filled with ash and clay helps durability by acting as a shock absorber if the stove is dropped or struck. The inner lining of the firebox and ashbox is fired clay. The grate inside is of clay and is replaceable. The whole stove costs $2–3 and in tests in Tanzania have proved to have twice the efficiency of the metal charcoal stoves traditionally used there.* (N.D. Vietmeyer)

* Information supplied by K. Openshaw, Division of Forestry, University of Dar es Salaam, Box 643, Morogoro, Tanzania.

Improved baking oven. In Ghana over 19,000 women operate small, wood-fired bake ovens and produce the bulk of the nation's bread. The traditional ovens are small, dome-shaped, built of sun-dried or fired-clay bricks, and take one layer of about 50 loaves. The University of Ghana Department of Nutrition and Food Science has recently found a simple way to double the stove's efficiency. Its oven has broken glass inside the base (to retain more heat) and is insulated with additional bricks. The most important modification, however, was a new construction technique that allowed the addition of a shelf so that the oven can accommodate a total of 100 loaves. This produces twice the quantity of bread for the same amount of fuelwood. (G. Campbell-Platt)

Sawdust stove. In areas where sawdust or other forestry and agricultural wastes are available, stoves can be fabricated from discarded drums and cans. Sawdust is compacted into a can, with sticks used to maintain openings for the fire; a meal is being prepared over this fire (note the sheet metal chimney). (E. Simon)

The Family Cooker. Based on efficient cooking stoves developed in The Netherlands during World War II, the Family Cooker was developed at Eindhoven University. Efficiency is optimized by reducing heat losses and controlling combustion. The housewife has easy and flexible control over the rate of cooking. This is because air for the fire is sucked in through the air-inlet pipes and controlled by the damper in the chimney. Heat transfers efficiently because the fire plays directly on the cooking pot and cannot escape sideways. Flue gases then pass downwards, heating the inlet air and a "hotbox" that can be used to preheat food or water or to keep cooked food warm. (For more information contact Faculty of Industrial Engineering, Eindhoven University of Technology, Postbus 513, 5600 MB Eindhoven, The Netherlands)

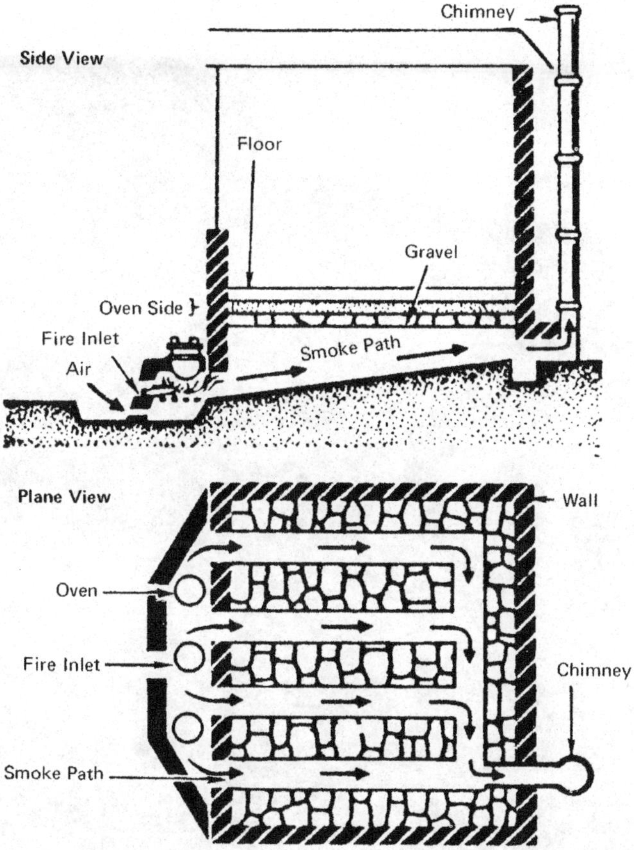

Ondol system of cooking and heating. In the Korean countryside, houses are traditionally built over a labyrinth of clay pipes packed in sand that honeycomb the foundations. Heat and smoke from the kitchen fire, or from an outside furnace, passing through these "horizontal chimneys" warms the floor and thus the rooms above. Ondols are often fired with leaves, twigs, chopped wood, brush, or hay. The double use of the heat could serve as a model for use in other locations where both cooking and heating are needed.

Pottery kiln near Seoul, South Korea. In Korea and northern China it is traditional to build pottery kilns taking advantage of the natural slope. Pots to be fired are placed on shelves within the kiln through side doors, which are sealed with bricks and mud. The kiln can be "filled to the brim." The incline creates a draft, the shelves give maximum use of space, every level is used, and the kiln is its own chimney. The kilns are usually fired in the evenings, left to burn all night, then sealed and allowed to cool slowly. (J.A. Bannigan)

Improved brick kilns. Working from ideas suggested by an unsuccessful experimental lime kiln, a Honduran brick manufacturer designed and built this kiln, described as an "inclined chimney kiln." Bricks are placed throughout the kiln, which is fired from below. The kiln acts as a chimney, drawing the hot gases through itself. The fuel used is sawdust, which would otherwise be wasted. (R. Fera)

Tobacco curing. The drying of tobacco leaves is a major user of fuelwood in some developing countries. The traditional flue-curing barns are not very efficient. An engineering team of the Tropical Products Institute recently boosted the efficiency of barn furnace/flue systems at the Kasungu Flue Curing Tobacco Authority in Malawi. The improved system uses shiloh flues (arched sheets of galvanized iron that are sealed to the floor beneath which pass the hot fumes) as heat exchangers. It saves up to 27 percent of the wood formerly used and cures the tobacco leaves 20 percent faster. (Tropical Products Institute)

Tabarka region, Tunisia. Charcoal is traditionally made in a pit or earth-covered mound. The use of a kiln reduces waste caused by the contamination of the charcoal with soil. It also produces a better product, as the rate of carbonization can be controlled and wasteful burning reduced. (M. Benaissa, FAO photo)

The mini-CUSAB bin with lower plugs in. This is a smaller version of the CUSAB bin. Basically, it is an oil drum or tar barrel fitted with air ducts. Portability is its great advantage; it can readily be taken to the source of fuel. Its size makes it convenient for construction in a village and for an individual to produce charcoal for his family's use from small wood scraps.

Mini-CUSAB bins are being used in Tonga to convert coconut shell and coconut wood into charcoal. In the South Pacific, coconut logs must be disposed of because rotting coconut wood breeds the rhinoceros beetle, a serious pest. The wood is too wet for firewood but makes useful charcoal in the Mini-CUSAB bin. (E.C.S. Little)

CUSAB kiln. Acronym for charcoal from useless scrub and brush, the CUSAB kiln can rapidly carbonize scrap wood, twigs, branches, and shrubs that cannot be utilized by traditional charcoaling methods. It provides a means for exploiting woody material that otherwise would go to waste. Its capacity is about 1 ton of charcoal. The kiln is provided with a vertical row of air ducts that pierce the wall. Looking through these holes, operators can see wood burning. As charcoaling proceeds up the kiln, they see the flames change to the red glow of charcoal burning and they plug up the air ducts as the level of charcoal rises. In the model shown here the kiln can be tipped over for easy emptying. (E.C.S. Little)

Case Study: Ethiopia

Addis Ababa owes its existence and its primary source of fuel to a tree, and the tree owes its existence to the shrewdness and determination of the Ethiopian peasant "forester."

When Addis Ababa was founded as Ethiopia's capital in 1890, the large population it attracted made heavy demands on the scattered remnants of indigenous native forest that existed in the area. In less than a decade, lumber and fuelwood were being hauled in from a distance of over 20 km, and soon all the land within "donkey and mule reach" of the city had been denuded. The shortages then became so serious that the city's foreign residents were certain the capital would be moved to another source of wood somewhere else in the country, as had occured several times previously. It was in this critical period that the first eucalyptus trees (the *bahir zaf,* "sea-tree" or "tree from across the sea" in Amharic) were introduced. They are now grown in many parts of the country.

ADDIS ABABA'S EUCALYPTUS FOREST

Eucalyptus assured a supply of firewood, permitting the peripatetic government to come to rest, and Addis Ababa, the "New Flower," became the permanent capital and not just another way station. At least 15 eucalyptus species have been brought into Ethiopia since the early years of the century, but *E. globulus* has always been the one cultivated most extensively, followed by *E. camaldulensis.*

Photographs taken in 1906 reveal only a few scattered clumps of eucalyptus trees, but by 1910 some landowners had densely planted a few hectares with the tree in hopes of making their fortunes. Imperial incentive, including tax relief and the distribution of free seeds, encouraged the tree's spread in the early years. By 1920, the streets and paths of Addis Ababa began to look like clearings in a vast, continuous forest, and it was even suggested that the city's name might appropriately be changed to Eucalyptopolis. By 1935, just before the Italian invasion, the city, seen from Entoto, appeared to be a green mass of forest.

The Italians were apparently the first to seriously study the forest, which they estimated covered at least 4,000 hectares. The European portion of

*This appendix is adapted from articles by Ronald J. Horvath (Addis Ababa's eucalyptus forest, *Journal of Ethiopian Studies*, 1968, 6:13–19) and Gunnar Poulsen (Eucalypts for people, *Sylva Africana*, 1978, 7:4–8).

Addis Ababa was relatively free of trees, but the Ethiopian section was merely an extension of the forest.

Aerial photographs taken in 1957 showed that the forest (excluding the trees in the builtup area of Addis Ababa) covered 100 km^2. The furthest limit extended 10 km along the Ambo road from the heart of Addis Ababa.

By 1964 the forest had expanded to cover 135 km^2. Thus, in the intervening years, there has been an increase of almost 5 km^2 per year. The furthest extension along the Ambo road had reached 27 km.

PLANTING

Above 2,000 m sea level, Ethiopian farmers plant mainly *E. globulus*, and at lower altitudes, *E. camaldulensis. E. camaldulensis* is grown not only under arid conditions where its tolerance makes it a seemingly obvious choice, but also in areas where the annual rainfall exceeds 1,000 mm. Ethiopia's success with this tree in heavy rainfall areas confounds most foresters, who would opt for other species known to thrive and produce straighter stems under such conditions.

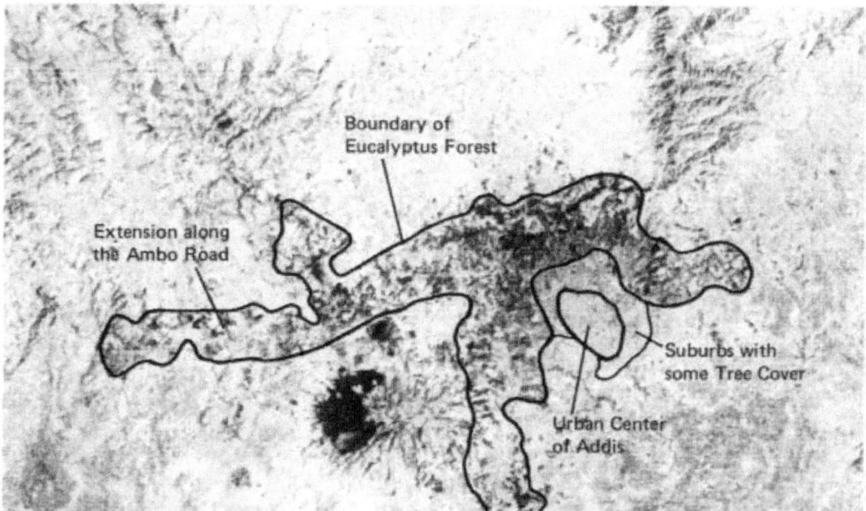

Addis Ababa's eucalyptus forest, as photographed by the Landsat satellite, February 12, 1976. The tree-covered area embraces the urban center of the city and stretches westward along the Ambo road. The suburban area with its light tree cover also can be seen. (L. Pettinger)

The explanation for the survival of eucalyptus, however, is simple. The two species chosen, better than any other, combine good adaptability to climate and soil with the necessary toughness that enable them to survive the crude planting techniques applied by Ethiopian farmers. It may be all wrong scientifically—but it works.

Ethiopian farmers usually grow the eucalyptus plants themselves or buy them from an enterprising neighbor. Nursery beds are carefully prepared, tended, and protected from harsh sunlight, rainfall, and wind. After about a year, the crowded seedlings are approximately 0.75 m high, very slender, and have few side branches.

At the onset of the rainy season, the farmer prepares his land for planting

The lightly forested suburban area of Addis Ababa. (G. Poulsen)

and waits for a cloudy, windless day, preferably with a slight drizzle. The plants are then lifted from the nursery bed with a small forked hoe, damaging the delicate root systems as little as possible. The long, slender stems are neither pruned nor stripped of leaves. The farmer simply bundles them together and carries them to the planting area. From a forestry perspective, the plants are doomed. The shoots are too long for the roots, especially as the roots often become seriously mutilated during lifting and transportation. A professional forester would expect that the roots would be too short to sustain evapo-transpiration of the long stem.

The seedlings are planted very close together, with a density of between 40,000 and 100,000 seedlings per hectare. The textbook forester would insist that even 10,000 seedlings per hectare is too dense, and if one visits a plantation a couple of days after planting, he would seem to be right. The top-shoots of the plants hang limply and many leaves wither. Surprisingly, however, as many as one-quarter—sometimes even half—of the plants survive the rough treatment. Farmers do not expect survival rates higher than that. If the final result of their efforts is a plantation containing 10,000-25,000 plants per hectare, they are pleased.

The dense plantation closes canopy within a few months, eliminating the problem of severe grass competition and making weeding unnecessary. If the trees remain small and slender, so much the better—most Ethiopian houses, both in town and in the country, are of a mud-plastered wattle type that need small stems for their construction. After thinning, the farmers' stands produce larger stems that are marketed as fuelwood and building poles.

HARVESTING

The eucalypts grow very rapidly on the fertile, volcanic soils surrounding Addis Ababa, where rainfall is plentiful year-round and usually well distributed. Once a tree is planted, it can be cut every 2-3 years for over 40 years without being replanted, because new sprouts grow from the stump. Trimmed poles 5 m long and 8 cm in diameter can be produced in this time.

Wood is transported to Addis Ababa by porters, pack animals, or trucks. According to one survey, 15 percent of all trucks entering Addis Ababa carry wood. However, an astonishingly large quantity of wood is carried into the city on people's backs. On almost any day, one may observe a steady stream of porters carrying loads of leaves, faggots, and poles down the slopes of Entoto. Perhaps one-third of all the pack animals and porters entering Addis Ababa carry wood or leaves. Many of the women seen carrying loads are toting their own fuel, which they buy from woodcutters in the forest. They walk as far as 14 km from the center of Addis Ababa just to buy leaves and twigs and to collect leaves and bark along the tree-lined roads. The trees shed large amounts of leaves, twigs, woody fruits, and particularly bark and are thus a constant source of fuel. The ground around the trees appears as if it were swept clean because every leaf, twig, fruit, and piece of bark is scavenged for fuel.

MARKETING

The Addis Ababa woodshed now extends far beyond the city-forest; eucalyptus wood comes in from distant farmlands.

Most of the wood is sold from small stores located all over the city. There, wood is piled on the ground and is sold either as poles for house building or as cut-up firewood. Large trees are sold directly to the government for telephone and electricity poles.

EUCALYPTUS TODAY

Residents of Addis Ababa are so dependent on the eucalyptus trees that the city probably could not exist without them. Eucalyptus wood is used for fuel and building materials. Eucalyptus lumber frames the majority of its structures. Much of the cooking in the city is done with the wood and the leaves. Eucalyptus leaves are now thought to be essential for making good quality *enjera* (the unleavened bread that is a traditional Ethiopian staple) because they burn with a quick, hot flame. Eucalyptus wood is also used for tool handles, furniture, telegraph and telephone poles, fences, particle board, and for numerous other products. Moreover, the trees are important for the beautification they provide.

The eucalyptus tree was introduced to Ethiopia at a time when residents were desperate for a solution to the wood-shortage problem.* Even today, no replacement for the eucalyptus is in sight. Now, the people have developed a rapport with the tree, which is not a burdensome crop and does not require special handling. Today, the "tree from across the sea" has become an integral part of the Ethiopian landscape and has clearly become one of Ethiopia's important resources.

*The most widely believed story is that the introduction was not intentional but that the seeds fell off the boots of a visiting Australian.

Case Study: South Korea*

by Erik Eckholm

Passing through the South Korean countryside today, a visitor cannot help being aware that something important having to do with trees is underway and that the government, known for its tight control over public activities, must be terribly interested in forestry. Sprinkled about the hillsides are billboards bearing slogans such as "Love Trees, Love Your Country." Other signs forbid unauthorized entry into a patch of forest, while still others simply sport a number–"1975" or "1976"–indicating the year in which that hill was planted. That a date, alone on a billboard, carries a recognized message for passers-by is testimony in itself of the pervasive awareness of forestry that has, one way or another, been instilled in the populace. And everywhere the trees, the proof that something real has come of all this: row upon row of young trees that in parts of the country seem to cover virtually every spot of land not being farmed.

Large areas of South Korea have been transformed from lands of barren hills into lands of young pines. According to government figures, more than one-third of the national land area is stocked with trees less than 10 years old. While official statistics here, as anywhere, should be taken with a grain of salt, the country is clearly in the process of changing its face.

The historical obstacles to sound forest management have not withered away; South Korea's forestry sector faces plenty of technical, economic, and managerial problems. Planting failures on long-abused lands now bereft of organic matter are not rare; in some areas pests are killing trees. But a critical about-face has occurred, and the country's challenge is increasingly one of instituting optimal management rather than of halting degradation.

While desultory efforts at village forestry had been tried since the 1950s, the really dramatic turnabout occurred in 1973, when the government devised a new forestry policy. Before then, the national emphasis had been on reducing the industrial timber deficit, relying mainly on the work of the government's forestry department. Village-level forestry groups– which had emerged spontaneously in some areas over the previous several centuries–were encouraged in principle but not given the strong support that would later result in their establishment throughout the country. Since 1973, in contrast, priority has been placed on meeting the needs of the rural population by enlisting their energies and unused lands. That a community-based approach could simultaneously provide more wood for the forest-products industry was also recognized.

*Adapted from Erik Eckholm's Worldwatch paper *Planting for the Future: Forestry for Human Needs*, Worldwatch Paper 26, February 1979.

In South Korea, reforestation efforts have been spectacularly successful. Formerly bare hills, now forest-clad, surround and benefit the agricultural bottomlands throughout the country. (E. Eckholm)

The new interest in forestry for rural needs was a logical extension of the much broader shift in national development policy that had occurred 2 years earlier. While boosting the gross national product very swiftly, South Korea's rapid industrial growth during the 1960s pulled more people to the cities, widened disparities between urban and rural incomes, and caused general social and economic dislocation in the villages. In 1971, a major new program known as the *saemaul*, or "new community," movement was launched. Billed as self-help by which villages could advance economically and build local institutions, the *saemaul* campaign has mobilized villages to construct irrigation works, roads, bridges, water supplies, and electrical facilities.

Following in the same spirit, the new forestry campaign has tried to mobilize villages to plant public and private lands, to form cooperatives to produce and market assorted products such as mushrooms and valuable leaves, and, above all, to establish firewood lots to meet local needs. The program has been implemented through an unusual combination of private and governmental organizations. Building on Korea's long history of village cooperative societies and the scattering of forest associations already in existence, the government encouraged the establishment of Village Forestry Associations (VFAs) in nearly every village. Nominally a private body (a local association consists of a representative from every household in the village), membership is mandatory and is directed by an elected chief. The VFAs are all part of a nationwide nongovernmental network, the Korea National Federation of Forest Association Unions, which is headquartered in Seoul.

South Korean villages are administrative entities with precise boundaries. Often they comprise a natural geographical unit, such as a farmed valley and its surrounding hills. As a first step in carrying out a firewood plantation project, officials from the government, federation, and village association together calculate wood requirements for the community and identify suitable lands–usually hillsides of little agricultural potential--for meeting this need. Most of the chosen lands are privately owned; the owners are given the option of either reforesting the areas themselves, or turning them over to the VFA in return for one-tenth of future proceeds from their plots. Though this share of eventual profits is not great, most landowners, who have received virtually no income from their plots in the past, give the land to the VFA.

Through this profit-sharing mechanism the Koreans have managed to co-opt private land for public purposes, overcoming the constraints that private land ownership had previously placed on forest improvement. The landowners have no choice about participating in one way or another, but they also receive tangible benefits as a result. Such a combination of latent stick with visible carrot seems to characterize many of South Korea's rural development programs.

The VFA plants, tends, and harvests the woodlots without pay. As wood is harvested, it is distributed among households; the proceeds from any marketable surplus are put into a cooperative fund for further community development projects. By the end of 1977, 643,000 hectares of village woodlots–which are primarily for fuel, though many also include trees planted for commercial purposes–had been established in this manner. According to Bong Won Ahn, a federation official who helped plan this enterprise, "The fuelwood component of our forestry program is essentially finished. We calculated the needs and set planting targets, and now these have been met. By the early 1980s, when increasing amounts of wood will be harvested from the new plantations, our rural fuel problems will be largely solved."

Critics of the program have argued that popular participation in the village associations, and in the *saemaul* movement generally, is as much a consequence of an authoritarian government's heavy hand as it is a genuine outpouring of civic spirit. Peace Corps workers living in villages recently told a visiting journalist that the regimentation of the *saemaul* movement is resented by many villagers.

Yet the physical and economic achievements of the village forestry and *saemaul* campaigns are there to see, and the benefits have been well distributed among the peasantry. It seems unlikely that such widespread cooperative behavioral changes could be long sustained by compulsion alone. Korea's

South Korea's success with forestry had been largely because it involved rural people themselves in the planning, planting, and stewardship of the forest. Here villagers harvest forage and firewood. (K.B. Yim Bin)

Confucian tradition, with its emphasis on obedience to hierarchical authority and on social cohesion, undoubtedly helps explain the success of the program. But beyond that, the glue binding people together in these efforts may be the genuine personal benefits they receive from their participation. For the many families who had been forced by wood scarcity into buying coal for home heating, for example, the switch to locally grown wood has meant an average 15 percent increase in income.

The combination of a strong political commitment at the top with broad public participation and shared benefits at the bottom, which characterizes forestry efforts in South Korea, is a pattern whose possibilities have yet to be explored in many countries. Yet logic, experience, and economic realities suggest that only the rural people themselves will be able to plant, protect, and harvest trees on the scale required in coming decades. Unless institutions are created that give them the means to do so, and that ensure they receive the fruits of their own labor, the interlocking crises of wood scarcity and land degradation can only get worse.

APPENDIX 4

Master List of Firewood Species

The following species received the highest rating in replies to the inquiry sent to several hundred plant scientists and foresters before the panel met to write this report. Species chosen by the panel for inclusion in the report are marked with an asterisk.

Humid Tropics

*Acacia auriculiformis**
A. aulacocarpa
A. crassicarpa
A. flava
A. koa
A. leucophloea
A. polyacantha
A. siamensis
A. tomentosa
Acrocarpus fraxinifolius
Adansonia digitata
Adina cordifolia
Afzelia africana
A. xylocarpa
Aglaia spp.
Albizia falcata
*A. lebbek**
A. moluccana
A. odoratissima
A. procera
Aleurites moluccana
*Alnus jorullensis**
Alstonia spp.
Anacardium occidentale
*Anogeissus latifolia**
*A. leiocarpus**
Anthocephalus cadamba
Antidesma ghaesembilla
Artocarpus spp.
Aspidosperma spp.
Astronium urundeuva
Aucoumea spp.
Avicennia spp.*
*Azadirachta indica**
Bambusa spp.
Baphia kirkii
Bauhinia malabarica
B. tomentosa
Bischoffia javanica
Bocageopsis multiflora
Bombax spp.
Bruguiera spp.*
Caesalpinia sappan
*Cajanus cajan**
*Calliandra calothyrsus**
C. surinamensis
Callicarpa arborea
Caloncoba gilgiana
Cananga odorata
Capparis spp.
Carapa guineensis
Cariniana pyriformis
Casearia spp.
Cassia macrantha
*C. siamea**
*C. spectabilis**
*Casuarina cunninghamiana**
*C. equisetifolia**
*C. lepidophloia**
C. nobile
Cecropia spp.
Cedrela spp.
Ceiba pentandra
Celtis spp.
Ceriops spp.
Chilopsis linearis
Chlorophora tinctoria
C. excelsa
Chloroxylon swietenia
Citrus spp.
Coccoloba sp.
Cocos nucifera
Coffea spp.
Combretum spp.
Conocarpus erectus
Cordia spp.
C. alliodora
Cratoxylon spp.
Crescentia cujete
Croton spp.
Cupressus lusitanica
Cynometra cauliflora
Daniella oliveri
Dendrocalamus strictus
Derris microphylla
Detarium senegalense.
Dialium guineensis
D. ovoideum
Dichrostachys glomerata
Dillenia spp.
Diospyros spp.
Diphysa robinioides
Dinizia excelsa
Duabanga grandiflora
D. moluccana
Elateriospermum spp.
Enterolobium cyclocarpum
Erythrina spp.
Erythrophleum spp.
Eschweilera mexiana
Eucalpytus alba
E. botryoides
E. brassiana
*E. camaldulensis**
*E. citriodora**
E. cloeziana
E. deglupta
*E. grandis**
*E. microtheca**
E. moluccana
E. pellita
E. resinifera
E. robusta
*E. saligna**
*E. tereticornis**
E. tornelliana
E. urophylla
Eugenia jambos
Ficus benghalensis
Garuga pinnata
*Gliricidia maculata**
*G. sepium**
*Gmelina arborea**
*Grevillea robusta**
Grewia spp.
Guatteria ferruginea
*Guazuma ulmifolia**
Haematoxylon campechianum
Hevea brasiliensis
Holoptelea integrifolia
Hymenocardia acida
Inga spp.
I. alba
*I. edulis**
I. laurina
*I. vera**
Inocarpus edulis
Intsia bijuga
Iryanthera hostmani
Khaya senegalensis
Kydia calycina
Laguncularia spp.

Lantana spp.
*Leucaena leucocephala**
Libidibia corymbosa
Licania spp.
Lindackeria maynensis
Lumnitzera racemosa
Macaranga spp.
Machaerium nictitans
Madhuca latifolia
Malmea spp.
Mammea americana
Mangifera indica
Mangroves*
Melaleuca leucadendron
Melastoma spp.
Melia azedarach
M. composita
Michelia champaca
Moringa oleifera
Morus mesozygia
Muntingia calabura *
Murraya paniculata
Musanga cecropioides
Myristica spp.
Nauclea diderrichii
Nectandra spp.
Ocotea spp.
Octomeles sumatrana
Olea africana
Ouratea calophylla
Parinari excelsa
Parkia spp.
*Parkinsonia aculeata**
Peltophorum pterocarpum
Pentaclethra macrophylla
Pentadesma butyracea
Persea spp.
Phyllanthus discoideus
Pinus caribaea
P. insularis
P. kesiya
P. merkusii
Piptadenia spp.
*Pithecellobium dulce**
P. jiringa
P. lobatum
Platonia insignis
*Pongamia glabra**
Populus euphratica
Pourouma spp.
Pseudosamanea guachapele
Psidium guajava
P. cattleianum
Pterocarpus erinaceus
P. indicus
Pterygota alata
Quercus spp.
Q. oocarpa
Q. penduncularis
Q. sapotaefolia
Rhamnus spp.
Rhizophora apiculata
R. candelaria
*R. mangle**
*R. mucronata**
Salix humboldtiana
Salvadora persica
Samanea saman
Schleichera oleosa
Schizolobium parahyba
Securinega virosa
Serialbizzia splendens
Sesbania aegyptica
*S. grandiflora**
Sterculia urens
Swartzia sp.
S. fistuloides
S. madagascariensis
Sweetia brachystachya
Swietenia macrophylla
S. mahogani
Symphonia globulifera
*Syzygium cummii**
S. guineense
Tamarindus indica
Tamarix passerinoides
Tectona grandis
Terminalia spp.*
T. paniculata
T. tomentosa
Tetragastris altissima
Tetrameles nudiflora
Thespesia populnea
*Trema guineensis**
*T. micrantha**
*T. orientalis**
other *Trema* spp.*
Trichilia hirta
Triplaris guayaquilensis
Triplochiton scleroxylon
Tristania obovata
Vitex spp.
Ximenia americana
Xylia kerrii
Xylocarpus spp.
Zanthoxylum spp.
Z. xanthoxyloides
Zizyphus spp.
Z. thyrsiflora

Tropical Highlands

Acacia acuminata
A. baileyana
A. cavenia
*A. dealbata**
*A. decurrens**
A. elata
A. macracantha
*A. mearnsii**
A. melanoxylon
A. pycnantha
A. visco
Acer negundo
A. obtusifolium
A. pseudoplatanus
Ailanthus glandulosa
Alnus formosana
*A. glutinosa**
*A. jorullensis**
*A. nepalensis**
A. nitida
A. orientalis
*A. rubra**
Altingia excelsa
Amorpha fruticosa
Aristotelia chilensis
Araucaria spp.
Aspidosperma quebracho-blanco
Baeckea frutescens
Bambusa sp.
Bauhinia retusa
Brachychiton populneum
Buddleia spp.
Callitris macleayana
Calycophyllum multiflorum
Carya spp.
Castanopsis spp.
C. acuminatissima
*Casuarina cunninghamiana**
*C. equisetifolia**
*C. junghuhniana**
*C. luehmannii**
Ceanothus spp.
Cedrela spp.
Cercocarpus
Cestrum spp.
Cinnamomum camphora
Citrus spp.
Coffea arabica
Commiphora spp.
Croton glabellus
Cupressus benthamii
C. cashmeriana
C. forbesii
C. goveniana
C. lusitanica
C. macnabiana
C. macrocarpa
C. sempervirens
C. torulosa
Dendrocalamus strictus
Didymopanax morototoni
Drimys winteri
Elaeagnus angustifolia
Escallonia spp.
Eucalyptus albens
*E. bicostata**
E. blakelyi
E. botryoides
E. calophylla
*E. camaldulensis**
*E. citriodora**
E. cladocalyx
E. cloeziana
E. deanei
E. delegatensis
E. diversicolor
*E. globulus**
*E. gomphocephala**
*E. grandis**
E. gummifera
E. largiflorens
E. leucoxylon
*E. macarthuri**
E. maculata
*E. maidenii**
E. melanoxylon
E. melliodora
E. microcorys
E. neglecta
E. nova-anglica
E. odorata
E. ovata
E. paniculata
E. resinifera
E. robusta
*E. saligna**

E. tereticornis
E. trabutii
*E. viminalis**
E. wandoo
Eugenia sp.
Ficus palmata
F. salicifolia
Fraxinus sp.
Gleditsia triacanthos
*Grevillea robusta**
Grewia spp.
Leptospermum spp.
Lespedeza bicolor
L. cyrtobotrya
L. maximowiczii
Ligustrum lucidum
Liquidambar formosana
L. styraciflua
Liriodendron tulipifera
Lithocarpus spp.
Maclura pomifera
Maytenus boaria
Melaleuca leucadendron
M. pubescens = M. preissiana
Melia azedarach
Nyssa aquatica
Olea africana
O. chrysophylla
O. cuspidata
O. europaea
Peumus boldus
Pinus canariensis
P. caribaea
P. elliottii
P. excelsa
P. kesiya
P. merkusii
P. nigra
P. oocarpa
P. pinea
P. pseudostrobus
P. radiata
P. rigida
Platanus occidentalis
P. orientalis
Podocarpus oleifolius
Polylepis spp.
P. tomentella
Populus balsamifera
P. betulifolia x *P. trichocarpa*
P. deltoides
P. grandidentata
P. nigra
P. tremuloides
Quercus sp.
Q. dilatata
Q. incana
Q. virginiana
Robinia pseudoacacia
Salix babylonica
S. caprea
S. humboldtiana
Schinopsis spp.
Schinus molle
Sophora japonica
Styrax sp.
Tecoma spp.
T. stans
Teijsmanniodendron ahernianum
Tetraclinis articulata
*Trema orientalis**
Tipuana tipu
Trevoa trinervis
Ulmus pumila
U. wallichiana
Vernonia baccharoides
Wendlandia spp.

Arid and Semiarid Regions

Acacia spp.
A. acuminata
A. albida
A. aneura
*A. arabica**
*A. auriculiformis**
A. baileyana
*A. brachystachya**
A. caffra
*A. cambagei**
A. catechu
A. cibaria
A. concinna
*A. cyanophylla**
*A. cyclops**
A. dealbata
*A. decurrens**
A. drepanolobium
A. elata
A. excelsa
A. farnesiana
A. giraffae
A. greggii
A. harpophylla
A. heteracantha
A. heterophylla
A. hockii
*A. holosericea**
A. homalophylla
A. karroo
A. kempeana
A. lasiopetala
A. leenthamii
A. leucophloea
A. litakunensis
A. longifolia
A. macracantha
A. melanoxylon
A. modesta
*A. mollissima**
*A. nilotica**
A. nilotica subsp. *adansonii*
A. nilotica var. *tomentosa*
A. oswaldii
A. pallacantha
A. pendula
A. peuce
A. planifrons
A. polyacantha subsp. *campylacantha*
A. pycnantha
*A. raddiana**
A. senegalensis
*A. seyal**
A. siamensis
A. tomentosa
*A. tortilis**
A. victoriae
*Albizia lebbek**
*Anogeissus leiocarpus**
*A. pendula**
Argania sideroxylon
Artemisia herba-alba
A. monosperma
A. scoparia
Aspidosperma quebracho-blanco
Atriplex bracteosa
A. canescens
A. leucoclada
*Azadirachta indica**
Balanites aegyptiaca
Bauhinia reticulata
B. thonningii
Bombacopsis quinata
Brasilettia mollis
Bunchosia armeniaca
Burkea africana
Caesalpinia paraguariensis
*Cajanus cajan**
Calliandra spp.
Calligonum comosum
Callistemon sp.
Carapa guineensis
Cassia garrettiana
*C. siamea**
C. sturtii
*Casuarina cristata**
*C. decaisneana**
*C. equisetifolia**
*C. glauca**
*C. stricta**
Cedrela odorata
Celtis integrifolia
C. spinosa
Ceratonia siliqua
Chloroxylon swietenia
*Colophospermum mopane**
Combretum ghasalense
C. glutinosum
Commiphora spp.
C. africana
Cordeauxia edulis
Cupressus arizonica
Cybistax donnell-smithii
Dalbergia sissoo
Diospyros spp.
Dodonaea viscosa
Erythrina senegalensis
Erythrophleum africanum
Eucalyptus alba
E. astringens
E. bicolor
E. blakelyi
E. brockwayi
E. calycogona
*E. camaldulensis**
E. cambageana
*E. citriodora**
E. crebra
E. flocktoniae
E. gardneri
E. glaucina
*E. gomphocephala**
E. gracilis
E. intertexta
E. melliodora
*E. microtheca**
*E. occidentalis**

E. oleosa
E. pilularis
E. platypus
E. populnea
E. pyriformis subsp. *youngiana*
E. robusta
E. rudis
E. salmonophloia
E. salubris
E. stricklandii
*E. tereticornis**
E. tetrodonta
E. torquata
*E. viminalis**
Ficus spp.
Geoffraea decorticans
Gleditsia triacanthos
*Gmelina arborea**
Grevillea pterosperma
Hakea leucoptera
Haloxylon spp.
*H. aphyllum**
*H. persicum**
Heterotheca abaxillaris
Hyphaene thebaica
Inga feyillei
Isoberlinia dalzielii
I. doka
Jacaranda acutifolia
Juglans neotropica
Krugeodendron ferreum
Lannea coromandelica
L. schimperi
Leucodendron argenteum
Lophira lanceolata
Lucuma paradoxa
Lysiloma sabicu
Maerua cressefolia
Melaleuca leucadendron
Melia azedarach
Mitragyna africana
Monotes kerstingii
Morus nigra
Olea europaea
Olneya tesota
Parkia clappertoniana
*Parkinsonia aculeata**
*Pinus brutia**
P. canariensis
P. edulis
*P. eldarica**
*P. halepensis**
P. pinea
Pistacia lentiscus
P. palaestina
P. terebinthus
*Pithecellobium dulce**
Popehax macrantha
Prosopis africana
*P. alba**
P. blanca
*P. caldenia**
*P. chilensis**
*P. cineraria**
*P. farcta**
P. ferox
P. glandulosa
P. inermis
*P. juliflora**
P. nigra
*P. pallida**
P. palmeri
P. pubescens
P. spicigera
P. stephaniana
*P. tamarugo**
P. torquata
Prunus andersoni
Pterocarpus erinaceus
P. lucens
Quercus spp.
Q. coccifera
Q. farnetta
Q. pubescens
Retama roetam
Rhanterium epapposum
Salvadora persica
Schinus molle
Sclerocarya birrea
Sterculia setigera = *S. tomentosa*
Tamarix spp.*
*T. aphylla**
*T. articulata**
T. gallica
T. meyeri
T. passerinoides
T. stricta
*Terminalia glaucescens**
T. tomentosa
Zizyphus abyssinica
*Z. jujuba**
*Z. mauritiana**
*Z. nummularia**
*Z. spina-christi**
Z. vulgaris

APPENDIX 5

Selected Readings

I Wood as Fuel

Adams, J. A. S., M. S. M. Mantovani, and L. L. Lundell. 1977. Wood versus fossil fuel as a source of excess carbon dioxide in the atmosphere: a preliminary report. *Science* 196(4285):54–56.

Anonymous. 1948. Excellent fuel woods. *Indian Forester* 74:279.

Anonymous. 1976. Le deboisement en Haute Volte; les besoins de chauffe de Ouagadougou. *Le Developpement Voltaique* No. 40.

Arnold, J. E. M. 1978. Wood energy and rural communities. Prepared for the 8th World Forestry Congress, Jakarta. Food and Agriculture Organization of the United Nations, Rome, Italy.

Arnold, J. E. M., M. F. E. de Backer, and S. L. Pringle. 1962. *Present Wood Consumption and Future Requirements in Kenya*. Report No. TA1503. Food and Agriculture Organization of the United Nations, Rome, Italy.

de Backer, M. F. E., J. E. M. Arnold, and S. L. Pringle. 1962. *Present Wood Consumption and Future Requirements in Tanzania.* Report No. TA1536. Food and Agriculture Organization of the United Nations, Rome, Italy.

Bebee, C. 1979. *Wood as Fuel and an Alternative Energy Source*. A bibliography of 122 citations covering 1968–1978. Available from Reference Branch, Technical Information Systems, National Agricultural Library, Beltsville, Maryland 20705, USA.

Bene, J. G., H. W. Beal, and A. Cote. 1977. *Trees, Food, and People: Land Management in the Tropics*. International-Development Research Centre, Ottawa, Canada.

Bethel, J. S. 1977. *Wood for Fuel in a Tropical Forest Utilization System*. Research Reports Contribution No. 29. University of Washington, Institute of Forest Products, Seattle, Washington, USA.

Bofinger, P. 1973. Wood heat: participatory energy. *American Forester* 84(10):28–31, 58–60.

Burley, J. 1978. Selection of species for fuelwood plantations. Prepared for the 8th World Forestry Congress, Jakarta. Food and Agriculture Organization of the United Nations, Rome, Italy.

Chandola, L. P. 1976. A pilot survey of fuel consumption in rural areas. *Indian Forestry* 102(10):692–700.

Chow, P. 1977. Wood as fuel: forests can be utilized more fully as an energy source. *Illinois Research* 19(2):6–7.

Corder, S. E. 1973. Wood and bark as fuel. Forest Research Laboratory Research Bulletin No. 14, Oregon State University, Corvallis, Oregon, USA.

Earl, D. E. 1974. *A Report on Charcoal*. Food and Agriculture Organization of the United Nations, Rome, Italy.

Earl, D. E. 1975. *Forest Energy and Economic Development*. Clarendon Press, Oxford, England.

Earl, D. E. 1975. A renewable source of fuel. *Unasylva* 27:110.

Eckholm, E. P. 1975. *The Other Energy Crisis: Firewood*. Worldwatch Paper 1. Worldwatch Institute, Washington, D.C., USA.

Eckholm, E. P. 1976. *Losing Ground: Environmental Stress and World Food Prospects*. W. W. Norton and Co., New York, New York, USA.

Eckholm, E. P. 1979. *Planting for the Future: Forestry for Human Needs*. Worldwatch Paper 26. Worldwatch Institute, Washington, D.C., USA.

Ellis, T. H. 1975. Should wood be a source of commercial power? Fuel for steam-electric plants. *Forest Products Journal* 25(10):13–16.

Evans, I. 1978. Using firewood more efficiently. Prepared for the 8th World Forestry Congress, Jakarta. Food and Agriculture Organization of the United Nations, Rome, Italy.

Ferguson, I. S. 1973. *The Economics of Plantation Forestry in the Savanna Region*. No. FO:DP/NIR/64/516. Project Working Document. Food and Agriculture Organization of the United Nations, Samaru, Nigeria.

Fleuret, P., and A. Fleuret. 1978. Fuelwood use in a peasant community. *Journal of Developing Areas* 12(3):315–322.

Floor, W. M. 1978. Energy options in rural areas of the Third World. Prepared for the 8th World Forestry Congress, Jakarta. Food and Agriculture Organization of the United Nations, Rome, Italy.

Food and Agriculture Organization of the United Nations. 1977. Energy in agriculture. In *The State of Food and Agriculture 1976*. Food and Agriculture Organization of the United Nations, Rome, Italy.

Food and Agriculture Organization of the United Nations. 1978. *China: Forestry Support for Agriculture*. Report No. F1554. Food and Agriculture Organization of the United Nations, Rome, Italy.

Food and Agriculture Organization of the United Nations. 1978. *Forestry for Local Community Development*. Forestry Paper No. 7. Food and Agriculture Organization of the United Nations, Rome, Italy.

Food and Agriculture Organization of the United Nations. 1978. *Rice-Husk Conversion to Energy*. Report No. F1526. Food and Agriculture Organization of the United Nations, Rome, Italy.

Food and Agriculture Organization of the United Nations. 1979. *Forestry for Rural Communities*. Forestry Department, Food and Agriculture Organization of the United Nations, Rome, Italy.

French, D. 1975. Energy crisis, with all its strains,

boosts utility and value of wood fiber. *Forest Industry* 102(9):22–23.

French, D. 1978. Energy for Africa's future. *Africa Report* 23(3):9–14.

Grut, M. 1972. *The Market for Firewood, Poles and Sawnwood in the Major Towns and Cities in the Savanna Region.* No. FO:SF/NIR/16 and Technical Report 6. Food and Agriculture Organization of the United Nations, Rome, Italy.

Gupta, R. K. 1977. Energy forests in farm and community lands. *Indian Farming* 26:84–86.

Hammer, T. 1977. *Wood for Fuel–Energy Crisis Implying Desertification: The Case of Bara, the Sudan.* Department of Geography, University of Bergen, Bergen, Norway.

Hammond, A. L. 1977. Photosynthetic solar energy: rediscovering biomass fuels: wood, sugarcane, algae, and even material produced by artificial photosynthetic processes. *Science* 197(4305):745–746.

Harris, A. C. 1978. Charcoal production. Prepared for the 8th World Forestry Congress, Jakarta. Food and Agriculture Organization of the United Nations, Rome, Italy.

Hayes, D. 1977. *Energy for Development: Third World Options.* Worldwatch Paper 15. Worldwatch Institute, Washington, D.C., USA.

Hayes, D. 1977. *Rays of Hope: The Transition to a Post-Petroleum World.* W. W. Norton and Co., New York, New York, USA.

Houghton, J. E., and L. R. Johnson. 1976. Wood for energy. *Forest Products Journal* 26(4):15–18.

Iannazzi, F. D. 1978. Wastepaper–worth more for its fiber value or as a fuel source? *Pulp and Paper* 52(9):140–143.

International Labour Office. 1975. *Charcoal Making for Small-Scale Enterprises: An Illustrated Training Manual.* International Labour Organization, Geneva, Switzerland.

Johnson, R. C. 1975. Some apsects of wood waste preparation for use as fuel. *TAPPI* (Technical Association of the Pulp and Paper Industry) 58(7):102–106.

Jones, P. 1978. Choosing efficient fuelwood for wood burning. *American Christmas Tree Journal* 22(1): 12.

Kaul, R. N., and H. S. Mann. 1977. Tree planting and energy crisis: firewood. *Indian Farming* 26(11): 79–81.

Krishna, S., and S. Ramaswamy. 1932. Calorific value of some Indian woods. *Indian Forestry Bulletin* (N.S.) No. 79. Chemistry 1–27.

Laurie, M. V. 1974. *Tree Planting Practices in African Savannas.* FAO Forestry Development Paper No. 19. Food and Agriculture Organization of the United Nations, Rome, Italy.

Little, E. C. S. 1972. A kiln for charcoal-making in the field. *Tropical Science* 14(3).

Mabonga-Mwisaka, J. 1978. Wood energy and rural communities in Zambia. Prepared for the 8th World Forestry Congress, Jakarta. Food and Agriculture Organization of the United Nations, Rome, Italy.

Makhijani, A. 1976. *Energy Policy for the Rural Third World.* International Institute for Environment and Development, London, England.

Makhijani, A., and A. Poole. 1975. *Energy and Agriculture in the Third World.* Ballinger Publishing Co., Cambridge, Massachusetts, USA.

Mathur, R. S. 1975. Certain trends in the consumption of wood in India. *Indian Forester* 101(1).

National Academy of Sciences. 1977. *Leucaena: Promising Forage and Tree Crop for the Tropics.* National Academy of Sciences, Washington, D.C., USA.

Openshaw, K. 1971. *Tanzania: Present Consumption and Future Requirements of Wood in Tanzania.* Technical Report No. 3, SF/TAN 15, and Project Working Document. Food and Agriculture Organization of the United Nations, Rome, Italy.

Openshaw, K. 1974. Wood fuels the developing world. *New Scientist* 61:883.

Openshaw, K. 1978. Woodfuel–a time for re-assessment. *Natural Resources Forum* 3:35–51.

Poulsen, G. 1978. Wood-fuel and nutrition: the complementarity of tree cover and food supply. In *Man and Tree in Tropical Africa.* International Development Research Centre, Ottawa, Canada.

Powell, J.W. 1978. Wood waste as an energy source in Ghana. In *Renewable Energy Resources and Rural Applications in the Developing World* (AAAS Selected Symposium), Norman L. Brown, ed. Westview Press, Boulder, Colorado, USA.

Ratcliff, P. 1976. Wood disposal or wood harvesting. Wood chips as fuel. *Journal of Arboriculture* 2(4):79–80.

Richardson, S. D. 1966. *Forestry in Communist China.* The Johns Hopkins University Press, Baltimore, Maryland, USA.

Ripley, T. H., and R.L. Doub. 1978. Wood for energy: an overview. *American Forester* 84(10): 16-19, 42,44,46.

Rose, D. W. 1977. Cost of producing energy from wood in intensive cultures. *Journal of Environmental Management* 5(1):23–35.

Sanger, C. 1977. *Trees for People: An Account of the Forestry Research Program Supported by the International Development Research Centre.* International Development Research Centre, Ottawa, Canada.

Sarkanen, K. V. 1976. Renewable resources: wood for the production of fuels and chemicals. *Science* 191(4228):773–776.

Schneider, M. H. 1977. Energy from forest biomass, wood fuel. *Forestry Chronicle* 53(4):215–218.

Schob, D. E. 1977. Woodhawks, wood haulers and cordwood: steamboat fuel on the Ohio and Mississippi Rivers, 1820–1850. *Journal of Forest History* 21(3):124–132.

Seybold, W. H. 1978. Wood for home heating. The problem of moisture content. Cooperative Extension Programs of the University of Wisconsin, Madison, Wisconsin, USA.

Shafizadeh, F. 1977. Fuels from wood waste, pp. 141–159. In *Fuels from Waste.* L. L. Anderson and D. A. Tillman, eds. Academic Press, New York, New York, USA.

Singer, H. 1961. Improvement of fuelwood cooking stoves and economy in fuelwood consumption. Report No. TA 1315. Food and Agriculture Organization of the United Nations, Rome, Italy.

Smil, V. 1977. Energy solution in China. *Environment* 19(7).

Spurrell, R. M., and D. R. Moody. 1978. Integrated wood waste power plant moves towards energy independence. *TAPPI Engineering Conference Proceedings* (Technical Association of the Pulp and Paper Industry) 1:259–269.

Tatom, J. W. 1976. *Clean Fuels from Agricultural and Forestry Wastes.* Georgia Institute of Technology,

Engineering Experiment Station, Atlanta, Georgia, USA.

Troughton, J. H. 1976. Energy and chemicals from plants. *Forest Industries Review* 8(1):23–24.

Tschinkel, J., and H. Tschinkel. 1975. *Contribution à la Protection des Combustibles Ligneux: Performance et Economie de Quatre Types de Rechauds.* Project Working Document No. 4, TUN 71/540. Food and Agriculture Organization of the United Nations, Rome, Italy.

Uhart, E. 1975. *Charcoal in the Sahelian Zone.* United Nations Economic Commission for Africa, Addis Ababa, Ethiopia.

Uhart, E. 1976. *Charcoal Industry in the Sudan.* United Nations Economic Commission for Africa, Addis Ababa, Ethiopia.

Uhart, E. 1976. *Charcoal Problem in Somalia.* United Nations Economic Commission for Africa, Addis Ababa, Ethiopia.

Uhart, E. 1976. *Le Charbon de Bois à Madagascar.* United Nations Economic Commission for Africa, Addis Ababa, Ethiopia.

Watson, G. 1976. Kiln heating and heat recovery from wood waste. *Australian Forest Industry Journal* 42(2):6–7.

Wayman, M. 1977. High-yield wood: a promising fuel. *Canadian Forest Industries* 97(12):27–29, 31.

Weber, F. R. 1977. *Reforestation in Arid Lands.* Volunteers in Technical Assistance, Mt. Rainier, Maryland, and ACTION/Peace Corps, Washington, D.C., USA.

White, E. W. 1978. Charcoal from wood: fuel for thought. *American Forester* 84(10):20–23, 56, 58.

Whitworth, D. A. 1976. Forest and wood waste utilisation: conversion to fuel alcohol-a. *Forest Industries Review* 8(1):20, 22.

Winklemann, H. 1958. *Wood Burning.* Forestry Occasional Paper No. 1. Food and Agriculture Organization of the United Nations, Rome, Italy.

World Bank. 1978. *Forestry.* Sector Policy Paper. World Bank, Washington, D.C., USA.

Zerbe, J. 1978. The many forms of wood as fuel. *American Forester* 84(10):32–35, 52–54.

Zumbo, J. 1976. Impacts of the wood for energy push. *American Forester* 84(10):38–41.

Fuelwood Species

Acacia auriculiformis

Banerjee, A. K. 1973. Plantations of *Acacia auriculaeformis* (Benth.) A. Cunn. in West Bengal. *Indian Forester* 99:553–540.

Lamb, D. 1975. *Kunjingini Plantations 1965–1975.* Tropical Forestry Research Note Sr. 24. Department of Forests, P.O. Box 5055, Boroko, Papua New Guinea.

National Academy of Sciences. 1979. *Tropical Legumes: Resources for the Future.* National Academy of Sciences, Washington, D.C., USA.

Nicholon, E. I. 1965. A note on *Acacia auriculiformis* A. Cunn. ex Benth. in Sabah. *Malayan Forester* 28(3):243–244.

Parry, T. S. 1954. Tree planting in Tanzania, IV: Species for coastal areas. *East African Agriculture Journal* 20(1):49–53.

Sastroamidjojo, J. S. 1964. *Acacia auriculiformis* A. Cunn. *Rimba Indonesia* 9(3):214–225. (Indonesian with English summary.)

Streets, R. J. 1962. *Exotic Forest Trees in the British Commonwealth.* Clarendon Press, Oxford, England.

Soekarpi. 1955. Some notes on *Acacia auriculiformis. Forest Research Institute Report* No. 71. Bogor, Indonesia. (In Indonesian.)

Acacia brachystachya

Maiden, J. H. 1917. *Acacia brachystachya* Benth. *Forest Flora of New South Wales* 7:9–14.

Pedley, L. 1973. Taxonomy of the *Acacia aneura* complex. *Tropical Grasslands* 7(1):3–8.

Pedley, L. 1978. A revision of *Acacia* Mill. in Queensland. *Austrobaileya* 1(2):75–234 and 1(3):235.

Acacia cambagei

Hall, N., J. W. Turnbull, and M.I. H. Brooker. 1975. *Acacia cambagei* R. T. Bak. *Australian Acacias.* Forestry and Timber Bureau, Australian Department of Agriculture, Canberra, A.C.T., Australia.

Acacia cyclops

LeRoux, P. J. 1974. Establishing vegetation in saline soil to stabilize aolian sand at Walvis Bay, South West Africa. *Forestry in South Africa* 15:43–46.

Ross, J. H. 1975. Fabaceae Mimosoideae. *Flora of Southern Africa* 16(1). (Government Printer, Pretoria, South Africa.)

Roux, E. R. 1965. Salt tolerance in four invasive exotic acacias of the Cape Peninsula. *South African Journal of Science* 61:438.

Roux, E. R., and J. L. Warren. 1963. Symbiotic nitrogen fixation in *A. cyclops* A. Cunn. *South African Journal of Science* 59(6):294–295.

Seddon, G. 1972. *Sense of Place.* University of Western Australia Press, Nedlands, Western Australia 6009, Australia. (Distributed in the United States by International Scholarly Book Services, Inc., Forest Grove, Oregon, USA.)

Stirton, C. H., ed. 1978. *Plant Invaders: Beautiful but Dangerous,* pp. 40–43. Public Department of Nature and Environmental Conservation of the Cape Provincial Administration, Cape Town, South Africa.

Acacia mearnsii

Alphen van Veer, E. J. 1949. Controlling erosion by regeneration of *Acacia decurrens* stands on Java. *Tectona* 39:389–392. (In Dutch.)

Coster, C. 1939. The importance of *Acacia decurrens* in the Dutch Indies. *Tectona* 32:368-388. (In Dutch.)

Food and Agriculture Organization of the United Nations. 1975. *Pulping and Papermaking Properties for Fast-Growing Plantation Wood Species.* No. FO:MISC/75/31. Food and Agriculture Organization of the United Nations, Rome, Italy.

Kennedy, G. G. 1969. Wattle–an alternative to wheat or maize. *Kenya Farmer* 152:18–19.

Poynton, R. J. n.d. *Characteristics and Uses of Trees and Shrubs Obtainable from the Forest Department.* Bulletin No. 39. The Government Printer, Pretoria, South Africa.

Streets, R. J. 1962. *Exotic Forest Trees in the British Commonwealth.* Clarendon Press, Oxford, England.

Wiersum, K. F. 1976. Some Notes on *Acacia decurrens* (var. Mollis). Mimeographed report. (Copies available from author; see Research Contacts.)

Acacia nilotica* subsp. *indica

Anonymous. 1948. Excellent fuel woods. *Indian Forester* 27:280.

Chaturvedi, M. D. 1955. No tree fit to compare with Babul. *Indian Farming* 5:15.
Gamble, J. S. 1972. *A Manual of Indian Timbers: Mimosae*, pp. 292–294. Second reprint. Bishen Singh Mahendra Pal Singh, 23-A New Connaught Place, Dehra Dun, India.
Khan, M. A. W. 1970. Phenology of *Acacia nilotica and Eucalyptus microtheca* at Wad Medani (Sudan). *Indian Forester* 96(3):226–248.
Krishna, S., and S. Ramaswamy. 1932. Calorific value of some Indian wood. *Indian Forestry Bulletin* (N.S.) No. 79. Chemistry 1–27.
Shetty, K. A. B. 1977. Social forestry in Tamil Nadu. *Indian Farming* 26(11):82.
Singh, R. 1976. *A National Programme for Rodent and Pest Management, Weed Control and Planting and Saving of Trees*. Indian Council of Agricultural Research, New Delhi, India.
Wealth of India: Raw Materials. 1948. *Acacia*. Vol. I, pp. 5–20. Council of Scientific and Industrial Research, New Delhi, India.
Yadav, J. S. P. 1977. Tree growth on salt-affected lands. *Indian Farming* 26(11):43–45.

Acacia saligna
Aveyard, J. M. 1968. The effect of seven pre-sowing seed treatments on total germination rate for six *Acacia* species. *Journal of the Soil Conservation Service of New South Wales* 24(1):43–54.
Barr, D. A., and W. J. Atkinson. 1970. Stabilization of coastal sands after mining. *Journal of the Soil Conservation Service of New South Wales* 26(2): 89–107.
Hall, N., and W. Turnbull. 1976. *Acacia saligna* (Labill.) H.Wendl. *Australian Acacias*. Division of Forest Research, Commonwealth Scientific and Industrial Research Organization, Canberra, Australia.
Maslin, B. R. 1974. Studies in the genus *Acacia*–3–The Taxonomy of *A. saligna* (Labill.)H.Wendl. *Nuytsia* 1(4):332–340.
Michaelides, E. D. 1979. *Acacia cyanophylla*. Mimeo report prepared for the Technical Consultation on Fast-Growing Plantation Broadleaved Trees for Mediterranean and Temperate Zones, Lisbon, Oct 1979. (Copies available from the author; See Research Contacts.)
Ross, J. H. 1975. The naturalized and cultivated exotic *Acacia* species in South Africa. *Bothalia* 11(4): 463–470.
Roux, E. R. 1961. History of the introduction of Australian *Acacias* on the Cape Flats. *South African Journal of Science* 57(4):99–102.

Acacia senegal
Dalziel, J. M. 1937. *Useful Plants of West Tropical Africa*. Crown Agents, London, England.
Hammer, T. 1977. *Wood for Fuel: Energy Crisis Implying Desertification, the Case of Bara, The Sudan*. Department of Geography, University of Bergen, Norway.
Kaul, R. N. 1970. *Afforestation in Arid Zones*. Monographiae Biologicae. Vol. 20. Dr. W. Junk N.V. Publishers, The Hague, Netherlands.
National Academy of Sciences. 1979. *Tropical Legumes: Resources for the Future*. National Academy of Sciences, Washington, D.C., USA.
Streets, R. J. 1962. *Exotic Forest Trees in the British Commonwealth*. Clarendon Press, Oxford, England.

Acacia seyal
Brenan, J. P. M. 1970. *Flora Zambesiaca*. Vol. 3., part 1, pp. 88–89. Crown Agents, London, England.
Keay, R. W. J., C. F. A. Onochie, and D. P. Stanfield. 1960–64. *Nigerian Trees*. Vol. II. Department of Forest Research, Ibadan, Nigeria.

Acacia tortilis
Brenen, J. P. M. 1957. Notes on Mimosoideae: III. *Kew Bulletin* 12:86–89.
Carr, J. D. 1976. *The South African Acacias*. Conservation Press (Pty) Ltd., P.O. Box 2107, Johannesburg, South Africa.
El Amin, H. M. n.d. *Sudan Acacias*. Forest Research Institute Bulletin No. 1. Khartoum, Sudan.
Gupta, R. K., and G. S. Balara. 1972. Comparative studies on the germination, growth and seedling biomass of two promising exotics in Rajasthan desert: *Prosopis juliflora* (Swartz) DC and *Acacia tortilis* (Forsk.) Hayne spp. *tortilis*. *Indian Forester* 98(5):280–285.
Karschon, R. 1975. *Seed germination of* Acacia raddiana *Savi and* A. tortilis *Hayne as related to infestation by Bruchids*. Leaflet No. 52. Agricultural Research Organization, Division of Scientific Publications, Bet Dagan, Israel.
Kaul, R. N. 1970. Indo-Pakistan. In *Afforestation in Arid Zones*, pp. 155–210. Monographiae Biologicae. Vol. 20. Dr. W. Junk N.V. Publishers, The Hague, Netherlands.
Mariaux, A. 1975. A dendroclimatology trial on *Acacia raddiana* in the Sahelian climate. *Bois et Forêts des Tropiques* 163:27–35.
Muthana, K. D., and G. D. Arora. 1979. *Acacia tortilis* (Forsk)–a promising fast growing tree for Indian arid zones. Mimeographed report. Central Arid Zone Research Institute, Jodhpur, India.
National Academy of Sciences. 1979. *Tropical Legumes: Resources for the Future*. National Academy of Sciences, Washington, D.C., USA.
Puri, D. N., K. D. Muthana, D. P. Handa, and M. Singh. 1973. Study on the comparative growth and establishment of *Acacia tortilis* (Forsk.) and *Acacia senegal* in rocky habitats. *Annals of the Arid Zone* 12:167–171.
Roy, A. D., R. N. Kaul, and Gyanchand. 1973. Israeli babool–a promising tree for arid and semiarid lands. *Indian Farming* 23(8):19–20.

Adhatoda vasica
Chopra, R. N., J. C. Chopra, K. L. Handa, and L. D. Kapur. 1958. *Chopra's Indigenous Drugs of India*. Second edition. U.N. Dhur & Sona, Calcutta, India.
Chopra, R. N., S. L. Nayar, and I. L. Chopra. 1956. *Glossary of Indian Medicinal Plants*. Council of Scientific and Industrial Research, New Delhi, India.
Dakshini, K. M. M. 1972. Indian subcontinent. In *Wildland Shrubs: Their Biology and Utilization*, pp. 3–15. C. H. Mekell, J. P. Blaisdell, and J. R. Goodin, eds., USDA Forest Service, General Technical Report INT-1. Ogden, Utah, USA.
Duthie, J. F. 1973. *Flora of the Upper Gangetic Plain and of the Adjacent Siwalik and sub-Himalayan Tracts. Acanthaceae*. p. 207. Reprint edition. Bishen Singh Mahendra Pal Singh, 23-A New Connaught Place, Dehra Dun, India.
Gamble, J. S. 1972. *A Manual of Indian Timbers: Acanthaceae*, p. 523. Second reprint. Bishen Singh

Mahendra Pal Singh, 23-A New Connaught Place, Dehra Dun, India.
Jain, S. K. 1968. *Medicinal Plants*, pp. 11–13. National Book Trust, New Delhi, India.
Watt, G. 1966. *The Commercial Products of India*. Today and Tomorrow's Printers and Publishers, New Delhi, India.
Wealth of India. Raw Materials. 1948 *Adhatoda*. Vol. 1, pp. 31–32. Council of Scientific and Industrial Research, New Delhi, India.

Ailanthus altissima
Cozzo, D. 1972. Comportamiento inicial de *Ailanthus altissima* en una plantacion experimental. *Revista Forestal Argentina* 16:47–52.
Goor, A. Y., and C. W. Barney. 1976. *Forest Tree Planting in Arid Zones*. Second edition. The Ronald Press Company, New York, New York, USA.
Guhathakurta, P., and R. C. Ghosh. 1972. *Ailanthus grandis* Prain–its prospect in forestry. *Indian Forester* 98:261–270.
Moslemi, A. A., and S. G. Bhagquat. 1970. Physical and mechanical properties of the wood of tree-of-heaven. *Wood and Fiber* 1:319–323.
Richardson, S. D. 1966. *Forestry in Communist China*, pp. 77–80, 114. Johns Hopkins University Press, Baltimore, Maryland, USA.

Albizia lebbek
Abraham, P. 1957. Karana is excellent as shade tree for cardamom. *Indian Farming* 7:14–16.
Gamble, J. S. 1972. *A Manual for Indian Timbers: Mimosaceae*, pp. 303–304. Second reprint. Bishen Singh Mahendra Pal Singh, 23-A New Connaught Place, Dehra Dun, India.
Guha, S. R. D., and B. D. Prasad. 1961. Chemical pulps for writing and printing papers from *Albizzia lebbek* Benth. (Siris). *Indian Pulp and Paper* 15:487–489.
Krishna, S., and S. Ramaswamy. 1932. Calorific value of some Indian woods. *Indian Forestry Bulletin* (N.S.) No. 79. Chemistry 1–27.
Shetty, K. A. B. 1977. Social forestry in Tamil Nadu. *Indian Farming* 26(11):82.
Wealth of India: Raw Materials. 1948. *Albizia lebbek*. Vol. 1, p. 43. Council of Scientific and Industrial Research, New Delhi, India.
Yadav, J. S. P. 1977. Tree growth on salt-affected lands. *Indian Farming* 26(11):43–45.

Alnus acuminata
Balza Viloria, M. 1959. Estudios sobre Aliso (*Alnus jorullensis*) y su regeneracion natural en el Valle del Alto Chama (Tesis de grado). Universidad de Los Andes, Merida, Venezuela.
Bond, G. 1976. The results of the IBP survey of root-nodule formation in non-leguminous angiosperms. In *Symbiotic Nitrogen Fixation in Plants*, P. S. Nutman, ed. International Biological Programme 7. Cambridge University Press, Cambridge, England.
Hunt, I. S. 1967. Las ptopiedades y usos de la madera de *Alnus. Instituto Forestal Latinoamericano de Investigacion y Capacitacion. Boletin* 23:29–42.
Rosero, P. 1975. *Alnus jorullensis* HBK. Conservation of Gene Resources Data Sheet, IUFRO Working Party S2.02.2. Copies available from Centro Tropical de Enseñanza Investigación (see Research Contacts).
Sicco Smit, G. 1971. Notas silviculturales sobre el *Alnus jorullensis* de Caldas, Colombia. *Turrialba* 21(1):83–88.
A Data sheet on this species is available from Professor L. Roche, Chairman, FAO Working Party on Conservation of Gene Resources, Department of Forestry and Wood Science, University College of North Wales, Bangor, Gwynedd LL57 2UW, United Kingdom.

Alnus nepalensis
Gerhards, C. C. 1964. Limited evaluation of the physical and mechanical properties of Nepal alder grown in Hawaii. Research Note FPL-036. Forest Products Laboratory, U.S. Forest Service, Madison, Wisconsin, USA.
Lutz, J. F., and G. G. Roessler. 1964. Veneer and plywood characteristics of Nepal alder. Unpublished report. Forest Products Laboratory, U.S. Forest Service, Madison, Wisconsin, USA.
Peters, C. C., and J. F. Lutz. 1966. Some machining properties of two wood species grown in Hawaii, Molucca albizzia and Nepal alder. Research Note FPL-0117. Forest Products Laboratory, U.S. Forest Service, Madison, Wisconsin, USA.
Tanner, R. M. 1964. Clones of Nepal alder in Hawaii. *Journal of Forestry* 62(9):636–637.

Alnus rubra
Bergstron, D. 1979. Let's harness the energy of red alder. Copies available from Pacific Northwest Forest and Range Experiment Station, Forest Service, U.S. Department of Agriculture, 809 NE Sixth Avenue, Portland, Oregon 97208, USA.
Briggs, D. G., D. S. DeBell, and W. A. Atkinson, eds. 1978. *Utilization and Management of Alder*. Pacific Northwest Forest and Range Experiment Station, Forest Service, U.S. Department of Agriculture, Portland, Oregon, USA. For sale by Government Book Store, 194 Federal Building, 915 Second Avenue, Seattle, Washington 98174.
Trappe, J. M., J. F. Frnklin, R. F. Tarrant, and G. M. Hansen, eds. 1968 (reprinted 1978). *Biology of Alder*. Pacific Northwest Forest and Range Experiment Station (address above).
U.S. Forest Service. 1974. *Seeds of Woody Plants in the United States*, pp. 206–211. Agriculture Handbook No. 450. For sale by the Superintendent of Documents, U.S. Government Printing Office, Washington, D.C. 20402, USA.

Anogeissus latifolia
Ghosh, R. C. 1977. *Handbook on Afforestation Techniques*. Controller of Publications, Delhi, India.
Guha, S. R. D., Y. K. Sharma, and R. P. Goyal. 1974. Pulping of *Anogeissus latifolia* (Roxb.) Bedd. *Indian Forester* 100(5):351–356.
Pearson, R. S. 1932. *Commercial Timbers of India*. Vol. 1, pp. 537–543. Government of Calcutta, Calcutta, India.
Verma, S. K. 1972. Observation on the mortality in the forests of *Anogeissus* in Rajasthan. *Indian Forester* 93(3):199–205.
Vyas, L. N., R. K. Garg, and M. P. S. Ranawat. 1973. Plant biomass and net production of *Anogeissus latifolia* Wall. in forests of semiarid zone of Rajasthan (India). *Biologia Plantarum* 15(4):280–285.
Vyas, L. N., R. K. Garg, and M. P. S. Ranawat, 1974. Branch dimensions and estimation of branch

productions in *Anogeissus latifolia* Wall. *Geobios* (Jodhpur) 1(1):3-6.

Azadirachta indica

Gravsholt, S. 1967. *Provisional Tables for Growth and Yield of Neem (*Azadirachta indica*) in Northern Nigeria.* Research Paper No. 1. Savanna Forestry Research Station, Samaru, Zaria, Nigeria.

Kemp, R. H. 1969. *Trials of Exotic Tree Species in the Savanna Region of Nigeria. Part 1. Aims, Procedure and Summary of Results.* Research Paper No. 4. Savanna Forestry Research Station, Samaru, Zaria, Nigeria.

Ketkar, C. M. 1976. *Utilization of Neem (*Azadirachta indica *Juss.) and its By-products.* Final Technical Report. Directorate of Non-Edible Oils and Soap Industry, Khadi and Village Industries Commission, Gramodaya, 3 Irla Road, Vile Parle (W), Bombay 400 056, Maharashtra, India.

Mackay, J. H. 1952. Notes on the establishment of Neem plantations in Bornu Province of Nigeria. *Farm and Forest* 11:9–13.

Mitra, C. R. 1963. *Neem.* Indian Central Oilseeds Committee, Himayatnagar, Hyderabad, Andhra Pradesh, India.

Radwanski, S. A. 1969. Improvement of red acid sands by Neem tree (*Azadirachta indica*) in Sokoto, North-Western State of Nigeria. *Journal of Applied Ecology* 6:507–511.

Radwanski, S. A. 1977. Neem tree. *World Crops and Livestock.* 29:62–66, 111–113, 167–168, 222–224.

Troup, R. S. 1921. *The Silviculture of Indian Forest Trees.* Vol. 1. Clarendon Press, Oxford, England.

Cajanus cajan

Akinola, J. O., P. C. Whiteman, and E. S. Wallis. 1975. The agronomy of pigeonpea (*Cajanus cajan* (L.) Millsp). Review Series No. 1/1975. Commonwealth Bureau of Pastures and Field Crops, Hurley, Maidenhead, Berks SL6 5LR, England.

Commonwealth Bureau of Pastures and Field Crops. n.d. Pigeon peas (*Cajanus cajan*). In *Annotated Bibliography 1253, 1959–1970.* Commonwealth Bureau of Pastures and Field Crops, Hurley, Maidenhead, Berks, SL6 5LR, England.

Gooding, M. J. 1962. The agronomic aspects of pigeonpeas. *Field Crop Abstracts* 15:1–5.

Kay, D. E. 1979. *Food Legumes*, pp. 322–347. Tropical Products Institute, 56/62 Gray's Inn Road, London WC1X 8LU, England.

Morton, J. F. 1976. The pigeon pea (*Cajanus cajan* Millsp.), a high protein tropical bush legume. *Horticultural Science* 11:11–19.

Pathak, G. N. 1970. Red gram. In *Pulse Crops of India*, pp. 14–53. Indian Council of Agricultural Research, New Delhi, India.

Theisen, A. A., E. G. Knox, and F. L. Mann. 1978. *Feasibility of Introducing Food Crops Better Adapted to Environmental Stress.* Vol. 1. No. NSF/RA-780289. National Science Foundation, Washington, D.C., USA.

Calliandra calothyrsus

Anonymous. 1977. *The Possibility of Kaliandra Wood as a Source of Energy.* Special report of the Forest Products Research Institute, Bogor, Indonesia.

Suyono. 1975. *Calliandra* as fuelwood and forest protector. *Duta Rimba* 3:3–6, 11 and 4:9–12, 33. (In Indonesian.)

Verhoef, L. 1941. Preliminary results of species trials with some legumes from tropical America. *Tectona* 34:711–736. (In Dutch.)

Cassia siamea

Little, E. L., and F. H. Wadsworth. 1964. *Common Trees of Puerto Rico and the Virgin Islands.* Agriculture Handbook No. 249. Forest Service, U.S. Department of Agriculture, Washington, D.C., USA.

Razzaque, M. A., A. B. Siddique, and P. Das. 1970. Pulping and paper making studies of minjri (*Cassia siamea*) wood. *Science and Industry* 7(1 and 2): 26–31.

Venkateremany, S. P. 1968. *The Silviculture of the Species of the Genus* Cassia *Linn.* Forest Research Institute and College, Dehra Dun, India.

Weber, F. 1977. *Reforestation in Arid Lands.* Volunteers in Technical Assistance, Mt. Rainier, Maryland, USA.

Casuarina equisetifolia

Anonymous. 1973. The properties and potential uses of velau (*Casuarina* sp.). In *Fiji Timbers and Their Uses.* Department of Forestry, Fiji.

Bond, G. 1976. The results of the IBP survey of root-nodule formation in non-leguminous angiosperms. In *Symbiotic Nitrogen Fixation in Plants*, P. S. Nutman, ed. International Biological Programme 7. Cambridge University Press, Cambridge, England.

Hall, N., R. W. Boden, C. S. Christian, R. W. Condon, F. A. Dale, A. J. Hart, J. H. Leigh, J. K. Marshall, A. G. McArthur, V. Russell, and J. W. Turnbull. 1972. *The Use of Trees and Shrubs in the Dry Country of Australia.* Australian Government Publishing Service, Canberra, Australia.

Hall, N., R. D. Johnston, and G. M. Chippendale. 1970. *Forest Trees of Australia.* Australian Government Publishing Service, Canberra, Australia.

Olson, D. F., Jr., and E. Q. P. Petteys. 1974. *Casuarina* L. *Agricultural Handbook* No. 450, pp. 278–280. U.S. Department of Agriculture, Washington, D.C., USA.

Rao, E. V. 1968. *A Study of* Casuarina equisetifolia *Forst.* Commonwealth Forestry Institute, South Parks Road, Oxford OX1 3RB, England.

Ray, M. P. 1971. Plantations of *Casuarina equisetifolia* in the Midnapore district of West Bengal. *Indian Forester* 97:443–457.

Somasundaram, T. R., and S. Jagadees. 1977. Propagation of *Casuarina equisetifolia* Forst. by planting shoots. *Indian Forester* 103(11):735–739.

Streets, R. J. 1962. *Exotic Forest Trees in the British Commonwealth*, pp. 211–218. Clarendon Press, Oxford, England.

Wealth of India: Raw Materials. 1950. *Casuarina* Linn. Vol. II, pp. 101–103. Council of Scientific and Industrial Research, New Delhi, India.

Colophospermum mopane

Eyles, P. A. 1971. *The Effect of Soil Conditions and Rhizobium Treatment on the Growth and Nitrogen Content of* Colophospermum mopane. B.Sc. Honours dissertation, University of the Witwatersrand, Johannesburg, South Africa.

Henning, A. C. 1976. *A Study of Edaphic Factors Influencing the Growth of* Colophospermum mopane *(Kirk ex Benth.) Kirk ex J. Leonard.* Ph.D. dissertation, University of the Witwatersrand, Johannesburg, South Africa.

Jarman, P. J., and P. I. Thomas. 1969. Observation on the distribution and survival of mopane. *Kirkia* 7:103–107.

Lawton, R. M. 1967. The value of browse in the dry tropics. *East African Agriculture and Forestry Journal* 33:227–230.

Palmer, E., and N. Pitman. 1972. *Trees of Southern Africa*. Vol. 2, pp. 842–845. A. A. Balkema, Cape Town, South Africa.

Pardy, A. A. 1953. Notes on indigenous trees and shrubs of Southern Rhodesia. *Rhodesia Agricultural Journal* 50:152–153.

Porter, R. N. 1968. *The Distribution and Ecology of* Colophospermum mopane. B.Sc. Honours dissertation, University of Pretoria, South Africa.

Derris indica

Brown, W. H. 1954. *Useful Plants of the Philippines*. Vol. 2, pp. 157–160. Department of Agriculture and Natural Resources, Manila, Philippines.

Mandal, L., and G. C. Banerjee. 1975. Extracted karanja cake (*Pongamia glabra*)—a new feed ingredient for poultry. *Indian Poultry Review* 6(12):480–481.

Emblica officinalis

Das, S. B. 1949. A note on the economic importance of *Phyllanthus emblica. Indian Forester* 75(7): 243–245.

Gamble, J. S. 1972. *A Manual of Indian Timbers: Euphorbiaceae*. Second reprint, pp. 599–600. Bishen Singh Mahendra Pal Singh, 23-A New Connaught Place, Dehra Dun, India.

Ghosh, R. C. 1977. *Handbook on Afforestation Techniques*. Controller of Publications, Delhi, India.

Lynch, S. J., and F. J. Fuchs, Sr. 1955. A note on the propagation of *Phyllanthus emblica* L. *Proceedings of the Florida State Horticulture Society* 68:301–302.

Morton, J. F. 1960. The emblic (*Phyllanthus emblica* L.). *Economic Botany* 14(2):119–128.

Singh, L. B. 1956. A new technique for propagating aonla (*Phyllanthus emblica*). *Science and Culture* 17:345–346.

Watt, G. 1966. *The Commercial Products of India*, p. 886. Today and Tomorrow's Printers and Publishers, New Delhi, India.

Wealth of India: Raw Materials. 1952. *Emblica*. Vol. III, pp. 168–170. Council of Scientific and Industrial Research, New Delhi, India.

Eucalyptus camaldulensis

Barrett, R. L., and D. T. Carter. 1976. *Eucalyptus camaldulensis* provenance trials in Rhodesia (later results). *Rhodesian Bulletin of Forestry Research* No. 2(2).

Eldridge, K. G. 1975. *An Annotated Bibliography of Genetic Variation in* Eucalyptus camaldulensis. Tropical Forestry Paper No. 8. Commonwealth Forestry Institute, Oxford, and Division of Forest Research, Commonwealth Scientific and Industrial Research Organization, Canberra, Australia.

Hall, N., R. W. Boden, C. S. Christian, R. W. Condon, F. A. Dale, A. J. Hart, J. H. Leigh, J. K. Marshall, A. G. McArthur, V. Russell, and J. W. Turnbull. 1972. *The Use of Trees and Shrubs in the Dry Country of Australia*. Australian Government Publishing Service, Canberra, Australia.

Hall, N., R. D. Johnston, and G. M. Chippendale. 1970. *Forest Trees of Australia*, pp. 100–101. Australian Government Publishing Service, Canberra, Australia.

Hillis, W. E., and A. G. Brown, eds. 1978. *Eucalypts for Wood Production*. Commonwealth Scientific and Industrial Research Organization, Melbourne, Australia.

Karschon, R. 1974. The relation of seed origin to growth of *Eucalyptus camaldulensis* Dehn. in Israel. *Israel Journal of Agriculture Research* 23:159–173.

Turnbull, J. W. 1973. The ecology and variation of *Eucalyptus camaldulensis*, pp. 32–40. Forestry Occasional Paper 1973/2. Forest Genetic Resources Information, Food and Agriculture Organization of the United Nations, Rome, Italy.

Weber, F. 1977. *Reforestation in Arid Lands*. Volunteers in Technical Assistance, Mt. Rainier, Maryland, USA.

Eucalyptus citriodora

Hall, N., R. D. Johnston, and G. M. Chippendale. 1970. *Forest Trees of Australia*, pp. 44–45. Australian Government Publishing Service, Canberra, Australia.

Hillis, W. E., and A. G. Brown, eds. 1978. *Eucalypts for Wood Production*. Commonwealth Scientific and Industrial Research Organization, Melbourne, Australia.

Jacobs, M. 1976. *Eucalypts for Planting*. No. FO: MISC/76/10. Forest Management Branch, Food and Agriculture Organization of the United Nations, Rome, Italy.

Larsen, E. 1965. A study of the variability of *Eucalyptus maculata* Hook. and *Eucalyptus citriodora Hook.* Leaflet No. 95. Forestry and Timber Bureau, Department of Agriculture, Canberra, A.C.T., Australia.

Eucalyptus globulus

Brown, A., and N. Hall. 1968. *Growing Trees on Australian Farms*. Forestry and Timber Bureau, Department of Agriculture, Canberra, A.C.T., Australia.

Hall, N., R. D. Johnston, and G. M. Chippendale. 1970. *Forest Trees of Australia*. Government Publishing Service, Canberra, Australia.

Hillis, W. E., and A. G. Brown, eds. 1978. *Eucalypts for Wood Production*. Commonwealth Scientific and Industrial Research Organization, Melbourne, Australia.

Jacobs, M. 1976. *Eucalypts for Planting*. No. FO: MISC/76/10. Forest Management Branch, Food and Agriculture Organization of the United Nations, Rome, Italy.

Kirkpatrick, J. 1975. *Geographical Variation in* Eucalyptus globulus. Bulletin No. 47. Forestry and Timber Bureau, Department of Agriculture, Australian Government Publishing Service, Canberra, Australia.

Turnbull, J. W. 1974. *Gene Resource Conservation IUFRO (International Union of Forest Research Organizations) Working Party* S2.02.2. Data sheet, *Eucalyptus globulus*. Forestry Occasional Paper No. 1. Forestry and Forest Products Division, Food and Agriculture Organization of the United Nations, Rome, Italy.

A Data sheet on this species is available from Professor L. Roche, Chairman, FAO Working Party on Conservation of Gene Resources, Department of Forestry and Wood Science, University College of North Wales, Bangor, Gwynedd LL57 2UW, United Kingdom.

Eucalyptus gomphocephala

Hall, N., R. D. Johnston, and G. M. Chippendale. 1970. *Forest Trees of Australia*. Australian Government Publishing Service, Canberra, Australia.

Hillis, W. E., and A. G. Brown, eds. 1978. *Eucalypts for Wood Production*. Commonwealth Scientific and Industrial Research Organization, Melbourne, Australia.

Jacobs, M. 1976. *Eucalypts for Planting*. No. FO: MISC/76/10. Forest Management Branch, Food and Agriculture Organization of the United Nations, Rome, Italy.

Eucalyptus grandis

Anonymous. 1972. *Handbook on Eucalypts Growing*. The Wattle Research Institute, P.O. Box 375, Pietermaritzburg 3200, Republic of South Africa.

Barrett, R. L., D. T. Carter, and B. R. T. Seward. 1975. *Eucalyptus grandis* in Rhodesia. *Rhodesian Bulletin of Forestry Research* No. 6.

Clarke, B. 1975. Establishment of eucalypts plantations, Coffs Harbour, New South Wales. *Science and Technology* 12(6).

Dyson, W. G. 1974. Experiments on growing *Eucalyptus* wood fuel in the semi-deciduous forest zone in Kenya. *East African Agricultural and Forestry Journal* 39(4):349-355.

Hall, N., R. D. Johnston, and G. M. Chippendale. 1970. *Forest Trees of Australia*. Australian Government Publishing Service, Canberra, Australia.

Hillis, W. E., and A. G. Brown, eds. 1978. *Eucalypts for Wood Production*. Commonwealth Scientific and Industrial Research Organization, Melbourne, Australia.

Howland, P. 1969. Effects of singling coppice in *Eucalyptus saligna* wood fuel crops at Muguga, Kenya. *East African Agriculture and Forestry Journal* 35(1):66-67.

Howland, P., and G. H. Freeman. 1970. Interim results of a fuel yield trial on eucalypts. *East African Agricultural and Forestry Journal* 35(3):257-264.

Streets, R. J. 1962. *Exotic Trees of the British Commonwealth*. Clarendon Press, Oxford, England.

Wimbush, S. H. 1946. *The Planting and Management of* Eucalyptus *for Fuel*. Pamphlet No. 9. Kenya Forest Department, Nairobi, Kenya.

Eucalyptus microtheca

Hall, N., R. W. Boden, C. S. Christian, R. W. Condon, F. A. Dale, A. J. Hart, J. H. Leigh, J. K. Marshall, A. G. McArthur, V. Russell, and J. W. Turnbull. 1972. *The Use of Trees and Shrubs in the Dry Country of Australia*. Australian Government Publishing Service, Canberra, Australia.

Hall, N., R. D. Johnston, and G. M. Chippendale. 1970. *Forest Trees of Australia*, pp. 220-221. Australian Government Publishing Service, Canberra, Australia.

Hillis, W. E., and A. G. Brown, eds. 1978. *Eucalypts for Wood Production*. Commonwealth Scientific and Industrial Research Organization, Melbourne, Australia.

Jacobs, M. 1976. *Eucalypts for Planting*. No. FO: MISC/76/10. Forest Management Branch, Food and Agriculture Organization of the United Nations, Rome, Italy.

Khan, W. M. A. 1966. *Direct Sowing of* Eucalyptus microtheca *in Polyethene Tubes*. Pamphlet No. 34. Forestry Research and Education Project, The Sudan.

Khan, W. M. A. 1966. *Growth of* Eucalyptus microtheca *in Conjunction with Agricultural and Fodder Crops*. Pamphlet No. 35. Forestry Research and Education Project, The Sudan.

Malrauda, G. S. 1959. Eucalypts for the dry zone. *Indian Forester* 85:211-224.

Eucalyptus occidentalis

Hillis, W. E., and A. G. Brown, eds. 1978. *Eucalypts for Wood Production*. Commonwealth Scientific and Industrial Research Organization, Melbourne, Australia.

Gliricidia sepium

Little, E. L., and F. H. Wadsworth. 1964. *Common Trees of Puerto Rico and the Virgin Islands*. Agriculture Handbook No. 249, Forest Service, U.S. Department of Agriculture, Washington, D.C.

Gmelina arborea

Adam, J. G. 1975. Un grand project de plantation de Gmelina (*Gmelina arborea* Roxb.) au Liberia. *Journal d'Agriculture Tropicale et de Botanique Appliquée* 22(1 and 3):59-66.

Anonymous. 1964. The silviculture of *Gmelina arborea* (Roxb.). *Nyasaland Farmer and Forester* 6:13-22.

Boulet, M.-G. 1977. Monographie du *Gmelina arborea*. *Bois et Forêts des Tropiques* 172:2-23.

Doat, J. 1976. Caracteristiques papetieres d'une essence tropicale de reboisement le *Gmelina arborea*. *Bois et Forêts des Tropiques* 168:47-63.

Greaves, A. 1979. Gmelina–large scale planting, Jarilandia, Amazon basin. *The Commonwealth Forestry Review* 58(4):267.

Lamb, A. F. A. 1968. Gmelina arborea. *Fast Growing Timber Trees of the Lowland Tropics*. Commonwealth Forestry Institute, University of Oxford, England. (Also available in Spanish translation: Especies maderables de crecimiento rapido en la Tierra Baja Tropical *Gmelina arborea*. *Instituto Forestal Latinoamiercano de Investigacion y Capacitacion. Boletin Bibliografico.* 33/34:21-51.)

Lauridsen, E. B. 1977. *Gmelina arborea*–international provenance trials study trees and soil collection in India, 1976. In *Forest Genetics Resources–Information*, No. 6. pp. 24-37. Food and Agricultural Organization of the United Nations, Rome, Italy.

Mitchell, B. A. 1963. Possibilities for forest plantations. *Malayan Forester* 26:259-286.

Palmer, E. R. 1973. *Gmelina arborea* as a potential source of hardwood pulp. *Tropical Science* 15(3): 243-260.

Sharma, R. P., and R. C. Jain. 1977. Standard volume tables for *Gmelina arborea* Roxb. *Indian Forester* 103(8):536-539.

Grevillea robusta

Food and Agriculture Organization of the United Nations. 1977. Report of the FAO panel of experts on forest gene resources, Canberra, Australia, 9-11 March 1977. Food and Agriculture Organization of the United Nations, Rome, Italy.

Nelson, R. E. 1960. *Silk-Oak in Hawaii–Pest or Potential Timber?* Misc. Paper 47. Pacific Southwest Forest and Range Experiment Station, U.S. Forest Service, Berkeley, California, USA.

Troup, R. S. 1921. *Silviculture of Indian Trees*. Clarendon Press, Oxford, England.

Youngs, R. L. 1964. *Hardness, Density, and Shrinkage*

Characteristics of Silk-Oak from Hawaii. Research Note FPL-074. Forest Products Laboratory, U.S. Forest Service, Madison, Wisconsin, USA.

Guazuma ulmifolia

Little, E. L., and F. H. Wadsworth. 1964. *Common Trees of Puerto Rico and the Virgin Islands.* Agriculture Handbook No. 249. Forest Service, U.S. Department of Agriculture, Washington, D.C., USA.

Haloxylon aphylla

Nechaeva, N. T., V. K. Vasilevskaia, and K. G. Antonova. 1973. *Life Forms of Plants of the Karakumy Desert.* Nauka Science Publishing House, Moscow, USSR.

Nikitin, S. A. 1966. *Tree and Bush Vegetation of the Deserts of the USSR.* Nauka Science Publishing House, Moscow, USSR.

Haloxylon persicum

Clor, M. A., T. A. al-Ani, and F. Charchafchy. 1976. Germinability and seedling vigor of *Haloxylon salicornicum* as affected by storage and seed size. *Journal of Range Management* 29(1):60–62.

Karschon, R. 1969. Contributions to the arboreal flora of Israel: *Haloxylon persicum* Bge. *La-Yaaran* 19(1):17–23.

Ishankuliyev, M. 1975. Effect of *Haloxylon ammodendron* and *Haloxylon persicum* on sandy desert soils. *Soviet Soil Science* 5:519–530.

Nechaeva, N. T., V. K. Vasilevskaia, and K. G. Antonova. 1973. *Life Forms of Plants of the Karakumy Desert.* Nauka Science Publishing House, Moscow, USSR.

Nikitin, S. A. 1966. *Tree and Bush Vegetation of the Deserts of the USSR.* Nauka Science Publishing House, Moscow, USSR.

Ovezliev, A. O., M. I. Frolov, and O. R. Kurbanov. 1978. Forests of Turkmenistan, their conservation and utilization. *Nature Conservation of Turkmenistan*, Issue 4. Ashkhabad, Turkmenistan.

Inga vera

Little, E. L., and F. H. Wadsworth. 1964. *Common Trees of Puerto Rico and the Virgin Islands.* Agriculture Handbook No. 249, Forest Service, U.S. Department of Agriculture, Washington, D.C.

Leucaena leucocephala

Philippine Council for Agriculture and Resources Research. 1978. *International Consultation on Ipil-ipil Research: Papers and Proceedings.* Philippine Council for Agriculture and Resources Research, College, Laguna, Philippines.

National Academy of Sciences. 1977. *Leucaena: Promising Forage and Tree Crop for the Tropics.* National Academy of Sciences, Washington, D.C., USA.

Mangroves

Clough, B. F., ed. In Press. *The Structure and Function of Mangrove Ecosystems in Australia.* Australian National University Press, Canberra, Australia.

Macnae, W. 1968. Fauna and flora of mangrove swamps. In *Advances in Marine Biology.* F. Russell and M. Yange, eds. Academic Press, New York, New York, USA.

Morton, J. F. 1965. Can the Red Mangrove provide food, feed, and fertilizer? *Economic Botany* 19: 113–123.

Morton, J. F. 1976. Craft industries from coastal wetland vegetation. In *Estuarine Processes.* Vol. 1, pp. 254–266. Academic Press, New York, New York, USA.

Walsh, G. E. 1974. Mangroves, a review. In *Ecology of Halophytes*, pp. 51–174. Academic Press, New York, New York, USA.

Walsh, G. E., S. C. Snedaker, and H. J. Teas. 1975. *Proceedings of International Symposium on Biology and Management of Mangroves.* Institute of Food and Agricultural Sciences, University of Florida, Gainesville, Florida, USA.

A geographical listing of the 50 mangrove species is available from V. J. Chapman; see Research Contacts.

Mimosa scabrella

Barrichelo, L. E. G., and C. E. B. Foelkel. 1975. Utilizacao de madeiras de essencias florestais nativas na obtencao de celulose. *Edicao Especial, Technologia de Celulose e Papel. Instituto de Pesquisas e Estudos Forestais (IPEF)* 10:43–56.

Hoehne, F. C. 1930. *A bracaatinga ou abaracaatinga.* Secretaria da Agricultura, Industria e Commercio de Estado de São Paulo, São Paulo, Brazil.

Muntingia calabura

Germek, B. E. 1964. Cultura da calabura. *Agronomico, Brazil* 16 (9 and 10):34–36.

Little, E. L., R. O. Woodbury, and F. H. Wadsworth. 1974. *Trees of Puerto Rico and the Virgin Islands.* Vol. 2, pp. 512–513. Agriculture Handbook No. 449. Forest Service, U.S. Department of Agriculture, Washington, D.C., USA.

Parkinsonia aculeata

Kaul, R. N., ed. 1970. *Afforestation in the Arid Zones.* Monographiae Biologica. Vol. 20. Dr. W. Junk N.V. Publishers, The Hague, Netherlands.

Pinus halepensis

Brown, A., and N. Hall. 1968. *Growing Trees on Australian Farms.* Forestry and Timber Bureau, Department of Agriculture, Canberra, A.C.T., Australia.

Calderon, E. F. 1973. Experiencias sobre propagacion del pino halepensis en Chihuahua. *Bosques Fauna* 10(3):20–29.

Fisher, H. H. 1971. The Aleppo pine, *Pinus halepensis. California Horticulture Journal* 32(4):129–132.

Karschon, R. 1961. Studies in nursery practice for pines. *La-Yaaran* 11(1):41–56.

Mirov, N. T. 1967. *The Genus* Pinus. The Ronald Press Co., New York, New York, USA.

Palmberg, C. 1975. Geographic variation and early growth in south-eastern semiarid Australia of *Pinus halepensis* Mill. and the *Pinus brutia* Ten. species complex. *Silvae Genetica* 24(5/6):150–160.

Poduje, L., and J. D. Lell. 1973. Informacion sobre el cultivo de pino de Alepo (*Pinus halepensis* Mill.) in la region semiarida y subhumeda de la Provincia de la Pampa. *Revista Forestal Argentina* 17(1):11–17.

Serre, F. 1976. Les rapports de la croissance et du climat chez le pin d'alep (*Pinus halepensis* Mill.). *Oecologia Plantarum* 11(2):143–171.

Souleres, G. 1975. Site classes and production of Aleppo pine *Pinus halepensis* in Tunisian forests. *Revue Forestiere Française* 27(1):41–49.

Pithecellobium dulce

Brown, W. H. 1954. *Useful Plants of the Philippines*. Vol. 2, pp. 154-157. Department of Agriculture and Natural Resources, Manila, The Philippines.

Martinez, M. 1959. *Plantas Utiles de la Flora Mexicana*, pp. 276-278. Ediciones Botas, Mexico, D.F. Mexico.

Wealth of India: Raw Materials. 1969. *Pithecellobium dulce Benth*. Vol. VIII, pp. 140-142. Council of Scientific and Industrial Research, New Delhi, India.

Prosopis alba

Goor, A. Y., and C. W. Barney. 1976. *Forest Tree Planting in Arid Zones*. Second edition. The Ronald Press Company, New York, New York, USA.

National Academy of Sciences, 1979. *Tropical Legumes: Resources for the Future*. National Academy of Sciences, Washington, D.C., USA.

Prosopis chilensis

Burkurt, A. 1974. Prosopis chilensis. *Darwiniana* 4:131.

Donoso, C., and A. Cabello. 1978. Antecedentes fenotogian y de germinacion de especies lens sas chilens. *Ciencias Forestales* 1(2).

National Academy of Sciences. 1979. *Tropical Legumes: Resources for the Future*. National Academy of Sciences, Washington, D.C., USA.

Prosopis cineraria

Bhimaya, C. L., R. N. Kaul, and B. N. Ganguli. 1964. Studies on lopping intensities of *Prosopis spicigera*. *Indian Forester* 90:19-23.

Ghosh, R. C. 1976. *Handbook on Afforestation Techniques*. Controller of Publications, Delhi-6, India.

Kaul, R. N., and M. B. Jain. 1967. Growth attributes: their relation to fuel yield in *Prosopis cineraria* Linn. McBridge (*P. spicigera* Linn.) *Commonwealth Forestry Review* 46:155-158.

Khan, M. I. R. 1955. Tropical thorn forest of West Pakistan. *Pakistan Journal of Forestry* 5:161-171.

Leakey, R. R. B., and F. T. Last. 1980. Biology and potential of *Prosopis* species in arid environments, with particular reference to *P. cineraria*. *Journal of Arid Environments* 3:9-24.

Shankar, V., N. K. Dadhich, and S. K. Saxena. 1976. Effect of Khejri tree (*Prosopis cineraria* Macbride) on the productivity of range grasses growing in its vicinity. *Forage Research* 2:91-96.

Sharma, B. M. 1966. Ecological study of desert plants: *Prosopis spicigera* Linn. *Tropical Ecology* 7:54-66.

Singh, K. S., and P. Lal. 1969. Effect of Khejari (*Prosopis spicigera* Linn.) and Babod (*Acacia arabica*) trees on soil fertility and profile characteristics. *Annals of Arid Zone* 8:33-36.

Wadhwani, B. B. 1953. *Prosopis spicigera* (Linn.). *Indian Forester* 79(8).

Prosopis juliflora

Burkart, A. 1976. *A Monograph of the Genus* Prosopis (*Mimosoideae*). *Journal of the Arnold Arboretum* 57:216-249, 450-525.

Griffith, A. L. 1961. Acacia *and* Prosopis *in the Dry Forests of the Tropics*. Food and Agricultural Organization of the United Nations, Rome, Italy.

Felker, P. J. 1976. *Potential utilization of leguminous trees for minimal energy input agriculture*. Ph.D. thesis. Michigan State University, East Lansing, Michigan, USA.

Felker, P. J., and G. Waines. 1977. Potential use of mesquite as a low energy water and machinery requiring food source. In *Proceedings of the Energy Farms Workshop, Sacramento, California July 14, 1977*. (Available from Publications Unit, Energy Resources Conservation and Development Commission, 1111 Howe Avenue, Sacramento, California 95825, USA.)

Felker, P. 1979. Mesquite: an all-purpose, leguminous, arid land tree. In *New Agricultural Crops*, G. A. Ritchie, ed. Published for American Association for the Advancement of Science by Westview Publishers, Boulder, Colorado, USA.

Gomez-Lorence, F., J. Signoret-Poillon, and M. del C. Atuin-Morciras. 1970. *Mezquites y Huizaches Algunos aspectos de la economia, ecologia y taxonomia de los generos* Prosopis *y* Acacia *en Mexico*. Ediciones del Instituto Mexicano de Recursos Naturales Renovables, Mexico D.F., Mexico.

Jurriaanse, A. 1973. *Are They Fodder Trees?* Pamphlet 116. Department of Forestry, Private Bag X93, Pretoria, South Africa.

National Academy of Sciences. 1979. *Tropical Legumes: Resources for the Future*. National Academy of Sciences, Washington, D.C., USA.

Schuster, J. L. 1969. *Literature on the Mesquite (*Prosopis *L.) of North America, An Annotated Bibliography*. Special Report No. 26. International Center for Arid and Semiarid Studies, Texas Tech University, Lubbock, Texas, USA.

Simpson, B. B., ed. 1977. *Mesquite–Its Biology in Two Desert Scrub Ecosystems*. US/IBP Synthesis Series. Dowden, Hutchinson and Ross, Inc., Stroundsberg, Pennsylvania 18360, USA.

Smith, J. R. 1953 (reprinted 1977). *Tree Crops–A Permanent Agriculture*. Devin-Adair Publishing Company, Old Greenwich, Connecticut, USA.

Prosopis pallida

Burkart, A. 1976. A monograph of the genus *Prosopis* (Leguminosae subfam. Mimosoideae). *Journal of the Arnold Arboretum* 57(3):219-525.

Smith, J. R. 1953 (reprinted 1977). *Tree Crops–A Permanent Agriculture*. Devin-Adair Publishing Company, Old Greenwich, Connecticut, USA.

Prosopis tamarugo

Elgueta-Salinas, H., and S. Calderon-Sanchez. 1971. *Estudio del Tamarugo como Production de Alimento del Ganado Lanar en la Pampa del Tamarugal*. Informe Tecnico No. 38, Seccion Silvicultura, Instituto Forestal, Departamento Forestal, Santiago, Chile.

Sesbania bispinosa

Abrol, I. P., and D. R. Bhumbla. 1971. Start with dhaincha on saline sodic soil *Sesbania aculeata* reclamation. *Indian Farming* 21(2):41-42.

Bhardwaj, K. K. K. 1974. Note on the distribution and effectiveness of Rhizobium of *Sesbania aculeata* Poir. in saline-alkali soils. *Indian Journal of Agricultural Science* 44(10):683-684.

Chandra, V., and M. I. H. Farooqi. 1979. Dhaincha for seed gum. *Extension Bulletin No. 1*. Economic Botany Information Service, National Botanical Research Institute, Lucknow, India.

Chela, K. S., and Z. S. Brar. 1973. Green-manuring

popular again. *Sesbania aculeata, Cyamopsis tetragonoloba, Crotalaria juncea. Progressive Farming* 10(3,9.3.4):11.

Farooqi, M. I. H., and V. N. Sharma. 1972. *Sesbania aculeata* Pers. seeds–new source for gum. *Research and Industry* 17:94–95.

Gillett, J. B. 1963. Sesbania in Africa. *Kew Bulletin* 17(1):91–158.

Hussain, A., and M. Ahmad. 1965. *Sesbania aculeata*–a promising new crop in West Pakistan. *World Crops* 17:28–31.

Hussain, A., and D. M. Khan. 1962. Jantar (*Sesbania aculeata*)–a source of protein supplement and industrial raw material. *West Pakistan Journal of Agricultural Research* 1(1):31–35.

Hussain, A., and D. M. Khan. 1962. Nutritive value and galactomannan content of jantar–*Sesbania aculeata* and *Sesbania aegyptica. West Pakistan Journal of Agricultural Research* 1(1):36–40.

Katiyar, R. C., and S. K. Ranjhan. 1969. Yield and chemical composition of dhaincha (*Sesbania aculeata*)–its nutritive value for sheep. *Indian Journal of Dairy Science* 22(1):33–36.

Khan, A. A., and A. H. Awan. 1967. Salinity tolerance character of dhaincha (*Sesbania aculeata*). *West Pakistan Journal of Agricultural Research* 5(3): 135–136.

Mazumdar, A. K., A. Day, and P. D. Gupta. 1973. Composition of dhaincha fibre (*Sesbania aculeata* Pers.) *Science and Culture* 39(10):473–474.

Razzaque, M. A., and A. B. Siddique. 1971. Pulping and paper making experiments on dhaincha (*Sesbania cannabina*). *Science and Industry* 8:315–319.

Wealth of India: Raw Materials. 1972. *Sesbania*. Vol. IX, pp. 293–295. Council of Scientific and Industrial Research, New Delhi, India.

Sesbania grandiflora

Bhat, A. S., M. M. Menon, T. N. Soundararajan, and R. L. Bhargave. 1971. *Sesbania grandiflora* (a potential pulpwood). *Indian Forester* 97(3):128–144.

Heyne, K. 1950. *De Nuttige Planten van Indonesie*. N.V. Uitgevery W. Van Hoeve–S. Gravenhage, Bandung, Indonesia.

National Academy of Sciences. 1979. *Tropical Legumes: Resources for the Future*. National Academy of Sciences, Washington, D.C., USA.

Syzygium cumini

Ghosh, R. C. 1976. *Handbook on Afforestation Techniques*. Controller of Publication, Delhi-6, India.

Morton, J. F. 1963. The Jambolan (*Syzygium cumini* Skeels)–its food, medicinal, ornamental and other uses. *Proceedings of the Florida State Horticultural Society* 76:328–338.

Troup, R. S. 1921. *The Silviculture of Indian Trees*. Vol. II. Clarendon Press, Oxford, England.

Wealth of India: Raw Materials. 1976. *Syzygium cumini*. Vol. X, pp. 100–104. Council of Scientific and Industrial Research, New Delhi, India.

Tamarix aphylla

Baum, B. R. 1967. Introduced and naturalized tamarisks in the United States and Canada (Tamaricaceae). *Baileya* 15:19–25.

Baum, B. R. 1978. *The Genus* Tamarix. Israel Academy of Sciences and Humanities, Jerusalem, Israel.

Richardson, A. M. 1954. Propagating the Athel tree. *Queensland Agricultural Journal* 79(6):335–337.

Sahni, K. C. 1968. *Important trees of the Northern Sudan*. United Nations Development Program and Food and Agriculture Organization of the United Nations, Forestry Research and Education Centre, Khartoum, Sudan.

Troup, R. S. 1921. *The Silviculture of the Indian Trees*. Vol. 1, pp. 18–20. Clarendon Press, Oxford, England.

Terminalia catappa

Kadambi, K. 1954. *Terminalia catappa*, Linn., its silviculture and management. *Indian Forester* 80(11): 718–720.

Trema spp.

Anonymous. 1953. *Trema orientalis*–fuelwood. *Unasylva* 7:26.

Bhat, R. V., and M. S. Jaspal. 1953. Chemical pulps and writing and printing papers from *Trema orientalis. Indian Paper Pulp* 8:175.

Calip, J. E. 1958. Possibility of coppice methods of propagating *Trema orientalis. Biological Abstracts* 40(3):932.

Gamble, J. S. 1972. *A Manual of Indian Timbers: Urticaceae*, pp. 630–631. Second reprint. Bishen Singh Mahendra Pal Singh, 23-A New Connaught Place, Dehra Dun, India.

Krishna, S., and S. Ramaswamy. 1932. Calorific value of some Indian woods. *Indian Forestry Bulletin* (N.S.) No. 79. Chemistry 1–27.

Singh, R. 1976. *A National Programme for Rodent and Pest Management, Weed Control and Planting and Saving of Trees*. Indian Council of Agricultural Research, New Delhi, India.

Wealth of India: Raw Materials. 1976. *Trema* Lour. Vol. X, pp. 277–279. Council of Scientific and Industrial Research, New Delhi, India.

Zizyphus mauritiana

Dalziel, J. M. 1937. *The Useful Plants of West Tropical Africa*, pp. 299–300. Crown Agents, London, England.

Hallam, G. M. 1977-79. Notes (with illustrations) on useful plants in the Gambia. (Unpublished and not yet fully completed. c/o Forestry Division, Yundum, W. Division, The Gambia.)

Singh, R., and R. C. Khanna. 1968. Some North Indian cultivars of Ber. *Indian Horticulture* 12:23–26, 36.

Troup, R. S. 1921. *Silviculture of Indian Trees*. Clarendon Press, Oxford, England.

Weber, F. 1977. *Reforestation in Arid Lands*. Volunteers in Technical Assistance, Mt. Rainier, Maryland, USA.

A set of notes on this species is available from the agroforestry project, Royal Tropical Institute, Mauritskade 63, Amsterdam, Netherlands.

Zizyphus spina-christi

Weber, F. 1977. *Reforestation in Arid Lands*. Volunteers in Technical Assistance, Mt. Rainier, Maryland, USA.

Appendix 1–Using Fuelwood Efficiently

Anonymous. *The Lorena Cookstove*. Estacion Experimental Choqui, Apartado Postal 159, Quezaltenango, Guatemala.

Anonymous. 1977. *The Haybox (A Fireless Cooker)*. Low Energy Systems, Larkfield Gardens, Dublin 6, Ireland.

Breag, G. R., and A. P. Harker. 1979. *The Utilization of Waste Heat Produced During the Manufacture of Coconut Shell Charcoal for the Centralised Production of Copra*. Report G127. Tropical Products Institute, 56-62 Gray's Inn Road, London, WC1X 8LU, England.

Campbell-Platt, G. 1979. Improvement of the traditional Ghanaian baking oven. *Appropriate Technology* 6:28–30.

Chanco, M. P. 1978. The rice husk stove. *Appropriate Technology* 5(3):18–19.

Draper, S. A. 1977. *Wood Processing and Utilization at the Village Level*. Third FAO/SIDA Expert Consultation on Forestry for Local Community Development. Food and Agriculture Organization of the United Nations, Rome, Italy.

Dutt, G. S. 1978. *Reducing Cooking Energy Use in Rural India*. Report PU/CES74. Princeton University Center for Environmental Studies, Princeton University, Princeton, New Jersey 08540, USA.

Evans, I., and D. Wharton. 1977. The Lorena mudstove–a wood-conserving cookstove. *Appropriate Technology* 4(2):8–10.

Evans, I., J. Kalin, and K. Darrow. 1979. *Lorena Owner-Built Stoves*. Volunteers in Asia, Stanford, California, USA.

Havens, D. 1973. *The Woodburner's Handbook*. Available from W.E.T.S., Harpswell Press, Sinpions Point Road, Brunswick, Maine 04011, USA.

International Labour Office. 1975. *Charcoal Making for Small Scale Enterprises: An Illustrated Training Manual*. International Labour Office, Geneva, Switzerland.

Little, E. C. S. 1978. The Mini-Cusab kiln for rapid small-scale manufacture of charcoal from scrub, coconut wood and coconut shells. *Appropriate Technology* 5(1):12–14.

Little, E. C. S. 1972. A kiln for charcoal making in the field. *Tropical Science* 14(3).

Paddon, A. R., and A. P. Harker. 1979. *The Production of Charcoal in a Portable Metal Kiln*. Report G119. Tropical Products Institute, 56-62 Gray's Inn Road, London WC1X 8LU, England.

Raju, S. P. 1953 (reprinted in 1966). *Smokeless Kitchens for the Millions*, The Christian Literature Society, Post Box 501, Park Town, Madras 3, India.

Rau, S. P. 1961. *Smokeless Kitchens for the Millions*. The Christian Literature Society, Post Box 501, Park Town, Madras 600 003, India.

Rhodes, D. 1968. *Kilns: Design, Construction, and Operation*. Available from W.E.T.S., Harpswell Press, Sinpions Point Road, Brunswick, Maine, 04011, USA.

Simon, E., and Solis, P. 1977. Economic stove that burns as fuel. *Appropriate Technology* 4(1).

Singer, H. 1961. *Improvement of Fuelwood Cooking Stoves and Economy in Fuelwood Consumption*. Report No. 1315. Report to the Government of Indonesia, Expended Technical Assistance Program, Food and Agriculture Organization of the United Nations, Rome, Italy.

Sussman, A., and R. Frazier. 1978. *Handmade Hot Water Systems*. Garcia River Press, P.O. Box 527, Point Arena, California 95468, USA.

Volunteers in Technical Assistance. 1979. *Designs of Woodburning Stoves/Ovens*. Volunteers in Technical Assistance, Mt. Rainier, Maryland, USA. USA.

Volunteers in Technical Assistance. 1979. *Wood Conserving Stoves: Two Stove Designs and Construction Techniques*. Volunteers in Technical Assistance, Mt. Rainier, Maryland, USA.

Volunteers in Technical Assistance. 1979. *Woodstove Design Manual*. Volunteers in Technical Assistance, Mt. Rainier, Maryland, USA.

Wartluff, J. 1975. *Double Drum Sawdust Stove*. Research Note No. NE-208 and Photo Story No. 30. Northeastern Forest Experiment Station, 6816 Market Mt., Upper Darby, Pennsylvania 19082, USA.

Wik, O. n.d. *How to Build an Oil Barrel Stove*. Alaska Northwest Publishing Co., Box 4-EEE, Anchorage, Alaska 99509, USA.

APPENDIX 6

Research Contacts

Acacia auriculiformis

A. K. Banerjee, Development Manager, The Titaghur Paper Mills Co., Ltd., Raw Material and Coal Section, Chartered Bank Buildings, Calcutta 700001, India

BIOTROP-SEAMEO, Regional Center for Tropical Biology, P.O. Box 17, Bogor, Indonesia

D. Cameron, Division of Forest Research, CSIRO, Northern Territory Regional Station, P.O. Box 39899, Winnellie, Darwin, Northern Territory 5789, Australia

Chief Research Officer, Forest Management Research Branch, Office of Forest, P.O. Box 5055, Boroko, Papua New Guinea

Division of Chemical Technology, CSIRO, P.O. Box 310, South Melbourne, Victoria 3205, Australia (A. F. Logan, F. H. Phillips)

C. R. Dunlop, Darwin Herbarium, Department of Industrial Development, Division of Primary Industry, P.O. Box 5160, Darwin, Northern Territory 5790, Australia

Forest Products Research Center, P.O. Box 1358, Boroko, Port Moresby, Papua New Guinea (D. J. Eddowes, J. F. U. Zieck)

Forest Research Institute, Jalan Gunung Batu, Bogor, Indonesia (R. Soerjono)

Forest Research Institute, Kepong, Selangor, Malaysia (F. S. P. Ng, T. B. Peh)

Tham C. Keong, Research Branch, Forest Department, P.O. Box 1407, Sandakan, Sabah, Malaysia

M. Ratnasabapathy, Associate Professor, Department of Botany, University of Malaya, Kuala Lumpur, Malaysia

Seed Section, Division of Forest Research, CSIRO, P.O.Box 4008, Canberra, A.C.T. 2600, Australia (J. C. Doran)

B. R. Thomson, Forest Research Officer, Forestry Division, Munda, New Georgia, Solomon Islands

K. F. Wiersum, Department of Tropical Silviculture, Agricultural University, Gen. Foulkesweg 64, 6700 AH Wageningen, The Netherlands

Acacia brachystachya

Conservator of Forests, Forests Department, 54 Barrack Street, Perth, Western Australia 6000, Australia

L. Pedley, Queensland Herbarium, Meirs Road, Indooroopilly, Queensland 4068, Australia

Seed Section, Division of Forest Research, CSIRO, P.O. Box 4008, Canberra, A.C.T. 2600, Australia (J. C. Doran)

Acacia cambagei

Department of Forestry, Division of Technical Services, GPO Box 944, Brisbane, Queensland 4001, Australia

Seed Section, Division of Forest Research, CSIRO, P.O. Box 4008, Canberra, A.C.T. 2600, Australia (J. C. Doran)

Acacia cyclops

P. Bridgewater, School of Environmental and Life Sciences, Murdoch University, Murdoch, Western Australia 6153, Australia

A. M. Gill, Division of Plant Industry, CSIRO, P.O. Box 1600, Canberra, A.C.T. 2601, Australia

P. J. LeRoux, Department of Forestry, Private Bag X93, Pretoria 0001, Republic of South Africa

J. D. Majer, Western Australian Institute of Technology, Hayman Road, South Bentley, Western Australia 6102, Australia

B. R. Maslin, Western Australian Herbarium, Department of Agriculture, Jarrah Road, South Perth, Western Australia 6151, Australia

S. J. Milton, Bolus Herbarium, University of Cape Town, Rondebosch 7700, Cape Town, Republic of South Africa

Acacia mearnsii

Chief Commissioner, Tasmanian Forestry Commission, GPO Box 207B, Hobart, Tasmania 7001, Australia

F. G. Craig, Chief, Division of Forestry Education and Research, Forests Commission, 601 Bourke St., Melbourne, Victoria 3002, Australia

Division of Forest Research, CSIRO, P.O. Box 4008, Canberra, A.C.T. 2600, Australia

Institute of Pacific Island Forestry, 1151 Punchbowl Street, Honolulu, Hawaii 96813, USA

Silviculturist, Forest Division, Silviculture Research Station, P.O. Box 95, Lushoto, Tanzania

South African Forestry Research Institute, P.O. Box 727, Pretoria 0001, Republic of South Africa

Wattle Research Institute, P.O. Box 375, Pietermaritzburg 3200, Republic of South Africa

K. F. Wiersum, Department of Tropical Silviculture, Agricultural University, Gen. Foulkesweg 64, 6700 AH Wageningen, The Netherlands

Woods and Forest Department, 135 Waymouth Street, Adelaide, South Australia 5000, Australia

Acacia nilotica

Arid Zone Forestry Research, Forestry Research Institute, P.O. Box 658, Khartoum, Sudan

Central Arid Zone Research Institute, Jodhpur, Rajasthan, India

A. M. Khan, Divisional Forest Officer, Silvicultural Research Division, 20-A, G.O.R. Colony, Hyderabad, Sind, Pakistan

K. P. Muniswamy, Forest Geneticist, Office of the Chief Conservator of Forests, Hyderabad 500004, India

K. M. Siddiqui, Forest Geneticist, Pakistan Forest Institute, Peshawar, N.W.F.P., Pakistan
G. E. Wickens, Royal Botanic Gardens, Kew, Richmond, Surrey TW9 3AE, England

Acacia saligna
Conservator of Forests, Forests Department, 54 Barrack Street, Perth, Western Australia 6000, Australia
Director of the Department of Forests, Ministry of Agriculture and Natural Resources, Nicosia, Cyprus (E. D. Michaelides)
Forest Department, Land Development Authority, P.O. Box 45, Kiriat Hayim, Haifa, Israel
Forestry Division, Agricultural Research Organisation, Ilanot, Doar Na Lev Hasharon, Israel
J. D. Majer, Western Australian Institute of Technology, Hayman Road, South Bentley, Western Australia 6102, Australia
Seed Section, Division of Forest Research, CSIRO, P.O. Box 4008, Canberra, A.C.T. 2600, Australia (J. C. Doran)

Acacia senegal
D. M. W. Anderson, Department of Chemistry, University of Edinburgh, West Mains Road, Edinburgh, EH9 3JJ, Scotland (Gum chemistry)
Arid Zone Forestry Research, Forestry Research Institute, P.O. Box 658, Khartoum, Sudan
J. P. M. Brenan, Royal Botanic Gardens, Kew, Richmond, Surrey TW9 3AE, England (Taxonomy)
Central Arid Zone Research Institute, Jodhpur, Rajasthan, India (H. S. Mann, R. S. Paroda)
Direction des Eaux, Forêts et Chasses, Ministère du Developpement Rural, Dakar, Sénégal
R. W. Fishwick, The World Bank, 1818 H Street, Washington, D.C. 20433, USA
Gum Research Officer, P.O. Box 302, El Obeid, Sudan (A. Rahman A. Dafei)
G. M. Khattak, Director General, Pakistan Forest Institute, Peshawar, N.W.F.P., Pakistan
M. Obeid, Gezira University, Wad Medani, Sudan
Tropical Products Institute, 56-65 Gray's Inn Road, London WC1X 8LU, England (Postharvest utilization)
J. Vassal, Laboratoire de Botanique, Université Paul Sabatieu, 39 Alleé J. Guesde, 31077 Toulouse Cedex, France
G. E. Wickens, Royal Botanic Gardens, Kew, Richmond, Surrey TW9 3AE, England

Acacia seyal
Division of Forestry, Faculty of Agriculture and Forestry, University of Dar es Salaam, P.O. Box 643, Morogoro, Tanzania
D. Dulier, Agro-pastoraliste, Laboratoire de Foscha, B.P. 433, N'djamena, Chad
Federal Department of Forest Research, Ministry of Agriculture and Natural Resources, P.M.B. 5054, Ibadan, Western State, Nigeria
J-P. Lebrun, Institut d'Elevage et de Médecine Vétérinaire des Pays Tropicaux, 10 rue Pierre Curie, 94700 Maisons Alfort, France
J. Vassal, Laboratoire de Botanique, Université Paul Sabatieu, 39 Alleé J. Guesde, 31077 Toulouse Cedex, France
G. E. Wickens, Royal Botanic Gardens, Kew, Richmond, Surrey TW9 3AE, England

Acacia tortilis
Arid Zone Forestry Research, Forestry Research Institute, P.O. Box 658, Khartoum, Sudan (Ahmed El Houri Ahmed)
M. Baumer, Clos des trois pins, 446 Combe caude, 34100 Montpellier, France
L. Boulos, National Research Centre, Tahrir Street, Dokki, Cairo, Egypt
Centre Experimental de Reboisement de M'Biddi, Direction des Eaux, Forêts et Chasses, Ministère du Developpement Rural, Dakar, Sénégal
Central Arid Zone Research Institute, Jodhpur, Rajasthan, India (H. S. Mann, K. D. Muthana, G. D. Arora)
M. Evenari, Botany Department, Hebrew University of Jerusalem, Jerusalem, Israel
Forest Department, Nakawa, P.O. Box 1752, Kampala, Uganda
M. Forti, Research and Development Authority, Ben-Gurion University, P.O. Box 1025, Beer-Sheva, Israel
Gum Research Officer, P.O. Box 302, El Obeid, Sudan (A. Rahman A. Dafei)
R. Karshon, Director, Forestry Division, Agricultural Research Organization, Ilanot, Doar Na Lev Hasharon, Israel
M. Kassas, Botany Department, Faculty of Science, Cairo University, Giza, Cairo, Egypt
H. F. Lamprey, UNEP-MAB Integrated Project in Arid Lands, UNESCO, Regional Office for Science and Technology for Africa, P.O. Box 30592, Nairobi, Kenya
H. N. Le Houérou, Coordinator, Range Development Projects, FAO/UNDP, P.O. Box 358, Tripoli, Libya
The Officer in Charge, National Herbarium and Botanic Gardens, P.O. Box 8100, Causeway, Salisbury, Zimbabwe
Royal Botanic Gardens, Kew, Richmond, Surrey TW9 3AE, England (J. P. M. Brenan, G. E. Wickens)
A. G. Seif el Din, Research Advisor, Forestry Cooperative Research Project, c/o International Development Research Centre, P.O. Box 30677, Nairobi, Kenya
Serengeti Research Institute, P.O. Seronera via Arusha, Tanzania
UNDP/FAO Improvement of Crop Production, Al-Kod/Giar, P.O. Box 1188, Aden, People's Democratic Republic of Yemen (M. A. Rehman Bazaraa)
K. K. Wachiira, Department of Geography, Kenyatta University College, P.O. Box 43844, Nairobi, Kenya

Adhatoda vasica
C. K. Atal, Director, Regional Research Laboratory, Council of Scientific and Industrial Research, Canal Road, Jammu-Tawi 180001, India
J. F. Morton, Morton Collectanea, University of Miami, Box 248204, Coral Gables, Florida 33124, USA
H. Y. Mohan Ram, Department of Botany, University of Delhi, Delhi 110007, India
P. Soderholm, USDA Subtropical Horticulture Research Unit, 13601 Old Cutler Road, Miami, Florida 33158, USA

Ailanthus altissima
Central Silviculturist, Forest Research Institute and

Colleges, P.O. New Forest, Dehra Dun, Uttar Pradesh, India
Dirección Forestal, Pargues y Fauna, Ministerio de Ganadería y Agricultura, Maldonado 1276, Montevideo, Uruguay
Director of the Department of Forests, Ministry of Agriculture and Natural Resources, Nicosia, Cyprus
Director, Ministry of Forests and Environmental Protection, rue Antime I, 17 Sofia, Bulgaria
P. Feret, Virginia Polytechnic Institute and State University, Department of Forestry, Blacksburg, Virginia 24061, USA
International Paper Co., Southlands Experiment Forest, Bainbridge, Georgia 31717, USA (R. L. Bryant, Siroos Jahromi)
Jefe, Sección de Coordinación y Relaciones, Instituto Nacional para la Conservación de la Naturaleza, Ministerio de Agricultura, Paseo Infanta Isabel 1, Madrid 2, Spain
G. M. Khattak, Director General, Pakistan Forest Institute, Peshawar, N.W.F.P., Pakistan
Ministry of Forestry, General Directorate of Afforestation, Orman Bakanligi, Agacladirma Genel Mudurlugu, Adakle Soka No. 85, Ankara, Turkey
Tree Breeding Group, Institute of Forestry and Pedology, Shenyang, Liaoning, The People's Republic of China
U.S. Forest Tree Seed Center, U.S. Forest Service, P.O. Box 819, Macon, Georgia 31202, USA

Albizia lebbek
Central Arid Zone Research Institute, Jodhpur, Rajasthan, India
Central Silviculturist, Forest Research Institute and Colleges, P.O. New Forest, Dehra Dun, Uttar Pradesh, India
Division of Silviculture, Taiwan Forestry Research Institute, Botanical Garden, 53 Nan-Hai Road, Taipei, Taiwan (Ta-Wei Hu)
S. K. Jain, Director, Botanical Survey of India, P.O. Botanic Garden, Howrah 711103, West Bengal, India
S. Kondas, Head, Department of Forestry, Agricultural College and Research Institute, Tamil Nadu Agricultural University, Coimbatore, Tamil Nadu, India
R. Lal Banik, Senior Research Officer (Forest Ecologist), Forest Department, Forest Research Institute, P.O. Box 273, Chittagong, Bangladesh
A. Lugo, Institute of Tropical Forestry, P.O. Box AQ, Rio Piedras, Puerto Rico 00928, USA (No seed)
K. P. Muniswami, Forest Geneticist, Office of the Chief Conservator of Forests, Hyderabad 500004, India
C. L. Russell, Nassau Botanic Garden, P.O. Box N3028, Nassau, Bahamas
Silviculturist, Forest Division, Silviculture Research Station, P.O. Box 95, Lushoto, Tanzania

Alnus acuminata
Banco Nacional de Semillas, División Forestal, Institute for the Development of Renewable Natural Resources, Avenida Caracas Nr. 25A66, Bogotá, Colombia
Centro Agronómico Tropical de Investigación y Enseñanza (CATIE), Turrialba, Costa Rica (P. Rosero)
Departamento de Capacitación y Experimentación Dirección de Desarrollo Forestal, Ministerio de Agricultura y Ganadería, Quito, Ecuador
Forest Research Institute, Private Bag, Rotorua, New Zealand
L. A. Fournier-Origgi, Forest Ecologist, School of Biology, University of Costa Rica, San Pedro de Montes de Oca, Costa Rica
J. C. Garzón, Subjefe, Centro de Germoplasma Forestal, Instituto Nacional de Investigaciones Forestales, Progreso 5, Coyoacán 21, D.F., Mexico
Instituto Forestal Latino-Americano, Apartado 36, Mérida, Venezuela
Jefe, Departamento de Protección Forestal, Dirección General Forestal, Ave. 1 Calle 1, San José, Costa Rica
T. Lang, Greenpeace Experimental Farm, R.R. #1, Denman Island, B.C. V0R 1T0, Canada
A. H. Nordmeyer, Forest and Range Experiment Station, New Zealand Forest Service, Box 31-011, Christchurch, New Zealand
F. M. Schlegel, Director, Institute Silviculture, University Austral, Casilla 567, Valdivia, Chile
O. von Borries, Forest Projects, Ministry of Planning, Casilla 3548, La Paz, Bolivia
C. D. Whitesell, Silviculturist, U.S. Forest Service, 1179 Punchbowl Street, Honolulu, Hawaii 96834, USA

Alnus nepalensis
Central Silviculturist, Forest Research Institute and Colleges, P.O. New Forest, Dehra Dun, Uttar Pradesh, India
Centro Agronómico Tropical de Investigación y Enseñanza (CATIE), Turrialba, Costa Rica
Chief, Afforestation Office, Hatti Sar, Naxal, Kathmandu, Nepal
Department of Forestry, College of Agriculture, Tank Road, Solan 173212, Himachal Pradesh, India
Institute of Pacific Island Forestry, 1151 Punchbowl Street, Honolulu, Hawaii 96813, USA
Professor of Forestry, Arts and Science University, University P.O., Rangoon, Burma

Alnus rubra
College of Forest Resources, University of Washington, Seattle, Washington 98195, USA (D. G. Briggs, R. M. Kenady, C. D. Oliver)
C. Ehrenberg, The Royal College of Forestry/SLU, Roslagsvagen 113, S-105 05 Stockholm 50, Sweden (For information on *Alnus glutinosa* and *A. incana*)
Faculty of Forestry, University of British Columbia, Vancouver, B.C., Canada V6T 1W5 (J. H. G. Smith)
Forestry Research Laboratory, Oregon State University, Corvallis, Oregon 97331, USA (J. C. Gordon, M. Newton, R. F. Tarrant)
F. D. Johnson, Department of Forest Resources, University of Idaho, Moscow, Idaho 83843, USA
Pacific Northwest Forest and Range Experiment Station, Forest Service, U.S. Department of Agriculture, Forestry Sciences Laboratory, 3200 Jefferson Way, Corvallis, Oregon 97331, USA (J. F. Frankin, J. M. Trappe)
Pacific Northwest Forest and Range Experiment Station, Forest Service, U.S. Department of Agriculture, Forestry Sciences Laboratory, 3625 93rd Avenue, S.W., Olympia, Washington 98502, USA (D. S. DeBell, C. A. Harrington, R. E. Miller)

Anogeissus latifolia

Central Arid Zone Research Institute, Jodhpur, Rajasthan, India

Central Silviculturist, Forest Research Institute and Colleges, P.O. New Forest, Dehra Dun, Uttar Pradesh, India

Conservator of Forests, Department of Forests, 17, Rana Pratap Marg, Lucknow, India

Kerala Forest Research Institute, Peechi, Trichur 680007, Kerala, India

S. Shyamsunder, Conservator of Forests, Research and Utilisation Circle, Bangalore 560020, Karnataka, India

Azadirachta indica

Conservator of Forests, I/C Planning Branch, Forestry Department, P.O. Box 1917, Kumasi, Ghana

Centre Technique Forestier Tropical, B.P. 303, Ouagadougou, Upper Volta

Federal Department of Forest Research, Ministry of Agriculture and Natural Resources, P.M.B. 5054, Ibadan, Western State, Nigeria

R. W. Fishwick, The World Bank, 1818 H Street, Washington, D.C. 20433, USA

Forest Products Research Institute, University P.O. Box 63, Kumasi, Ghana

C. M. Ketkar, Neem Cake Promotional Scheme, Khadi and Village Industries Commission, 1153, Ganeshkhind Road, Shivajinagar Pune 411016, Maharashtra, India

A. Y. Komla, Forestry Department, Bolgatanga, Ghana

E. D. Morgan, Department of Chemistry, Keele University, Keele, Staffordshire, ST5 5BG, England (Chemistry of neem extracts)

H.A. R. Musnad, Silviculturist, Arid Zone Forestry Research, Forestry Research Institute, P.O. Box 658, Khartoum, Sudan

V. R. Nanayakkara, Conservator of Forests, Forest Department, Colombo 2, Sri Lanka

Principal Research Officer, Savanna Forestry Research Station, Forestry Research Institute of Nigeria, P.M.B. 1039, Samaru-Zaria, Nigeria

S. Radwanski, International Neemtree Research and Development Programme, 4 rue Georges Berger, 75017 Paris, France

Silviculturist, Forest Division, Silviculture Research Station, P.O. Box 95, Lushoto, Tanzania

P. K. Soderholm, USDA Subtropical Horticultural Research Laboratory, 13601 Old Cutler Road, Miami, Florida 33158, USA

Cajanus cajan

Because this is a common crop throughout the tropics, seeds of locally adapted varieties should be available from government, university, and commercial seed suppliers. Some researchers and institutions specializing in pigeon pea research are listed below.

R. Abams, Director, Agricultural Experiment Station, University of Puerto Rico, College of Agricultural Sciences, Mayagüez, Puerto Rico, USA

International Institute of Tropical Agriculture (IITA), Oyo Road, P.M.B. 5320, Ibadan, Nigeria

J. F. Morton, Morton Collectanea, University of Miami, Box 248204, Coral Gables, Florida 33124, USA

D. Sharma, Senior Plant Breeder, International Crops Research Institute for the Semi-Arid Tropics (ICRISAT), 1-11-256 Begumpet, Hyderabad 500016, A.P., India

Calliandra calothyrsus

Forest Products Research Institute, Jalan Gunung Batu, Bogor, Indonesia

Perum Perhutani, Forest State Corporation, Jl. Jendral Gatot Subroto 17-18, Post Box 111, Jakarta, Indonesia (Soekiman Atmosoedarjo)

R. Reid, Davies Laboratory, Division of Tropical Crops and Pastures, CSIRO, Private Mail Bag, P.O. Townsville, Queensland 4810, Australia

K. F. Wiersum, Department of Tropical Silviculture, Agricultural University, Gen. Foulkesweg 64, 6700 AH Wageningen, The Netherlands

Cassia siamea

R. M. Bennett, Chief Forest Officer, Department of Agriculture, Vila, Republic of New Hebrides

Centro Agronómico Tropical de Investigación y Enseñanza (CATIE), Turrialba, Costa Rica

Chief Conservator of Forests, Forest Department, P.O. Box 30513, Nairobi, Kenya

Division of Forestry, Faculty of Agriculture and Forestry, University of Dar es Salaam, P.O. Box 643, Morogoro, Tanzania

Division of Silviculture, Taiwan Forestry Research Institute, Botanical Garden, 53 Nan-Hai Road, Taipei, Taiwan (Ta-Wei Hu, Wuu Jiun Yang)

Forest Research Institute, GPO Box 273, Chittagong, Bangladesh (M. A. Razzaque, Abu Bakr Siddique, Paritosh Das)

A. Lugo, Institute of Tropical Forestry, P.O. Box AQ, Rio Piedras, Puerto Rico 00928, USA

Casuarina equisetifolia

P. S. Ashton, The Arnold Arboretum, Harvard University, 22 Divinity Avenue, Cambridge, Massachusetts 02138, USA

Biology Laboratory, 80 Meiers Road, Indooroopilly, Queensland 4068, Australia

Central Silviculturist, Forest Research Institute and Colleges, P.O. New Forest, Dehra Dun, Uttar Pradesh, India

Centro Agronómico Tropical de Investigación y Enseñanza (CATIE), Turrialba, Costa Rica

L. Conde, University of Florida, School of Forest Resources and Conservation, Newins-Ziegler Hall, Gainesville, Florida 32611, USA

Conservator of Forests, Department of Forestry, Suva, Fiji

Department of Forestry, Division of Technical Services, GPO Box 944, Brisbane, Queensland 4001, Australia

Director, Forest Research Institute, College, Laguna 3720, Philippines

Director of the Department of Forests, Ministry of Agriculture and Natural Resources, Nicosia, Cyprus

Division of Silviculture, Taiwan Forestry Research Institute, Botanical Garden, 53 Nan-Hai Road, Taipei, Taiwan (Ta-Wei Hu)

Institute of Pacific Islands Forestry, 1151 Punchbowl Street, Honolulu, Hawaii 96813, USA

R. Lal Banik, Senior Research Officer (Forest Ecologist), Forest Department, Forest Research Institute, P.O. Box 273, Chittagong, Bangladesh

V. R. Nanayakkara, Conservator of Forests, Forest Department, Colombo 2, Sri Lanka

Office of Forestry, Division of Primary Industry, Department of Southern Highlands, Mendi, S.H.P., Papua New Guinea (*Casuarina oligodon*, high-altitude provenances)

Project Leader, USDA Forest Service, Forest Resources Laboratory, Lehigh Acres, Florida 33936, USA
R. Reid, Davies Laboratory, Division of Tropical Crops and Pastures, CSIRO, Private Mail Bag, P.O. Townsville, Queensland 4810, Australia
C. L. Russell, Nassau Botanic Garden, P.O. Box N3028, Nassau, Bahamas
Seed Section, Division of Forest Research, CSIRO, P.O. Box 4008, Canberra, A.C.T. 2600, Australia (J. C. Doran)
Silviculturist, Forest Division, Silviculture Research Station, P.O. Box 95, Lushoto, Tanzania
Surish Srivastava, Provident Tree Farms, Inc., P.O. Box 240, MCC, Makati, Metro Manila, Philippines

Colophospermum mopane
Central Arid Zone Research Institute, Jodhpur, Rajasthan, India (H. S. Mann, Director)
Department of Forestry, P.O. Box 30048, Lilongwe 3, Malawi
Forest Research Centre, P.O. Box HG595, Highlands, Salisbury, Zimbabwe (D. T. Carter, Chief of Research)
T. Muller, c/o National Herbarium and Botanic Gardens, P.O. Box 8100, Causeway, Salisbury, Zimbabwe
Senior Forestry Research Officer, Forest Research Institute of Malawi, P.O. Box 270, Zomba, Malawi
Tree Improvement Research Centre, National Council for Scientific Research, P.O.Box 1210, Kitwe, Zambia (Amar Singh Hans)

Derris indica
Central Silviculturist, Forest Research Institute and Colleges, P.O. New Forest, Dehra Dun, Uttar Pradesh, India
Chief Forest Research Officer, Division of Forest Research, Ministry of Lands and Natural Resources, P.O. Box 2099, Kitwe, Zambia
Department of Animal Nutrition, Bidhan Chandra Krishi Viswa Vidyalaya (University), Mohanpur, Nadia 741246, West Bengal, India (G. C. Banerjee, L. Mandal)
Department of Natural Resources, Forest Research Institute, College, Laguna 3720, Philippines
M. A. Latif, Divisional Officer, Forest Research Institute, GPO Box 273, Chittagong, Bangladesh
Seed Section, Division of Forest Research, CSIRO, P.O. Box 4008, Canberra, A.C.T. 2600, Australia (J. C. Doran)

Emblica officinalis
Andhra Pradesh Forest Development Corporation, 1-2-593/21 Domala Guda, Hyderabad, Andhra Pradesh, India
P. N. Bajpai, Professor and Head, Horticulture, C.S. Azad, Agriculture University, Kanpur, India
Deputy Director, Horticulturist, Gum Officer In Charge, Horticulture and Training Centre, Saharanpur, Uttar Pradesh, India
Kerala Forest Research Institute, Peechi, Trichur 680007, Kerala, India
J. F. Morton, Morton Collectanea, University of Miami, Box 248204, Coral Gables, Florida 33124, USA
A. S. Raturi, Research Assistant, Seed Testing Laboratory, Forest Research Institute and Colleges, P.O. New Forest, Dehra Dun, Uttar Pradesh, India

Eucalyptus camaldulensis
R. L. Barrett, Genetics Section, Wattle Research Institute, P.O. Box 375, Pietermaritzburg 3200, Republic of South Africa
Centro Agronómico Tropical de Investigación y Enseñanza (CATIE), Turrialba, Costa Rica
Directeur des Forêts, Direction des Forêts, Ministère de l'Agriculture, Avenue Hédi Chaker No. 86, Tunis, Tunisia
Director, Dirección de Bosques, Inderena, Carrera 14 No. 25A66 Piso 2, Bogotá, D.E., Colombia
Director of the Department of Forests, Ministry of Agriculture and Natural Resources, Nicosia, Cyprus
Federal Department of Forest Research, Ministry of Agriculture and Natural Resources, P.M.B. 5054, Ibadan, Western State, Nigeria
Forestry Department, Land Development Authority, P.O. Box 45, Kiriat Hayim, Haifa, Israel
Jefe, Departamento de Protección Forestal, Dirección General Forestal, Ave. 1 Calle 1, San José, Costa Rica
R. Karschon, Director, Forestry Division, Agriculture Research Organization, Ilanot, Doar Na Lev Hasharon, Israel
J. Schechter, Research and Development Authority, Ben Gurion University, P.O. Box 1025, Beer-Sheva, Israel
Seed Section, Division of Forest Research, CSIRO, P.O. Box 4008, Canberra, A.C.T. 2600, Australia (J. C. Doran)

Eucalyptus citriodora
R. L. Barrett, Genetics Section, Wattle Research Institute, P.O. Box 375, Pietermaritzburg 3200, Republic of South Africa
Centro Agronómico Tropical de Investigación y Enseñanza (CATIE), Turrialba, Costa Rica
Department of Forestry, Division of Technical Services, GPO Box 944, Brisbane, Queensland 4001, Australia
Director, Dirección de Bosques, Inderena, Carrera 14 No. 25A66 Piso 2, Bogotá, D.E., Colombia
Division of Silviculture, Taiwan Forestry Research Institute, Botanical Garden, 53 Nan-Hai Road, Taipei, Taiwan (Ta-Wei Hu)
Forest Department, West Bengal, Shankar Villa, Darjeeling 734101, India (S. Palit)
Institute of Pacific Island Forestry, 1151 Punchbowl Street, Honolulu, Hawaii 96813, USA
Jefe, Departamento de Protección Forestal, Dirección General Forestal, Ave. 1 Calle 1, San José, Costa Rica
Regeneration Section, Commonwealth Aluminum Corporation, Ltd., Weipa, North Queensland 4874, Australia
Seed Section, Division of Forest Research, CSIRO, P.O. Box 4008, Canberra, A.C.T. 2600, Australia (J. C. Doran)

Eucalyptus globulus
In many areas, good-quality planting materials can probably be obtained from well-established local plantations or government forest departments.

Centro Agronómico Tropical de Investigación y Enseñanza (CATIE), Turrialba, Costa Rica
Chief Commissioner, Tasmanian Forestry Commission, GPO Box 207B, Hobart, Tasmania 7001, Australia

Corporación Nacional Forestal, Ministerio de Agricultura, Avenida Bulnes 285, Santiago, Chile

F. G. Craig, Chief, Division of Forestry Education and Research, Forests Commission, 601 Bourke Street, Melbourne, Victoria 3002, Australia

Département de Foresterie, Institut Facultaire des Sciences Agronomiques (I.F.A.), B.P. 28, Yangambi, Republic of Zaire

Director, Dirección de Bosques, Inderena, Carrera 14 No. 25A66 Piso 2, Bogotá, D.E., Colombia

D. G. M. Donald, Professor of Silviculture, Faculty of Forestry, University of Stellenbosch, Victoria Street, Stellenbosch 7600, Republic of South Africa

The Forestry Commission, P.O. Box 415-E, Launceston, Tasmania 7250, Australia (R. K. Orme)

J. A. Lopez, Director, Forest Seeds Centre, National Forestry Corporation, Constitución 291 Casilla 5, Chillán, Chile

Oficina Nacional de Evaluación de Recursos Naturales, Calle 17 No. 355, Apartado 4992, Urb. El Palomar, San Isidro, Lima, Peru

Silviculturist, Forest Division, Silviculture Research Station, P.O. Box 95, Lushoto, Tanzania

W. Turner, Department of Forestry, P.O. Box 1067, Riverside, California 92502, USA

B. Zobel, School of Forest Resources, North Carolina State University, Raleigh, North Carolina 27607, USA

Eucalyptus gomphocephala

Conservator of Forests, Forests Department, 54 Barrack Street, Perth, Western Australia 6000, Australia

Dirección Forestal, Pargues y Fauna, Ministerio de Ganadería y Agricultura, Maldonado 1276, Montevideo, Uruguay

Director of the Department of Forests, Ministry of Agriculture and Natural Resources, Nicosia, Cyprus

Director, Forest Department, Land Development Authority, P.O. Box 45, Kiryat Hayim, Haifa, Israel

Jefe, Sección de Coordinación y Relaciones, Instituto Nacional para la Conservación de la Naturaleza, Ministerio de Agricultura, Paseo Infanta Isabel, 1, Madrid 2, Spain

R. D. Johnston, Division of Forest Research, CSIRO, P.O. Box 4008, Canberra, A.C.T. 2600, Australia

R. Karschon, Director, Forestry Division, Agricultural Research Organization, Ilanot, Doar Na Lev Hasharon, Israel

Eucalyptus grandis

R. L. Barrett, Genetics Section, Wattle Research Institute, P.O. Box 375, Pietermaritzburg 3200, Republic of South Africa

P. Burgess, Division of Forest Research, CSIRO, P.O. Box 4008, Canberra, A.C.T. 2600, Australia

E. Campinhos Jr., Aracruz Florestal S.A., 29190–Aracruz–E. Santo, Brazil

Centro Agronómico Tropical de Investigación y Enseñanza (CATIE), Turrialba, Costa Rica

Chief Research Officer, Forest Management Research Branch, Office of Forests, P.O. Box 5055, Boroko, Papua New Guinea

Departamento de Fitotecnia, Faculdade de Agronomia de Jaboticabal, Estrada de Barrinha Sn., Jaboticabal 14870, Estado de São Paulo, Brazil

D. R. DeWet Forestry Research Station, P/Bag X520, Sabie 1260, Republic of South Africa (T. J. Hodgson, Project Leader; G. van Wyk, Officer in Charge)

W. G. Dyson, Tropical Agricultural Research and Training Center, Turrialba, Costa Rica

Forest Department, Nakawa, P.O. Box 1752, Kampala, Uganda

Forest Department, West Bengal, Shankar Villa, Darjeeling 734101, India (S. Palit)

Forest Research Centre, P.O. Box H6 595, Highlands, Salisbury, Zimbabwe

Institute of Pacific Islands Forestry, 1151 Punchbowl Street, Honolulu, Hawaii 96813, USA (R. G. Skolmen)

Kerala Forest Department, Trivandrum 14, Kerala, India

Project Leader, USDA Forest Service, Forest Resources Laboratory, Lehigh Acres, Florida 33936, USA

Regional Director, Dirección Regional Santa Cruz, Centro de Desarrollo Forestal, Villa Gral Barrigntos Casilla 521, Santa Cruz, Bolivia

Seed Section, Division of Forest Research, CSIRO, P.O. Box 4008, Canberra, A.C.T. 2600, Australia (J. C. Doran)

B. Zobel, School of Forest Resources, North Carolina State University, Raleigh, North Carolina 27607, USA

Eucalyptus microtheca

Latif Alnajjar, College of Forestry, Haman Alil, Mosul, Iraq

Conservator of Forests, Forests Department, 54 Barrack St., Perth, Western Australia 6000, Australia

Department of Forestry, Division of Technical Services, GPO Box 944, Brisbane, Queensland 4001, Australia

Division of Forestry, Faculty of Agriculture and Forestry, University of Dar es Salaam, P.O. Box 643, Morogoro, Tanzania

S. A. Kandeel, Head, Forest and Wood Technology Department, Faculty of Agriculture, Alexandria University, El-Shatby, Alexandria, Egypt

Seed Section, Division of Forest Research, CSIRO, P.O. Box 4008, Canberra, A.C.T. 2600, Australia (J. C. Doran)

Eucalyptus occidentalis

B. J. Beggs, Conservator, Forests Department, 54 Barrack Street, Perth, Western Australia 6000, Australia

Director des Forêts, Direction des Forêts, Ministère de l'Agriculture, Avenue Hédi Chaker No. 86, Tunis, Tunisia

Director of the Department of Forests, Ministry of Agriculture and Natural Resources, Nicosia, Cyprus

Faculté des Sciences, Institut Scientifique Cherifien, Université Mohammed V., Avenue Moulay Cherif, Rabat, Morocco

Research and Development Authority, Ben Gurion University, P.O. Box 1025, Beer-Sheva, Israel (M. Forti, Dov Pasternak)

Gliricidia sepium

Planting materials for this species can be readily obtained throughout tropical America, as well as in Southeast Asia, India, Mauritius, and some African countries, for example, Uganda, Tanzania, and Kenya.

Bureau of Forest Development, Bantay, Ilocos Sur, Philippines

Department of Natural Resources, Forest Research Institute, College, Laguna 3720, Philippines (M. R. Reyes, R. T. Cadiz)

El Departamento Forestal, Dirección de Recursos Naturales y Renovables, Ministerio de Agricultura y Ganadería, Managua, Nicaragua

Forestry Department, Centro Universitario de Nor-Occidente "CUNOROC," Universidad de San Carlos de Guatemala, A.P. 68, Huehuetenango, Guatemala

A. Lugo, Institute of Tropical Forestry, P.O. Box AQ, Rio Piedras, Puerto Rico 00928, USA

Programa de Recursos Naturales Renovables, Centro Agronómico Tropical de Investigación y Enseñanza (CATIE), Turrialba, Costa Rica (G. Budowski, N. Gewald)

H. Zambrana, Servicio Forestal y de Fauna, MAG, Apto 2265, San Salvador, El Salvador

Gmelina arborea

International interest in *Gmelina arborea* has caused the FAO Panel on Forest Gene Resources to include this species as one of top priority in the list of forest trees whose gene resources needed exploration, utilization, and conservation. The *Gmelina* exploration/collection scheme is developing with the collection of some 30 seed samples from widely scattered areas in India, Thailand, and Africa. Coordination of the project has been delegated to the Danish/FAO Tree Seed Centre, Denmark, with close collaboration with the Central Silviculturist, Forest Research Institute and Colleges, Dehra Dun, India. The purpose of this provenance collection is to obtain a knowledge of genetical variation on *Gmelina arborea* and, accordingly, to select the seed sources best suited for plantation establishment in different environments.

Gmelina arborea has been selected for early attention in the Commonwealth Forestry Institute project for investigation as a species for large-scale planting in the lowland tropics.

R. M. Bennett, Chief Forest Officer, Department of Agriculture, Vila, Republic of New Hebrides

S. Boonkird, Deputy Managing Director, Forest Industry Organization, Rajadamnern Nok Avenue, Bangkok 1, Thailand

E. F. Brewer, Wildlife Conservation Department, President's Office, The Quadrangle, Banjul, The Gambia

C. B. Briscoe, Consulting Forester, Rt. 1, Box 49, Wayne, Oklahoma 73095, USA

J. Brookman-Amissah, Silviculturist, Forest Products Research Institute, University P.O. Box 63, Kumasi, Ghana

Centre Technique Forestier Tropical, B.P. 8033, Abidjan, Ivory Coast

Centro Agronómico Tropical de Investigación y Enseñanza (CATIE), Turrialba, Costa Rica (N. Gewald, J. Palmer)

Chief Forest Officer, Forest Department, Ministry of Trade and Industry, Belmopan, Belize

Chief Forest Officer, Forestry Division, Ministry of Natural Resources, Honiara, Solomon Islands

Conservator of Forests, Forestry Division, Agricultural Department, Yundum Agricultural Station, Yundum, Western Division, The Gambia

Danish/FAO Forest Tree Seed Centre, DK-3050 Humlebaek, Denmark (E. B. Lauridsen, H. Keiding)

Department of Forestry, P.O. Box 30048, Lilongwe 3, Malawi

Director, Forest Research Institute, Kepong, Selangor, Malaysia

Director General, Forest Department, Nos. 25-26, Randeria Building, Phayre Street, Rangoon, Burma

I. L. Domingo, Paper Industries Corporation of the Philippines, P.O. Box 502, Commerical Center, Makati, Metro Manila, Philippines

Federal Department of Forest Research, Ministry of Agriculture and Natural Resources, P.M.B. 5054, Ibadan, Western State, Nigeria

Forest Department, West Bengal, Shankar Villa, Darjeeling 734101, India (S. Palit)

M. Gopal, Coordinator, Indo-Danish Project on Seed Procurement and Tree Improvement, State Forest Building, Saifabad, Hyderabad 500004, Andhra Pradesh, India

T. Greaves, Commonwealth Forestry Institute, South Parks Road, Oxford OX1 3RB, England

R. Lal Banik, Senior Research Officer (Forest Ecologist), Forest Department, Forest Research Institute, P.O. Box 273, Chittagong, Bangladesh

P. V. D. Ledoux, Research Scientist, Department Biol. Universidade Federal do Pará, Ministerio da Educação e Cultura, Caixa Postal 691, 66000 Belém, Pará, Brazil

N. Jones, Tree Improvement Officer, FAO Forestry Project BGD/72/005, Forest Research Institute, GPO Box 597, Chittagong, Bangladesh

Somphern Kittinanda, Deputy Director-General, Royal Forest Department, Bangkok 9, Thailand

Manager, Technical Forest Services Department, JARI Florestal e Agropecuaria Ltds., Caixa Postal 263, Belém, Pará, Brazil

F. F. Ordinario, Science Research Associate III, Forest Research Institute, College, Laguna 3720, Philippines

P. G. Pattanath, Research Officer, Indo-Danish Project on Seed Procurement and Tree Improvement, State Forest Building, Saifabad, Hyderabad 500004, Andhra Pradesh, India

J. O. Sargento, Forester, Central Forest Experiment Station, College of Forestry, University of the Philippines at Los Baños, College, Laguna 3720, Philippines

Senior Forestry Research Officer, Forest Research Institute of Malawi, P.O. Box 270, Zomba, Malawi

M. U. Slee, Lecturer, Department of Forestry, The Australian National University, P.O. Box 4, Canberra, A.C.T. 2600, Australia

Silviculturist, Forest Research Section, Forest Department Headquarters, Budruddin Road, Kuching, Sarawak, Malaysia (Hua Seng Lee)

A. J. Vincent, Chief Forestry Research Officer, Forestry Research Office, c/o Western Area Forestry Office, New England, Freetown, Sierra Leone

N. Q. Zabala, Associate Professor of Forest Tree Improvement, Forest Resources Management, College of Forestry, University of the Philippines at Los Baños, College, Laguna 3720, Philippines

Grevillea robusta

Centro Agronómico Tropical de Investigación y Enseñanza (CATIE), Turrialba, Costa Rica

Chief, Afforestation Office, Hatti Sar, Naxal, Kathmandu, Nepal

Chief Forest Research Officer, Division of Forest Research, Ministry of Lands and Natural Resources, P.O. Box 2099, Kitwe, Zambia

Department of Forestry, Faculty of Agriculture,

Makerere University, P.O. Box 7062, Kampala, Uganda

Department of Forestry, P.O. Box 30048, Lilongwe 3, Malawi

Division of Forestry, Faculty of Agriculture and Forestry, University of Dar es Salaam, P.O. Box 643, Morogoro, Tanzania

Division of Silviculture, Taiwan Forestry Research Institute, Botanical Garden, 53 Nan-Hai Road, Taipei, Taiwan (Ta-Wei Hu)

Forestry Department, 173 Constant Spring Road, Kingston 8, Jamaica, West Indies

Institute of Pacific Island Forestry, 1151 Punchbowl Street, Honolulu, Hawaii 96813, USA

Kenya Agricultural Research Institute, Forest Research Department, P.O. Box 30148, Nairobi, Kenya

Kenya Forest Department, Conservancy of Research Services, P.O. Box 74, Kikuyu, Kenya

K. A. Kushalappa, Silviculturist, Forest Department, Karnataka Government, 18th Cross, Malleshwaram, Bangalore 560003, Karnataka, India

The Officer in Charge, National Herbarium and Botanic Gardens, P.O. Box 8100, Causeway, Salisbury, Zimbabwe

Senior Forestry Research Officer, Forest Research Institute of Malawi, P.O. Box 270, Zomba, Malawi

South Africa Forestry Research Institute, P.O. Box 727, Pretoria 0001, Republic of South Africa

Guazuma ulmifolia

A. Lugo, Institute of Tropical Forestry, P.O. Box AQ, Rio Piedras, Puerto Rico 00928, USA

Oficina Nacional de Evaluación de Recursos Naturales, Calle 17 No. 355, Apartado 4992, Urb. El Palomar, San Isidro, Lima, Peru

Programa de Recursos Naturales Renovables, Centro Agronómico Tropical de Investigación y Enseñanza (CATIE), Turrialba, Costa Rica (G. Budowski, N. Gewald)

Haloxylon aphyllum* and *H. persicum

A. G. Babayev, President, Academy of Science of the Turkmen SSR, Ulitsa Gogolya 15, 744000 Ashkhabad, Turkmen SSR, USSR

M. Evenari, Botany Department, Hebrew University of Jerusalem, Jerusalem, Israel

R. Karschon, Director, Forestry Division, Agricultural Research Organization, Ilanot, Doar Na Lev Hasharon, Israel (*Haloxylon persicum* only)

M. A. Khasraghi, Research Unit, Forestry Office, Zanjan, Iran

Research Institute of Forests and Rangelands, Ministry of Agriculture and Rural Development, 569, Pahlavi Avenue, Tehran, Iran (B. Malekpour, Director)

Z. Sh. Shamsutdinov, All-Union Research Institute of Astra Khan Sheep Breeding, Samarkand, Karl Marx Street, 47, Uzbek SSR, 703032, USSR

A. L. Takhtajan, Director, Komarov Botanical Institute, 2 Prof. Popov Street, 197022 Leningrad, USSR

Inga vera

Centro Agronómico Tropical de Investigación y Enseñanza (CATIE), Turrialba, Costa Rica

Forestry Department, 173 Constant Spring Road, Kingston 8, Jamaica, West Indies

A. Lugo, Institute of Tropical Forestry, P.O. Box AQ, Rio Piedras, Puerto Rico 00928, USA

Y. P. de Pasztor, Research Scientist, Instituto Florestal, C.P. 1322, 00100 São Paulo, São Paulo, Brazil

Leucaena leucocephala

M. D. Benge, DS/AGR, SA 18 Room 420, ADCA, Agency for International Development, Washington, D.C. 20523, USA

J. L. Brewbaker, Department of Horticulture, University of Hawaii, Honolulu, Hawaii 96822, USA

Division of Tropical Crops and Pastures, CSIRO, Cunningham Laboratory, Mill Road, St. Lucia, Queensland 4067, Australia

J. Hesselink, c/o B.F. Goodrich Philippines, B.F. Goodrich Building, Makati, Metro Manila, Philippines

Philipinas Plant Center, P.O. Box 3350, Manila, Philippines (Miss Belen H. Abreu)

N. D. Vietmeyer, National Academy of Sciences, JH215, 2101 Constitution Avenue, Washington, D.C. 20418, USA

Mangroves

C. A. Arroyo, Coordinator, Mangrove Research Center, Forest Research Institute, College, Laguna 3720, Philippines

V. J. Chapman, Botany Department, Auckland University, Private Bag, Auckland, New Zealand

B. Christensen, Tjørnmarksvej 21, DK-4780 Stege, Denmark

B. Clough, Australian Institute of Marine Science, P.M.B. No. 3, Townsville M.S.D., Queensland 4810, Australia

A. M. Gill, Division of Plant Industry, CSIRO, P.O. Box 1600, Canberra, A.C.T. 2601, Australia

A. E. Lugo, Project Leader, USDA, Institute of Tropical Forestry, P.O. Box AQ, Rio Piedras, Puerto Rico 00928, USA

J. F. Morton, Morton Collectanea, University of Miami, Box 248204, Coral Gables, Florida 33124, USA

S. C. Snedaker, School of Marine and Atmospheric Science, 4600 Rickenbacker Causeway, Miami, Florida 33149, USA

H. J. Teas, Department of Biology, University of Miami, P.O. Box 249118, Coral Gables, Florida 33124, USA

Mimosa scabrella

Departamento de Silvicultura, ESALQ-USP, Cidade Universitaria, "Armando de Salles Oliviera," Caixa Postal 8191, São Paulo, Brazil

C. E. B. Foelkel, Riocell–Rio Grande Cia de Celulose do Sul, Caixa Postal 108, 92.500 Guaíba, Rio Grande do Sul, Brazil

Industrias Klabin do Paraná de Celulose S/A, Monte Alegre, Estado do Paraná, Brazil

Instituto de Pesquisas e Estudos Florestais (IPEF), Caixa Postal 9, 13.400 Piracicaba, São Paul, Brazil (L. E. G. Barrichelo, S. M. da Fonseca)

M. Reis, SON 20Z Bloco C, Apartamento 506, Brasilia, D.F., Brazil

Muntingia calabura

Department of Natural Resources, Forest Research Institute, College, Laguna 3720, Philippines

Departamento Silvicultura e Manejo, Caixa Postal 2959, 80.000 Curitiba, Paraná, Brazil

Director, Forest Research Institute, Kepong, Selangor, Malaysia

Director of Forests, Forest Department Headquarters, Badruddin Road, Kuching, Sarawak, Malaysia

A. Lugo, Institute of Tropical Forestry, P.O. Box AQ, Rio Piedras, Puerto Rico 00928, USA

J. F. Morton, Morton Collectanea, University of Miami, Box 248204, Coral Gables, Florida 33124, USA

Parkinsonia aculeata

Central Arid Zone Research Institute, Jodhpur, Rajasthan, India

Department of Botany, Panjab University, Chandigarth 160014, India

Director of the Department of Forests, Ministry of Agriculture and Natural Resources, Nicosia, Cyprus

Instituto Nacional de Investigaciones Forestales, Progreso 5, Coyoacan 21, D.F., Mexico

O. D. Knipe, Rocky Mountain Station, Forest Sciences, Arizona State University, Tempe, Arizona 85281, USA

A. Lugo, Institute of Tropical Forestry, P.O. Box AQ, Rio Piedras, Puerto Rico 00928, USA

H. A. R. Musnad, Silviculturist, Arid Zone Forestry Research, Forestry Research Institute, P.O. Box 658, Khartoum, Sudan

Silviculturist, Forest Division, Silviculture Research Station, P.O. Box 95, Lushoto, Tanzania

South African Forestry Research Institute, P.O. Box 727, Pretoria 0001, Republic of South Africa

Texas Forest Service, Texas A&M University, College Station, Texas 77843, USA

Pinus halepensis

R. Boardman, Senior Research Officer, Woods and Forests Department, 135 Waymouth Street, Adelaide, South Australia 5000, Australia

Centro Sperimentazione Agricola e Forestale, Casella Postale 9079, 00100 Rome, Italy (A. Eccher, E. Giordano)

Director General, Agricultural Research Centre, P.O. Box 2480, Sidi El-Masri, Tripoli, Libyan Arab Republic

Director-General, Forests and Soil Conservation, Ministry of Agriculture, P.O. Box 2179, Amman, Jordan

Forestry Division, Volcani Institute of Agricultural Research, Ilanot, Israel

Forest Research Institute, Ministry of Agriculture, Terma Alkmanos, Athens 615, Greece

A. Gradi, Director Ufficio Produzione Sementi Forestali, Azienda Stato Foreste Demoniali, Strada Tiberina, 52036 Piene Santo Stefano, Arezzo, Italy

J. L. B. Gutiérrez, Departamento Forestal, Apartado 8.111, Carretera La Coruña Km. 7, Madrid 35, Spain

Institut National Agronomique, Department de Forestiere, Alger-el-Harrach, Algeria

Institut National de Recherches Forestières, Route de la Soukra, B.P. 2, Ariana, Tunisia

Instituto Sperimentale per la Selvicoltura, Viale Santa Margherita 80, 52100 Arezzo, Italy (R. Morandini, Director; A. Tocci)

R. Karschon, Director, Forestry Division, Agricultural Research Organization, Ilanot, Doar Na Lev Hasharon, Israel

E. Magini, Prof. of Silviculture, Istituto di Silvicoltura, Universta 'degli Studi, Piazzale delle Cascine 18, 50144 Firenze, Italy

I. Nahal, Professor of Forestry and Ecology, Faculty of Agriculture, University of Aleppo, Aleppo, Syria

Station de Recherches Forestieres, B.P. 763, Rabat-Agdal, Morocco

U. Topak, Technical Officer, Institute of Forest Tree Seeds Improvement, Ministry of Forestry, P.O. Box 11, Ankara, Gazi, Turkey

Tree Improvement Section, Forest Research Institute, Rabat, Morocco

M. M. Versequy, Director, S.A.R.L. Versepuy, R. C. Le Puy 57 B/3, 43 Le Puy, France

P. J. Wood, Commonwealth Forestry Institute, South Parks Road, Oxford OX1 3RB, England

Pithecellobium dulce

Department of Natural Resources, Forest Research Institute, College, Laguna 3720, Philippines

Division of Forestry, Faculty of Agriculture and Forestry, University of Dar es Salaam, P.O. Box 643, Morogoro, Tanzania

Division of Silviculture, Taiwan Forestry Research Institute, Botanical Garden, 53 Nan-Hai Road, Taipei, Taiwan (Ta-Wei Hu)

Institute of Pacific Islands Forestry, 1151 Punchbowl Street, Honolulu, Hawaii 96813, USA

Instituto Forestal Latino-Americano, Apartado 36, Merida, Venezuela

A. Kabayanda, Director, Départment des Eaux et Forêts, Ministère de l'Agriculture de l'Elevage et du Developpement Rural, Bujumbura, Republic of Burundi

W. A. Kinnison, Department of Horticulture and Landscaping, Central Arizona College, Signal Peak Campus, Coolidge, Arizona 85228, USA

S. Kondas, Head, Department of Forestry, Agricultural College and Research Institute, Tamil Nadu Agricultural University, Coimbatore, Tamil Nadu, India

A. Lugo, Institute of Tropical Forestry, P.O. Box AQ, Rio Piedras, Puerto Rico 00928, USA

Prosopis alba

Cátedra de Botánica, Facultad de Agronomía, Avda. San Martín 4453, 1417 Buenos Aires, Argentina

Departamento de Ciencias Biológicas, Facultad de Ciencias Exactas y Naturales, Buenos Aires, Suc. 28, Argentina (C. A. Naranjo, R. A. Palacios)

P. J. Felker, College of Agriculture, Texas A&I University, Kingsville, Texas 78363, USA

Instituto Argentino de Investigaciones de las Zonas Aridas, CONICET, Casilla de Correo 507, 5500 Mendoza, Argentina

Instituto Forestal Nacional, Secretaria de Estado de Agricultura y Ganadería, Buenos Aires, Argentina (J. R. Ottone, F. Cersosimo, H. Mangieri)

W. A. Kinnison, Department of Horticulture and Landscaping, Central Arizona College, Signal Park Campus, Coolidge, Arizona 85228, USA

Laboratorio de Genética, Facultad de Agronomía, Universidad Nacional de Tucumán, Ayacucho 482, San Miguel de Tucumán, Argentina (A. B. Andrada)

M. Rolfo, Department of Botany, Universidad de la República, Ave. 18 de Julio 1824, Montevideo, Uruguay

Prosopis chilensis

Banco Nacional de Semillas Forestales, Servicio Forestal y de Caza, Ministerio de Agricultura, Natalio Sanchez 20-Jesús María, Lima, Peru

Departamento de Ciencias Biológicas, Facultad de Ciencias Exactas y Naturales, Buenos Aires, Suc. 28, Argentina (C. A. Naranjo, R. A. Palacios)

P. Dreckman V., División de Conservación de Recursos Naturales, Gobierno de Chile, Santiago, Chile

P. J. Felker, College of Agriculture, Texas A&I University, Kingsville, Texas 78363, USA
W. A. Kinnison, Department of Horticulture and Landscaping, Central Arizona College, Signal Peak Campus, Coolidge, Arizona 85228, USA
H. A. R. Musnad, Silviculturist, Arid Zone Forestry Research, Forestry Research Institute, P.O. Box 658, Khartoum, Sudan
M. Teresa Serra, Professor, Botánica Forestal, Departamento de Silvicultura, Facultad de Ciencias Forestales, Universidad de Chile, Casilla 9206, Santiago, Chile

Prosopis cineraria
Central Arid Zone Research Institute, Jodhpur, Rajasthan, India (H. S. Mann)
Forest Research Institute and Colleges, P.O. New Forest, Dehra Dun, Uttar Pradesh, India (R. C. Ghosh)
Institute of Terrestrial Ecology, Bush Estate, Penicuik, Midlothian EH26 0QB, Scotland, United Kingdom (R. B. B. Leakey and F. T. Last)
G. M. Khattak, Director General, Pakistan Forest Institute, Peshawar, N.W.F.P., Pakistan

Prosopis juliflora
Department of Animal Science, University of Stellenbosch, Stellenbosch 7600, Republic of South Africa (J. Kargaard, F. J. van der Merwe)
P. J. Felker, College of Agriculture, Texas A&I University, Kingsville, Texas 78363, USA
H. C. Johnston, University of Texas, Austin, Texas 78712, USA
B. B. Simpson, Department of Botany, University of Texas, Austin, Texas 78712, USA
O. T. Solbrig, Gray Herbarium, Harvard University, Cambridge, Massachusetts 02138, USA
C. W. Weber, Nutrition and Food Sciences, University of Arizona, Tucson, Arizona 85721, USA

Prosopis pallida
J. L. Brewbaker, Department of Horticulture, University of Hawaii, Honolulu, Hawaii 96822, USA
Institute of Pacific Island Forestry, 1151 Punchbowl Street, Honolulu, Hawaii 96813, USA
D. L. Plucknett, Consultative Group on International Agricultural Research, 1818 H Street N.W., Washington, D.C. 20433, USA
Y. N. Tamimi, University of Hawaii, Agricultural Experiment Station, 461 W. Lanikaula Street, Hilo, Hawaii 96720, USA

Prosopis tamarugo
Corporación de Fomento de la Producción, Chile (CORFO), Departamento de Ganadería, Santiago, Chile
Director Ejecutivo, Instituto Forestal, Valenquela Llanos 1061, Casilla 3085, La Reina, Santiago, Chile
M. Habit, Oficial Regional de Producción y Protección Vegetal para América Latina, FAO, Casilla 10095, Santiago, Chile
H. M. Hull, School of Renewable Natural Resources, 325 Biological Sciences East Building, University of Arizona, Tucson, Arizona 85721, USA
F. Sudzuki, Escuela de Agronomía, Departamento Producción Agrícola, Campus Antumapu, Santiago, Chile

Sesbania bispinosa
I. P. Abrol, Division of Soils and Agronomy, Central Soil Salinity Research Institute, Karnal, Haryana 132001, India
Ayub Agricultural Research Institute, Lyallpur, Pakistan
P. C. D. Gupta, Jute Technological Research Laboratory, 12 Regent Park, Tollygunge, Calcutta 40, West Bengal, India
L. Lintu, Pulp and Paper Branch, Forest Industries Division, FAO, Via delle Terme di Caracalla, 00100 Rome, Italy
L. Markila, 5404 Candlewick Wynd, Delta, British Columbia V4M 3T6, Canada
K. L. Mehra, Director, National Bureau of Plant Genetic Resources, Indian Agricultural Research Institute, New Delhi 110012, India
National Botanic Gardens, Rana Pratap Marg, Lucknow 226001, India (T. N. Khoshoo, M. I. H. Farooqi)
Punjab Agricultural University, Ludhiana, Punjab 141004, India (Kartar Singh Chela)
Tran Van Nao, Forest Resources Division, FAO, Via delle Terme di Caracalla, 00100 Rome, Italy
R. Whistler, Department of Biochemistry, Purdue University, Lafayette, Indiana 47907, USA

Sesbania grandiflora
Department of Horticulture, The University of Agricultural Sciences, Dharwar Campus, Krishingar, Dharwar 580005, Karnataka, India
C. R. Dunlop, Darwin Herbarium, Department of Industrial Development, Division of Primary Industry, P.O. Box 5160, Darwin, Northern Territory 5790, Australia
Forest Department, Nakawa, P.O. Box 1752, Kampala, Uganda
C. Huang, Animal Industry Division, Joint Commission on Rural Reconstruction, 37 Nan-Hai Road, Taipei 107, Taiwan
N. S. Kaikini, IFS (Retd) Forest Advisor, West Coast Paper Mills, Ltd., Dandeli (NK) Karnataka, India
G. C. Lugod, Department of Botany, College of Sciences and Humanities, University of the Philippines at Los Baños, College, Laguna 3720, Philippines
B. P. Patil, Director, Indian Grassland Fodder Research Institute, Jhansi, Uttar Pradesh, India
H. Suijdendorp, Department of Agriculture, District Office, Carnarvon, Western Australia 6701, Australia
K. F. Wiersum, Department of Tropical Silviculture, Agricultural University, Gen. Foulkesweg 64, 6700 AH Wageningen, The Netherlands

Syzygium cumini
Agricultural Research and Education Center, University of Florida, 18905 S.W. 280th Street, Homestead, Florida 33030, USA
Department of Natural Resources, Forest Research Institute, College, Laguna 3720, Philippines
Division of Silviculture, Taiwan Forestry Research Institute, Botanical Garden, 53 Nan-Hai Road, Taipei, Taiwan (Ta-Wei Hu)
Institute of Pacific Islands Forestry, 1151 Punchbowl Street, Honolulu, Hawaii 96813, USA
A. Kumar, Research Assistant, Seed Testing Laboratory, Forest Research Institute and Colleges, P.O. New Forest, Dehra Dun, Uttar Pradesh, India

Lancetilla Experimental Garden, Apto. 49, Tela, Honduras
J. F. Morton, Morton Collectanea, University of Miami, Box 248204, Coral Gables, Florida 33124, USA
National Botanic Gardens, Rana Pratap Marg, Lucknow 226001, India

Tamarix aphylla
Arizona Agricultural Experiment Station, University of Arizona, Tucson, Arizona 85721, USA
Botany Department, Faculty of Science, Cairo University, Giza, Cairo, Egypt
Chihuahua Desert Research Institute, P.O. Box 1334, Alpine, Texas 79830, USA
Department of Botany, Tel-Aviv University, Tel-Aviv, Israel (Y. Waisel, J. Friedman)
Directeur, E.N.F.V.A., B.P. 37, Kaédi, Mauritania
Forest Department, Land Development Authority, P.O. Box 45, Kiriat Hayim, Israel
S. A. Kandeel, Head, Forest and Wood Technology Department, Faculty of Agriculture, Alexandria University, El-Shatby, Alexandria, Egypt
R. Karschon, Director, Forestry Division, Agricultural Research Organization, Ilanot, Doar Na Lev Hasharon, Israel
Kuwait Institute for Scientific Research, P.O. Box 12009, Kuwait, State of Kuwait

Terminalia catappa
Assistant Conservator of Forests (Research), Forestry Division, P.O. Bag 30, Port-of-Spain, Trinidad, West Indies
Botanical Gardens, Department of Agriculture and Fisheries, P.O. Box 834, Hamilton 5, Bermuda
Centro Agronómico Tropical de Investigación y Enseñanza (CATIE), Turrialba, Costa Rica
Division of Silviculture, Taiwan Forestry Research Institute, Botanical Garden, 53 Nan-Hai Road, Taipei, Taiwan (Ta-Wei Hu)
Institute of Pacific Islands Forestry, 1151 Punchbowl Street, Honolulu, Hawaii 96813, USA (R. G. Skolmen)
S. Kondas, Head, Department of Forestry, Agricultural College and Research Institute, Tamil Nadu Agricultural University, Coimbatore, Tamil Nadu, India
A. Lugo, Institute of Tropical Forestry, P.O. Box AQ, Rio Piedras, Puerto Rico 00928, USA
M. R. Pastor, Director de Silvicultura, Ministerio de Agricultura, Calle Cuba 552, Lima, Peru
Professor of Forestry, Arts and Science University, University P.O., Rangoon, Burma
C. L. Russell, Nassau Botanic Garden, P.O. Box N3028, Nassau, Bahamas

***Trema* species**
T. Jaffie, Office de la Recherche Scientifique et Technique Outre-Mer, B. P. V51, Abidjan, Ivory Coast
Ken Newcombe, Energy Planner, Department of Minerals and Energy, P.O. Box 2352, Konedobu, Papua, New Guinea
Dr. E. Soepadmo, Department of Botany, University of Malaya, Kuala Lumpur, Malaysia
M. J. Trinick, Division of Land Resources Management, CSIRO, Private Bag, Post Office, Wembley, Western Australia 6014, Australia

Trema guineenis
Centre Technique Forestier Tropical, 08BP33, Abidjan 108, Ivory Coast
Département de Foresterie, Institut Facultaire des Sciences Agronomiques (I.F.A.), B.P. 28, Yangambi, Republic of Zaire
Division of Forestry, Faculty of Agriculture and Forestry, University of Dar es Salaam, P.O. Box 643, Morogoro, Tanzania
P. D. Hardcastle, Forest Research Institute of Malawi, Box 270, Zomba, Malawi
Kerala Forest Research Institute, Peechi, Trichur 680007, Kerala, India
National Centre of Forestry Research, B.P. 243, Tananarive, Madagascar
S. N. Rai, Divisional Forest Officer, Dharwad 580008, India

Trema orientalis
Department of Natural Resources, Forest Research Institute, College, Laguna 3720, Philippines (V. B. Mendoza, A. G. Cortiguerra, E. B. Ramoran)
Division of Silviculture, Taiwan Forestry Research Institute, Botanical Garden, 53 Nan-Hai Road, Taipei, Taiwan (Ta-Wei Hu)
Institute of Pacific Islands Forestry, 1151 Punchbowl Street, Honolulu, Hawaii 96813, USA

Trema politoria
Central Silviculturist, Forest Research Institute and Colleges, P.O. New Forests, Dehra Dun, Uttar Pradesh, India

Zizyphus mauritiana
Arid Zone Forestry Research, Forestry Research Institute, P.O. Box 658, Khartoum, Sudan
Forestry Division, Agricultural Department, Yundum Agricultural Station, Yundum, Western Division, The Gambia (G. M. Hallam)
J. P. Gaillard, IRFA/GERDAT, B.P. 5035, 34032 Montpellier, France
G. M. Khattak, Director General, Pakistan Forest Institute, Peshawar, N.W.F.P., Pakistan
T. N. Khoshoo, Director, National Botanical Research Institute, Lucknow 226001, India
G. Stanford, Greenhills Center, Rt. 1, Box 861, Cedar Hill, Texas 75104, USA
G. E. Wickens, Royal Botanic Gardens, Kew, Richmond, Surrey TW9 3AE, England

Zizyphus spina-christi
Directeur, E.N.F.V.A., B.P. 37, Kaédi, Mauritania
Forest Products Research Institute, University P.O., Box 63, Kumasi, Ghana
H. H. Hajara, Head, Department of Biology, King Abdulaziz University, P.O. Box 1540, Jeddah, Saudi Arabia
Ministry of Agriculture, Sana, Yemen Arab Republic
G. E. Wickens, Royal Botanic Gardens, Kew, Richmond, Surrey TW9 3AE, England

Using Fuelwood Efficiently
Application of Science and Technology to Rural Areas (ASTEA), The Indian Institute of Science, Bangalore 560012, India (H. Geller, A. K. N. Reddy, Director)
B. Agarwal, Research Fellow, Science Policy Research

Unit, Mantell Building, Falmer, Brighton, Sussex EN1 9RS, England.

Aprovecho Institute, 359 Polk St., Eugene, Oregon 97402, USA (I. Evans)

J. Arnold, Energy and Transport, P.O. Box 258, Kenmore Station, Boston, Massachusetts 02215, USA

J. Asare, National Council of Women and Development, P.O. Box 459, Accra Ghana

K. Bennett, Department of Chemical Engineering and Chemical Technology, Imperial College, Prince Consort Rd., London SW7 2BY, England

G. Campbell-Platt, Lecturer, Department of Nutrition and Food Science, University of Ghana, P.O. Box 134, Legon, Ghana

M. P. Chanco, P.O. Box 1365, Manila, Philippines (The Rice Husk Stove)

C. Christensen, Kapelvej 37, DK 2830 Virum, Denmark

Christian Literature Society, Post Box 501, Park Town, Madras 3 India

Department of Appropriate Technology, Eindhoven University of Technology, Postbus 513, 5600 MB Eindhoven, The Netherlands

Department of Mechanical Technology, Danish Technical University, Copenhagen, Denmark

Department of Social Welfare and Community Development, Accra, Ghana

D. G. Donovan, Forestry Research Fellow, Institute of Current World Affairs, P.O. Box 1615, Kathmandu, Nepal

G. S. Dutt, University Center for Energy and Environmental Studies, Princeton, New Jersey 08540, USA

The East African Industrial Research Organization, POB 30650, Nairobi, Kenya

R. Hill, University of Maine, Orono, Maine 04473, USA

Intermediate Technology Development Group (ITDG), Parnell House, 25 Wilton Road, London SW1V 1JS, England

Investigaciones Cientificas Asociadas del Altiplano (ICADA), Apartado Postal 159, Quezaltenango, Guatemala (L. Jacobs)

E.C.S. Little, "Aroha" No. 1 R.D., Kerikeri, North Auckland, New Zealand (CUSAB and Mini-CUSAB Bins)

K. P. Maynard, Research Assistant, Department of Mechanical Engineering, University of the West Indies, St. Augustine, Trinidad, West Indies

Mechanical Engineering Department, Faculty of Engineering, University of Science and Technology, Kumasi, Ghana (F.O. Kwami, Head)

Mechanical Wood Products Branch, Forest Industries and Trade Division, Food and Agricultural Organization of the United Nations, 00100 Rome, Italy

K. Openshaw, Division of Forestry, Faculty of Agriculture, Forestry and Veterinary Science, University of Dar es Salaam, Morogoro, Tanzania

K. Krishna Prasad, Visiting Professor, Department of Applied Physics, Technische Hogeschool, Eindhoven, Postbus 513, 5600MB, Eindhoven, The Netherlands

Research Centre for Applied Science and Technology (RECAST), Tribhuvan University, Kirtipur, Kathmandu, Nepal

Societe Africaine d'Etudes et de Developpement (SAED), B.P. 593, Ouagadougou, Upper Volta

Tropical Products Institute, 56/62 Gray's Inn Road, London WC1X 8LU, England (T. Paddon, G. Breag and M. Parry)

United National Industrial Development Organization, P.O. Box 707, A-1011 Vienna, Austria

Volunteers in Technical Assistance (VITA), 3706 Rhode Island Avenue, Mt. Rainier, Maryland 20822, USA

B. Yoder, Department of Agricultural Engineering, Riley Robb Hall, Cornell University, Ithaca, New York 14853, USA

Explanation of Terms

Amenity planting: The cultivation of trees to add to one's comfort and convenience or to the attractiveness of the area; shade trees, ornamentals, wind breaks, etc.

Anthracite: A hard type of coal that gives much heat but little flame or smoke.

Bole: The main tree trunk.

Budding: Grafting by inserting a bud into a slit in the bark or rootstock.

Canopy: The layer of tree crowns in a forest.

Cladode: Modified stem having appearance and function of a leaf.

Clearcut: A harvesting and regeneration technique in which in one operation all trees are removed from an area regardless of size. Most used with species that require full sunlight to reproduce and grow well.

Coppice: The ability to regenerate by shoots or root suckers, or a forest so established.

Cultivar: A variety of a plant species in cultivation.

Density or Specific Gravity: Where given, these are based on air-dry samples, as to both weight and volume, and (wherever possible) for plantation-grown specimens.

Diameter: All tree trunk diameters were measured at breast height (1.37 m).

Direct Seeding or Broadcast Seeding: Scattering seed over the area on which a forest stand is to be raised.

Ecotype: A group of plants within a species genetically adapted to a particular habitat.

Firewood: Although the terms "firewood" and "fuelwood" are generally used interchangeably, "firewood" connotes small-scale use such as for home cooking, whereas "fuelwood" implies large-scale industrial use.

Funicle: The slender stalk of a seed or ovule.

Inarch: To graft by uniting a shoot to another plant while both are growing on their own roots.

Inoculation: The deliberate introduction of organisms, usually microorganisms, into a new environment. Used here especially for the introduction of beneficial rhizobia bacteria into soils to improve growth of leguminous plants.

Insolation: Solar radiation, as received by the earth.

Laterite: A red, residual soil containing large amounts of aluminum and ferric hydroxides, found especially in well-drained tropical rain forests.

Lignotuber: Woody tuberous root that resprouts when the main stem (trunk) is damaged (found notably in some eucalypts).

Naturalize: To adapt a plant to a new environment; acclimate.

Nitrogen Fixing: The conversion of elemental nitrogen (N_2) from the atmosphere to organic forms utilizable in biological processes.

Nodulation: The quality of having small knots or lumps on the roots, especially ones containing nitrogen-fixing bacteria.

Phenology: The study of natural phenomena that recur periodically (such as blossoming) and their relation to climate and seasons.

Phyllode: Flattened leaf stalks that look like, and function as, leaves (found notably in *Acacia* species).

Pollarding: Cutting back the crown of a tree to produce a close head of shoots (a pollard) at a height that puts it beyond the reach of browsing animals.

Propagule: Any part of a plant–e.g., seed, cutting, spore–capable of growing into a new organism.

Provenance: The original geographic source of seed, seedlings, or cuttings.

Provenance Trial: A planting of populations of different provenances of the same species, usually to identify those most suitable for silvicultural use in the test region.

Pulse Crop: Leguminous crop with edible seeds such as peas, beans, and peanuts.

Ratoon: A shoot growing from the root of a plant that has been cut down, used especially for sugarcane.

Root Sucker: A shoot arising from *below* the ground level either from a root or a rhizome.

Saw Timber: Trees of size and quality for producing sawn wood.

Scarify: To wear down by abrasion or by acid treatment the outer, more-or-less impervious, seedcoat in order to assist or hasten germination.

Shade Intolerance: The characteristic of some species to regenerate naturally only in open sunlight.

Silvics: The study of the life history and general characteristics of trees and forests.

Silviculture: The theory and practice of controlling the establishment, composition, and growth of forests.

Stratification: The operation of storing seeds in, and often in alternate layers with, a moist medium such as sand or peat, in order to maintain viability or overcome dormancy.

Ton: The symbol "t" is used for metric ton (2,200 lb. avoirdupois).

Truncheon: A long, thick stem-cutting, containing much old wood, used in propagating some species.

Yield: Figures quoted for each of the species in this report are merely rough estimates. Yields and densities vary greatly with age of the tree and the locale (the wood of *Eucalyptus grandis,* for example, has a specific gravity of 0.82 in natural Australian stands and 0.55 in plantations in South Africa). Some of the species do not produce usable construction timber and have not previously attracted the attention of foresters, so that the figures quoted (if any) may be based on a single sampling.

APPENDIX 8

Biographical Sketches of Panel Members

EDWARD S. AYENSU, Director of the Office of Biological Conservation, Smithsonian Institution, Washington, D.C., is currently the Secretary-General of the International Union of Biological Sciences. He received his B.A. in 1961 from Miami University in Ohio, M.Sc. from George Washington University in 1963, and his Ph.D. in 1966 from the University of London. His research interests are in comparative anatomy and phylogeny of flowering plants, commercial timbers, histology of monocotyledons, economic botany, and tropical biology. An internationally recognized expert on tropical plants, he has published extensively in these areas and on topics relating to science, technology, and development, especially in developing countries. Dr. Ayensu was co-chairman of the panel on underexploited tropical plants of the Advisory Committee on Technology Innovation (ACTI) and chairs and serves as a member of many international conferences.

JOHN BENE, after three decades as chief executive of one of the major integrated forest industries of Canada with affiliations in several tropical countries, joined the Canadian International Development Agency as its first forestry adviser in 1968 and subsequently became Director General of the Special Advisers Branch of that agency. He served on the first Board of Governors of the International Development Research Centre of Canada (IDRC) and now serves as adviser. In 1975, as the Senior Adviser to the President, he headed a team that identified agroforestry research as the priority gap in the effort to improve and sustain productivity on marginal tropical lands. This led to the establishment of the International Council for Research in Agroforestry in Nairobi, of which he is now chairman.

JAMES S. BETHEL, Dean of the College of Forest Resources of the University of Washington in Seattle, received a B.S.F. from that university in 1937 and an M.F. in 1939 and D.F. in 1947 from Duke University. His research has been concentrated in wood science, especially tropical wood utilization and wood energy technology. He was chairman of a panel on wood as a renewable resource for the Board on Agriculture and Renewable Resources of the National Academy of Sciences (NAS) and a member of the NAS Committee on the Effects of Herbicides in Vietnam. He also served as director of a study, "Energy from Wood," conducted for the Office of Technology Assessment, U.S. Congress.

LOUTFY BOULOS, Professor of Botany at the National Research Center, Dokki, Cairo, Egypt, received a B.Sc. in 1954, and M.Sc. in 1960 from Cairo University and D.Sc. from the University of Montpellier, France, in 1963. He joined the staff of several universities and international organizations: Cairo University, University of Libya, University of Jordan (with UNESCO), and International Livestock Centre for Africa in Addis Ababa. He established the first national herbaria in Libya and Jordan, a specialized herbarium for livestock research in Ethiopia, and contributed to the herbaria of Cairo University, Agricultural Museum, and Desert Institute of Cairo. His works are mainly on the flora of Egypt, Libya, and Jordan, and a worldwide systematic revision of the genus *Sonchus*, Compositae. He is a member of the Board of Advisers for preparing a checklist of Mediterranean flora.

GERARDO BUDOWSKI, Head of the Natural Renewable Resources Program at Centro Agronómico Tropical de Investigación y Enseñanza, Turrialba, Costa Rica, received a B.S. in agriculture in 1948 from Central, Caracas, Venezuela. He received an M.S. in forestry in 1953 from the Inter-American Institute of Agriculture Sciences (IAIAS), Turrialba, Costa Rica, and a Ph.D. in 1962 from Yale University School of Forestry. He was with IAIAS from 1953 to 1967 and has served internationally as an ecologist with UNESCO (1967-1970) and as Director General of the International Union for Conservation of Nature and Natural Resources in Morges, Switzerland (1970-1976). He is an internationally recognized forester specializing in tropical forestry and environmental sciences, with experience in Africa and Southeast Asia as well as in Latin America.

JEFFREY BURLEY, University Lecturer with the Department of Forestry, Commonwealth Forestry Institute, Oxford University, England, received his B.A. in forestry in 1961 from Oxford and his M.F. and Ph.D. in 1965 from Yale University. In addition to lecturing undergraduates and graduates on wood structure and properties, forest genetics, and plant breeding, he researches and supervises research students in the genetic variation of anatomical, chemical, and morphological properties of tropical trees. He spends much time touring tropical developing countries, advising on forestry problems on behalf of several international aid agencies.

ERIK P. ECKHOLM, a Senior Researcher with Worldwatch Institute, was formerly an Associate Fellow at the Overseas Development Council. He is coauthor of *By Bread Alone* and author of *Losing Ground: Environmental Stress and World Food Prospects* (W. W. Norton, 1976). Mr. Eckholm can be credited with bringing the firewood crisis to world attention. His booklet *The Other Energy Crisis: Firewood* (Worldwatch Paper 1, September 1975) provided incentive for this study.

HANS M. GREGERSEN, Professor of Forestry at the University of Minnesota, Minneapolis, received his M.S. in 1963 from the University of Washington and his Ph.D. in forest resource economics in 1969 from the University of Michigan. He has worked for the U.S. Department of Agriculture and FAO and has been a consultant to the World Bank, the Organisation for Economic Co-operation and Development, and the Inter-American Development Bank. His research has been concentrated in forest economics, benefit-cost analysis, project analysis, and economic development. He has spent much time traveling through developing countries, analyzing their needs and potentials including those related to fuelwood. At present he is writing a book for FAO on economic analysis of forestry projects in developing countries.

ANTHONY VINCENT HALL, Senior Lecturer in Botany and Assistant Curator of the Bolus Herbarium at the University of Cape Town, received a B.Sc. in 1955, an M.Sc. in 1959, and a Ph.D. in 1963 from the University of Cape Town. His field of research has spread from plant taxonomy to related interests in ecology and computer-based classification. He directs research groups on endangered species, on invasive plants, and on the integration of nature conservation with land-use practices. He is a Fellow of the Royal Society of South Africa and has played a leading role in a number of conservation bodies.

ROBERT DOUGLAS JOHNSTON, a Principal Research Scientist with the Division of Forest Research, Commonwealth Scientific and Industrial Research Organisation, Canberra, received a B.Sc. in 1951 from Sydney University and joined the Forestry and Timber Bureau as a botanical research officer. He received a Ph.D. in science (botany) in 1951 from Imperial College, University of London, and transferred to the Commonwealth Scientific and Industrial Research Organisation (CSIRO) when the Forest Research Institute became the CSIRO Division of Forest Research. His main interest has been the factors affecting the natural distribution of Australian trees, particularly eucalypts. He has extensive field experience and has coauthored several publications including *Forest Trees of Australia* and *Eucalypts.*

STANLEY L. KRUGMAN, Principal Research Forest Geneticist, Timber Management Research, U.S. Department of Agriculture (USDA), Forest Service, received a B.S. in 1955 from the University of Missouri and an M.S. in 1956 and Ph.D. in 1961 from the University of California, Berkeley. Following one year of postgraduate research at the University of California, he joined the Pacific Southwest Forest and Range Experiment Station, USDA Forest Service, as a plant physiologist. He became Project Leader of the Institute of Forest Genetics in 1965 and served in that role until his transfer in 1971 to his current position. His studies have included the identification of forest-tree pigments and naturally occurring seed hormones and investigations into the reproductive physiology of pines. Recent activities have included the establishment of tree species tests in a number of different countries for both fiber and fuel production.

B. F. KUKACHKA, Pioneering Research Unit, Forest Products Laboratory, Forest Service, U.S. Department of Agriculture, Madison, Wisconsin, received a B.S. in 1937 and Ph.D. in wood technology in 1942 from the University of Minnesota. His research career of over 33 years has been in the area of wood structure and properties and identification of tropical timbers. Current research is in the systematic wood anatomy of the neotropical Sapotaceae. He has traveled extensively in the American tropics and has served as a consultant for FAO and AID in this area. In 1978 he received the coveted Outstanding Achievement Award of the University of Minnesota Regents for his efforts in fostering tropical forestry and wood utilization in the American tropics.

HANS-JÜRGEN VON MAYDELL, Senior Research Scientist, Federal Research Centre for Forestry and Forest Products, Hamburg-Reinbek, Institute for World Forestry, received his Diplom-Forstwirt degree in 1957 and his Ph.D. in 1958 from the University of Freiburg, Germany. He has worked for FAO in many developing countries and has served as a consultant to the Government of the West German Federal Republic and to international organizations. He is a member of the Board of Trustees, International Council for Research in Agroforestry (ICRAF), leader of a working group of the International Union of Forestry Research Organizations (IUFRO), and teaches the geography of forest resources, industries, and international timber trade at the University of Hamburg. His special research emphasis is on forestry in semiarid regions (Sahel) and agroforestry systems, tropical forestry, and ecomonics.

FRANÇOIS MERGEN, Pinchot Professor of Forestry, Yale University, was Dean of the School of Forestry and Environmental Studies at Yale during 1965–1975. He received a B.A. from Luxembourg College and B.Sc.F. from the University of New Brunswick in 1950 and an M.F. in ecology in 1951 and a Ph.D. (forest genetics) in 1954 from Yale. He is especially knowledgeable about Francophone Africa and is currently chairman of the Sahel program of the Board on Science and Technology for International Development (BOSTID) and a member of the Advisory Committee on Technology Innovation (ACTI). He was research collaborator at the Brookhaven National Laboratory, 1960–65. He was the recipient of the Award for Outstanding Achievement in Biological Research by the Society of American Foresters in 1966 and was Distinguished Professor (Fulbright-Hays Program) in Yugoslavia, 1975. Before joining the Yale faculty, he served as project leader in forest genetics for the U.S. Forest Service in Florida. Has served as a consultant to FAO, foreign governments, and private forestry companies.

JULIA F. MORTON is Research Professor of Biology and Director of the Morton Collectanea, University of Miami, a research and information center devoted to economic botany. She received her D.Sc. from Florida State University in 1973 and was elected a Fellow of the Linnean Society of London in 1974. She has conducted extensive field studies for the U.S. National Institutes of Health and the Department of Defense, has served as horticultural development consultant in Florida and tropical America, and, since 1954, has been consultant for the Poison Control Centers in Florida. In 1978, she was selected as the First Distinguished Economic Botanist by the International Society for Economic Botany. She is currently President of the Florida State Horticultural Society.

H. Y. MOHAN RAM, Professor of Botany, University of Delhi, India, received his B.Sc. from Mysore University in 1950, his M.Sc. from Agra University in 1953, and his Ph.D. from the University of Delhi in 1959. He was a visiting Fulbright and Smith-Mundt Fellow at Cornell University from 1958 to 1960 and was a UNESCO Fellow from 1970 to 1971 at the Laboratoire de Physiologie Pluricellulaire, Fig-sur-Yvette, France. His interests include plant growth and development, tropical economic botany, pollination biology, and ethnobotany. He has been Secretary-Treasurer of the International

Society of Plant Morphologists since 1964, and is Fellow and Editor of Publications of the Indian National Science Academy, New Delhi.

JOZEF SWIDERSKI is Chief, Mechanical Wood Products Branch, Food and Agricultural Organization (FAO), Rome. Dr. Swiderski was educated in forest products engineering in Poland where he also worked in several forest products mills. Before joining FAO he was Director, Department of Technique and Development, Ministry of Forestry and Forest Industry, Warsaw. In his present position he has been sending forestry advisers all over the world to advise on firewood and charcoal production.

FRANK H. WADSWORTH, Acting Project Leader of the Institute of Tropical Forestry, U.S. Department of Agriculture Forest Service in Rio Pierdras, Puerto Rico, received a Ph.D. in forestry from the University of Michigan in 1950. For the past 35 years his work has been concentrated on the silviculture of moist forests in Puerto Rico and elsewhere in tropical America.

K. FREERK WIERSUM is staff member of the Department of Silviculture, Section on Tropical Silviculture, Agricultural University, Wageningen, the Netherlands. After completing his *ingenieurs* degree (M.Sc. equivalent) in tropical forest ecology and silviculture at Wageningen University in 1973, he worked with the UN Development Program/FAO Upper Solo Watershed Management and Upland Development Project in central Java, Indonesia. Here, among other activities, he was engaged in studies on the local fuelwood situation. After joining the Department of Silviculture in 1976, he was sent to the Institute of Ecology, Padjadjaran University, Bandung, Indonesia, to participate in a joint research project on vegetation and erosion, studying possibilities for integrating the productive and protective functions of forests in densely populated areas.

NOEL D. VIETMEYER, staff officer for this study, is a Professional Associate of the Board on Science and Technology for International Development, National Academy of Sciences. A New Zealander with a Ph.D. in organic chemistry from the University of California, Berkeley, he now works on innovations in science that are important for developing countries.

APPENDIX 9

Contributors

This report contains much information previously unavailable in the international literature. This is due to the generosity of scientists from 81 countries who have provided to the various chapters their personal observations as well as information from internal reports and limited-circulation journals. The following individuals and organizations participated by correspondence in the preparation of the report.

Argentina

R. H. Braun W., Director, Instituto Argentino de Investigaciones de las Zonas Aridas, Mendoza

C. A. Naranjo, Departamento de Ciencia Biológicas, Facultad de Ciencias Exactas y Naturales, Buenos Aires

J. R. Ottone, Instituto Forestal Nacional, Buenos Aires

R. A. Palacios, Departamento de Ciencia Biológicas, Facultad de Ciencias Exactas y Naturales, Buenos Aires

J. J. Valla, Cátedra de Botánica, Universidad de Buenos Aires, Buenos Aires

Australia

B. J. Beggs, Conservator, Forests Department, Perth, Western Australia

R. Boardman, Senior Research Officer, Woods and Forests Department, Adelaide, South Australia

P. Bridgewater, School of Environmental and Life Sciences, Murdoch University, Murdoch, Western Australia

P. Burgess, Division of Forest Research, CSIRO, Canberra, A.C.T.

D. Cameron, Division of Forest Research, CSIRO, Winnellie, Darwin, Northern Territory

B. Clough, Australian Institute of Marine Science, James Cook University, Townsville, Queensland

J. C. Doran, Seed Section, Division of Forest Research, CSIRO, Canberra, A.C.T.

C. R. Dunlap, Darwin Herbarium, Division of Primary Industry, Darwin, Northern Territory

K. Felton, Forestry Commission of Tasmania, Hobart, Tasmania

Forestry Commission of New South Wales, Wood Technology Division, Beecroft, New South Wales

J. J. Fox, The Research School of Pacific Studies, The Australian National University, Canberra, A.C.T.

A. M. Gill, Division of Plant Industry, CSIRO, Canberra, A.C.T.

N. Hall, Hopetoun Village, Castle Hill, New South Wales

A. C. Harris, Consultant in Forestry and Forest Industries, Perth, Western Australia

W. Jehne, Division of Mathematics and Statistics, CSIRO, The Cunningham Laboratory, Townsville, Queensland

J. P. Jessop, State Herbarium of South Australia, Adelaide Botanic Gardens, Adelaide, South Australia

R. D. Johnston, Division of Forest Research CSIRO, Canberra, A.C.T.

R. W. Johnson, Director, Queensland Herbarium, Indooroopilly, Queensland

A. F. Logan, Division of Chemical Technology, CSIRO, South Melbourne, Victoria

J. R. Maconochie, Curator, Herbarium of the Northern Territory, Department of the Northern Territory, Alice Springs, Northern Territory

J. D. Majer, Western Australian Institute of Technology, South Bentley, Western Australia

B. R. Maslin, Department of Agriculture, Western Australian Herbarium, South Perth, Western Australia

R. K. Orme, The Forestry Commission, Launceston, Tasmania

L. Pedley, Queensland Herbarium, Indooroopilly, Queensland

R. A. Perry, Chief, Division of Land Resources Management, Wembley, Western Australia

F. H. Phillips, Division of Chemical Technology, CSIRO, South Melbourne, Victoria

L. D. Pryor, Forestry Department, The Australian National University, Canberra, A.C.T.

J. P. Quirk, Director, Waite Agricultural Research Institute, The University of Adelaide, Glen Osmond, South Australia

Regeneration Section, Commonwealth Aluminium Corporation, Ltd., Weipa, North Queensland

R. Reid, Davies Laboratory, Division of Tropical Crops and Pastures, CSIRO, Townsville, Queensland

P. C. Richmond, Forest Department, Kalgoorlie, Western Australia

W. M. Robinson, Director, Division of Technical Services, Department of Forestry, Brisbane, Queensland

M. U. Slee, Department of Forestry, The Australian National Universtity, Canberra, A.C.T.

Soil Conservation Authority, Kew, Victoria

H. Suijdendorp, Department of Agriculture, Carnarvon, Western Australia

M. J. Trinick, Division of Land Resources, Management, CSIRO, Wembley, Western Australia

P. T. Unwin, Forestry Commission of Tasmania, Hobart, Tasmania

Austria
Bundesministerium für Land-undforstwirtschaft, Vienna

Bahamas
C. L. Russell, Nassau Botanic Garden, Nassau

Bangladesh
P. Das, Forest Research Institute, Chittagong
N. Jones, Tree Improvement Officer, FAO Forestry Project, Chittagong
R. Lal Banik, Forest Ecologist, Forest Research Institute, Chittagong
M. A. Latif, Forest Research Institute, Chittagong
S. R. Rahman, Assistant Chief Conservator of Forests, Dacca
M. A. Razzaque, Forest Research Institute, Chittagong
A. B. Siddique, Forest Research Institute, Chittagong

Belize
Chief Forest Officer, Forest Department, Ministry of Trade and Industry, Belmopan

Benin
The Director, Service des Eaux, Forêts et Chasse, Cotonou

Bermuda
Botanical Gardens, Department of Agriculture and Fisheries, Hamilton
D. Leighton, Botanical Gardens, Department of Agriculture and Fisheries, Paget

Bolivia
M. Sachtler, Project Manager, Proyecto FAO Forestal, Santa Cruz

Brazil
R. P. Bosshart, IRI Research Institute/Brazil, Belém, Pará
E. Campinhos Jr., Aracruz Florestal S.A., Aracruz, Espírito Santo
J. Dubois, Projeto de Desenvolvimento et Pesquisa Florestal, Batista Campos, Belém, Pará
C. E. B. Foelkel, Riocell, Rio Grande Cia de Celulose do Sul, Guaíba, Rio Grande do Sul
W. Gunk, Instituto Nacional de Pesquisas de Amazonia, Manaus
M. T. Inowe, Departamento Silvicultura e Manejo, Curitiba, Paraná
W. Kerr, Director General, Instituto Nacional de Pesquisas de Amazonia, Manaus, Amazonas
P. V. D. Ledoux, Departamento de Ciencias Biológicas, Universidade Federal do Pará, Belém, Pará
P. M. Mayer, Companhia Agrícola e Florestal Santa Bárbara, Belo Horizonte, Minas Gerais
A. von der Pahlen, Instituto Nacional de Pesquisas de Amazonia, Manaus
Y. P. de Pasztor, Instituto Florestal, São Paulo, São Paulo
H. Schubart, Instituto Nacional de Pesquisas de Amazonia, Manaus
Dr. William, Instituto Nacional de Pesquisas de Amazonia, Manaus

Bulgaria
M. Marinov, Forest Research Institute, Sofia

Burma
R. L. Sum, Department of Forestry, Arts and Science University, Rangoon

Burundi
A. Kabayanda, Directeur, Départment des Eaux et Forêts, Ministère de l'Agriculture de l'Elevage et du Développement Rural, Bujumbura

Canada
J. Bene, International Development Research Centre, Ottawa
T. Lang, Greenpeace Experimental Farm, Denman Island, British Columbia
L. Markila, (Pulp and Paper Branch, FAO, retired), Delta, British Columbia
J. H. G. Smith, Faculty of Forestry, University of British Columbia, Vancouver, British Columbia

Chad
B. Diguera, Direction des Eaux, Forêts et Chasses, Fort Lamy
D. Dulieu, Laboratoire de Foscha, N'djamena

Chile
Executive Director, Instituto Forestal, Santiago
R. Gajardo M., Botanica Forestal, Facultad de Ciencias Forestales, Santiago
M. Habit, Oficial Regional de Producción y Protección Vegetal para América Latina, FAO, Santiago
B. Husch, Project Manager, FAO Regional Office for Latin America, Santiago
J. A. Lopez, Director, Forest Seeds Centre, National Forestry Corporation, Chillán
L. R. Rocuant, Centro de Ciencias Forestales, Universidad de Concepción, Chillán
F. M. Schlegel, Director, Instituto Silviculture, University Austral, Valdivia
M. Teresa Serra, Departamento de Silviculture, Universidad de Chile, Santiago
F. Sudzuki, Escuela de Agronomía, Campua Antumapu, Santiago

Colombia
B. M. Arnolli, Forest Officer, FAO, Bogotá
K. O. Rachie, International Center of Tropical Agriculture, Cali

Costa Rica
G. Budowski, Centro Agronómico Tropical de Investigación y Enseñanza, Turrialba
W. G. Dyson, Tropical Agricultural Research and Training Center, Turrialba
L. A. Fournier-O., Forest Ecologist, University of Costa Rica, San Pedro de Montes de Oca
N. Gewald, Centro Agronómico Tropical de Investigación y Enseñanza, Turrialba
L. R. Holdridge, Tropical Science Center, San José
J. Leon, Regional Genetics Resources Program, Centro Agronómico Tropical de Investigación y Enseñanza, Turrialba
J. Palmer, Centro Agronómico Tropical de Investigación y Enseñanza, Turrialba
P. Rosero, Centro Agronómico Tropical de Investigación y Enseñanza, Turrialba
A. M. Vargas, Ministerio de Agricultura y Ganadería, Dirreción Forestal, San José

Cyprus
E. D. Michaelides, Director of the Department of Forests, Ministry of Agriculture and Natural Resources, Nicosia
A. H. Yiannis, Department of Forests, Ministry of Agriculture and Natural Resources, Nicosia

Denmark
E. B. Lauridsen, Danish/FAO Forest Tree Seed Centre, Humlebaek

Dominican Republic
The Director, Dirección General Forestal, Secretaria de Estado de las Fuerzas Armadas, Santo Domingo

Egypt
L. Boulos, National Research Centre, Dokki, Cairo
S. A. Kandeel, Head, Forest and Wood Technology, Alexandria University, El-Shatby, Alexandria
M. Kassas, Botany Department, Cairo University, Giza, Cairo

El Salvador
H. Zambrana, Servicio Forestal y de Fauna, MAG, San Salvador

Fiji
Conservator of Forests, Department of Forestry, Suva

France
M. Baumer, Montpellier
R. Catinot, Centre Technique Forestier Tropical, Nogent-sur-Marne
G. R. Ferlin, Les Cigales, Domaine de la Pinède, Sanary
J. P. Gaillard, IRFA/GERDAT, Montpellier
J-P. Lebrun, Institut d'Elevage et de Médecine Vétérinaire des Pays Tropicaux, Maisons-Alfort
S. Radwanski, International Neemtree Research and Development Programme, Paris
E. Uhart, Paris
J. Vassal, Laboratoire de Botanique, Université Paul Sabatieu, Toulouse
M. M. Versepuy, Directeur, S.A.R.L. Versepuy, Le Puy

French Guiana
J. P. Lescure, Office de la Recherche Scientifique et Technique Outre-Mer, Centre de Cayenne

The Gambia
E. F. Brewer, Wildlife Conservation Department, Banjul
G. M. Hallam, Forestry Division, Yundum Agricultural Station, Yundum

Germany
H-J. von Maydell, Institut für Weltforstwirtschaft, Hamburg
H. Ollmann, Bundersforschungsanstalt für Forst und Holzwirtschaft, Institut für Weltforstwirtschaft, Hamburg

Ghana
F. W. Addo-Ashong, Forest Products Research Institute, Kumasi
J. Brookman-Amissah, Forest Products Research Institute, Kumasi
D. E. Earl, UNDP/FAO Project, Forestry Department, Takoradi
A. Y. Komla, Forestry Department, Bolgatanga

Greece
Forest Research Institute, Ministry of Agriculture, Athens

Guatemala
F. Aguirre B., Deputy Director, Instituto Centroamericano de Investigación y Tecnología Industrial, Guatemala
W. G. López, Gerente General, Instituto Nacional Forestal, Guatemala
H. Tasaico, Fortalecimiento del Sector Forestal, Guatemala

Honduras
D. L. Hazlett, Lancetilla Experimental Garden, Tela

India
I. P. Abrol, Central Soil Salinity Research Institute, Karnal, Haryana
C. K. Atal, Director, Regional Research Laboratory, Council of Scientific and Industrial Research, Jammu-Tawi
C. R. Babu, Department of Botany, University of Delhi, Delhi
A. K. Banerjee, Development Manager, The Titaghur Paper Mills, Co., Ltd., Calcutta
G. C. Banerjee, Department of Animal Nutrition, Bidhan Chandra Krishi Viswa Vidyalaya, Mohanpur, Nadia, West Bengal
P. N. Bajpai, Head, Horticulture, C.S. Azad, Agriculture University, Kanpur
Z. S. Brar, Punjab Agricultural University, Ludhiana, Punjab
K. S. Chela, Punjab Agricultural University, Ludhiana, Punjab
Department of Forestry, College of Agriculture, Solan, Himachal Pradesh
Department of Horticulture, The University of Agricultural Sciences, Dharwar, Karnataka
Deputy Director, Horticulture Experiment and Training Center, Saharanpur, Uttar Pradesh
M. I. H. Farooqi, National Botanic Gardens, Lucknow
R. C. Ghosh, Forest Research Institute and Colleges, Dehra Dun, Uttar Pradesh
M. Gopal, Coordinator, Indo-Danish Project on Seed Procurement and Tree Improvement, Saifabad, Hyderabad, Andhra Pradesh
P. C. D. Gupta, Jute Technological Research Laboratory, Calcutta, West Bengal
S. K. Jain, Director, Botanical Survey of India, Howrah, West Bengal
N. S. Kaikini, West Coast Paper Mills, Ltd., Dandeli, Karnataka
Kerala Forest Research Institute, Peechi, Trichur, Kerala
C. M. Ketkar, Neem Cake Promotional Scheme, Shivajinagar Pune, Maharashtra
T. N. Khoshoo, National Botanic Gardens, Lucknow
S. Kondas, Head, Department of Forestry, Agricultural College and Research Institute, Tamil Nadu Agricultural University, Coimbatore, Tamil Nadu
A. Kumar, Seed Testing Laboratory, Forest Research Institute and Colleges, Dehra Dun, Uttar Pradesh
K. A. Kushalappa, Forest Department, Karnataka Government, Bangalore, Karnataka

D. N. Lohani, Chief Conservator of Forests, Lucknow, Uttar Pradesh
L. Mandal, Department of Animal Nutrition, Bidhan Chandra Krishi Viswa Vidyalaya, Mohanpur, Nadia, West Bengal
H. S. Mann, Director, Central Arid Zone Research Institute, Jodhpur
K. P. Muniswamy, Office of the Chief Conservator of Forests, Hyderabad
K. L. Mehra, Director, National Bureau of Plant Genetic Resources, Indian Agricultural Research Institute, New Delhi
B. P. Patil, Director, Indian Grassland Fodder Research Institute, Jhansi, Uttar Pradesh
S. B. Palit, Deputy Inspector General of Forests, Ministry of Agriculture and Irrigation, New Delhi
P. G. Pattanath, Indo-Danish Project on Seed Procurement and Tree Improvement, Saifabad, Hyderabad, Andhra Pradesh
S. N. Rai, Divisional Forest Officer, Dharwad
H. Y. Mohan Ram, Department of Botany, University of Delhi, Delhi
D. Sharma, International Crops Research Institute for the Semi-Arid Tropics, Hyderabad, Andhra Pradesh
S. Shyamsunder, Conservator of Forests, Research and Utilisation Circle, Bangalore, Karnataka
H. S. Sohi, Chairman, Department of Botany, Punjab University, Chandigarh
L. J. G. van der Maesen, International Crops Research Institute of the Semi-Arid Tropics, Hyderabad, Andhra Pradesh

Indonesia

BIOTROP-SEAMEO, Regional Center for Tropical Biology, Bogor
Setijati Sastrapradja, Director, National Biological Institute, Bogor
Toga Silitonga, Forest Products Research Institute, Bogor
Soekiman Atmosoedarjo, Perum Perhutani, Forest State Corporation, Jakarta
I. R. Soerjono, Director, Forest Research Institute, Agricultural Research and Development Agency, Bogor
W. C. Tappen, U.S. Agency for International Development, Jakarta
R. Voss, Agronomist and Soil Specialist, Weyerhaeuser (Far East), Jakarta

Iran

D. L. Goodwin, Team Leader, UNDP/FAO/Iran Watershed Management Project, Tehran
B. Malekpour, Director, Research Institute of Forests and Rangelands, Tehran

Iraq

L. Alnajjar, College of Forestry, Haman Alil, Mosul
S. A. Omar, Director, Botany Directorate, Ministry of Agriculture and Agrarian Reform, Abu-Ghraib

Israel

M. Evenari, Botany Department, Hebrew University of Jerusalem, Jerusalem
M. Forti, Research and Development Authority, Ben Gurion University, Beer-Sheva
J. Friedman, Department of Botany, Tel-Aviv University, Tel-Aviv
R. Karschon, Director, Forestry Division, Agricultural Research Organization, Ilanot
J. Schechter, Research and Development Authority, Ben Gurion University, Beer-Sheva
D. E. Tsuriell, International Cooperation Group on Sand Dune Reclamation, Kiryat Shmuel, Haifa
Y. Waisel, Department of Botany, Tel-Aviv University, Tel-Aviv

Italy

J. E. M. Arnold, Forestry Department, FAO, Rome
Centro di Sperimentazione Agricola e Forestale, Rome
R. Morandini, Director, Instituto Sperimentale per la Selvicoltura, Arezzo
L. Lintu, Pulp and Paper Branch, FAO, Rome
J. Prats-Llaurado, Agriculture Director, Forest Resources Division, FAO, Rome
J. Swiderski, Forest Industries Division, FAO, Rome

Ivory Coast

M. Alexandre, Office de la Recherche Scientifique et Technique Outre-Mer, Abidjan
F. Wencelius, Directeur, Centre Technique Forestier Tropical, Abidjan

Jamaica

K. C. Hall, Forestry Department, Kingston
W. Soegeng Reksodihardjo, Pioneer Hi-bred International Inc., Kingston

Kenya

Chief Conservator of Forests, Forest Industrial Development, Ministry of Natural Resources, Nairobi
Kenya Forest Department, Conservancy of Research Services, Kikuyu
H. F. Lamprey, UNEP-MAB Integrated Project in Arid Lands, UNESCO, Nairobi
G. Poulsen, IDRC-Regional Office for East and Central Africa, Nairobi
K. K. Wachira, Geography Department, Kenyatta University College, Nairobi
A. G. Seif el Din, Forestry Cooperative Research Project, IDRC, Nairobi

Korea, Republic of

Sohn, Soo Ik, Director-General, Office of Forestry, Ministry of Agriculture and Forestry, Seoul
K. B. Yim, College of Agriculture, Seoul National University, Suweon

Liberia

Assistant Secretary of Agriculture and Forestry, Department of Agriculture, Monrovia

Malawi

Department of Forestry, Lilongwe
P. D. Hardcastle, Forest Research Institute of Malawi, Malawi

Malaysia

Tham C. Keong, Research Branch, Forest Department, Sandakan, Sabah
H. Seng Lee, Forest Research Section, Forest Department Headquarters, Kuching, Sarawak
F. S. P. Ng, Forest Research Institute, Kepong, Selangor
T. B. Peh, Forest Research Institute, Kepong, Selangor
M. Ratnasabapathy, Department of Botany, University of Malaya, Kuala Lumpur
Sarawak Timber Industry Development Corporation, Kuching, Sarawak

Mauritania
Directeur, E.N.F.V.A., Kaédi

Mexico
President, Mexican Association of Professional Foresters, Mexico, D.F.

Nepal
Chief, Afforestation Office, Hatti Sar, Naxal, Kathmandu
K. Voegele, Swiss Association for Technical Assistance (SATA), Kathmandu

Netherlands
K. F. Wiersum, Department of Tropical Silviculture, Agriculture University, Wageningen

New Hebrides
R. M. Bennett, Chief Forest Officer, Department of Agriculture, Vila

New Zealand
V. J. Chapman, Botany Department, Auckland University, Auckland
Forest Research Institute, Rotorua
N. C. Lambrechtsen, Ministry of Works and Development, National Plant Materials Centre, Palmerston North
E. C. S. Little, Kerikeri, Bay of Islands
A. H. Nordmeyer, Forest and Range Experiment Station, New Zealand Forest Service, Christchurch

Nigeria
J. B. Ball, Silviculturist, FAO, Forest Development Project, Ibadan
P. Howland, Forest Research Institute of Nigeria, Ibadan
International Institute of Tropical Agriculture, Ibadan
Principal Research Officer, Forestry Research Institute of Nigeria, Samaru-Zaria

Pakistan
Ayub Agricultural Research Institute, Lyallpur
A. M. Khan, Silvicultural Research Division, Hyderabad, Sind
G. M. Khattak, Director General, Pakistan Forest Institute, Peshawar, N.W.F.P.
K. M. Siddiqui, Pakistan Forest Institute, Peshawar, N.W.F.P.

Papua New Guinea
D. J. Eddowes, Forest Products Research Center, Boroko, Port Moresby
Chief Research Officer, Forest Management Research Branch, Office of Forests, Boroko
K. J. White, Department of Primary Industry, Office of Forests, Boroko
J. F. U. Zieck, Forest Products Research Center, Boroko, Port Moresby

Peru
Oficina Nacional de Evaluación de Recursos Naturales, San Isidro, Lima
M. R. Pastor, Director de Silvicultura, Ministerio de Agricultura, Lima
Ministerio de Agricultura, Dirección General Forestal y de Fauna, Dirección de Investigación Forestal y de Fauna, Lima
A. A. Sobrevilla V., Department of Forestry, Universidad Nacional de la Amazonia Peruana, Iquitos

Philippines
I. L. Domingo, Paper Industries Corporation of the Philippines, Makati, Metro Manila
G. C. Lugod, Department of Botany, University of the Philippines at Los Baños, College, Laguna
V. B. Mendoza, Department of Natural Resources, Forest Research Institute, College, Laguna
F. S. Pollisco, Director, Forest Research Institute, College, Laguna
J. O. Sargento, Central Forest Experiment Station, University of the Philippines at Los Baños, College, Laguna
Surish Srivastava, Provident Tree Farms, Inc., Makati, Metro Manila
Dr. Tamolong, Forest Products Research and Industries Development Commission, Laguna
A. B. Velasco, Forest Research Institute, College, Laguna
N. Q. Zabala, Forest Resources Management, University of the Philippines at Los Baños, College, Laguna

Poland
H. A. Korzeniowski, Polskie Towarzystwo Lesne, Warsaw

Sénégal
A. O. Fall, Direction des Eaux, Forêts et Chasses, Ministère du Développement Rural, Dakar
El-Hadji Séne, Ingénieur des Eaux et Forêts, Service des Eaux et Forêts et Chasses, Dakar

Sierra Leone
Ministry of Agriculture and Natural Resources, Freetown
A. J. Vincent, Chief Forestry Research Officer, Forestry Office, New England, Freetown

Solomon Islands
Chief Forest Officer, Forestry Division, Ministry of Natural Resources, Honiara

South Africa
R. L. Barrett, Genetics Section, Wattle Research Institute, Pietermaritzburg
D. G. M. Donald, Faculty of Forestry, University of Stellenbosch, Stellenbosch
M. Grut, Faculty of Forestry, University of Stellenbosch, Stellenbosch
A. V. Hall, Bolus Herbarium, University of Cape Town, Cape Town
T. J. Hodgson, Project Leader, D.R. DeWet Forestry Research Station, Sabie
P. J. LeRoux, Department of Forestry, Pretoria
S. J. Milton, Bolus Herbarium, University of Cape Town, Cape Town
National Botanic Gardens of South Africa, Kirstenbosch, Botanic Gardens, Claremont
P. Reyneke, Department of Forestry, Cape Town
South African Forestry Research Institute, Pretoria
G. van Wyk, Officer in Charge, D.R. DeWet Forestry Research Station, Sabie

Spain
F. Barrientos, Sección de Coordinación y Relaciones, Instituto Nacional para la Conservación de la Naturaleza, Ministerio de Agricultura, Madrid

J. L. B. Gutierrez, Departamento Forestal, Madrid

Sri Lanka

V. R. Nanayakkara, Conservator of Forests, Forest Department, Colombo

Sudan

A. Rahman A. Dafei, Gum Research Officer, El Obeid

H. A. R. Musnad, Silviculturist, Arid Zone Forestry Research, Forestry Research Institute, Khartoum

M. Obeid, Vice-Chancellor, Gezira University, Wad Medani

Syria

I. Nahal, Professor of Forestry and Ecology, University of Aleppo, Aleppo

Taiwan

C. Huang, Joint Commission on Rural Reconstruction, Taipei

Ta-Wei Hu, Division of Silviculture, Taiwan Forestry Research Institute, Taipei

Tanzania

K. Openshaw, Division of Forestry, University of Dar es Salaam, Morogoro

Silviculturist, Forest Division, Silviculture Research Station, Lushoto

G. A. Sobuni, Serengeti Research Institute, Seronera

Thailand

S. Boonkird, Forest Industry Organization, Bangkok

J. K. Jackson, Mae Sa Integrated Watershed and Forest Land Use Project, Chiang Mai

Sompherm Kittinanda, Royal Forest Department, Bangkok

S. Pachotikarn, Director, Silviculture Division, Royal Forest Department, Bangkok

Tem Smitinand, Forest Botanist and Director, The Forest Herbarium, Royal Forest Department, Bangkok

Tunisia

H. Hamza, Directeur, Institut National de Recherches Forestières, Ariana

R. Levingston, FAO Forestry Officer, Reafforestation Project, Tunis

Turkey

U. Topak, Institute of Forest Tree Seeds Improvment, Ministry of Forestry, Ankara, Gazi

Uganda

Forest Department, Nakawa, Kampala

Ministry of Agriculture, Forestry and Cooperatives, Forest Department, Entebbe

Union of Soviet Socialist Republics

A. G. Babayev, President, Academy of Science of the Turkmen SSR, Ashkhabad, Turkmen SSR

M. I. Ishankuliyev, Repetek Reserve, Repetek, Chardzhou Region, Turkmen SSR

M. P. Petrov, Department of Geography, University of Leningrad, Leningrad

A. L. Takhtajan, Director, Komarov Botanical Institute, Leningrad

United Kingdom

D. M. W. Anderson, Department of Chemistry, University of Edinburgh, Edinburgh, Scotland

J. P. M. Brenan, Royal Botanic Gardens, Kew, Richmond, Surrey, England

A. H. Bunting, Plant Science Laboratories, University of Reading, Reading, England

J. Burley, University of Oxford, Commonwealth Forestry Institute, Oxford, England

R. W. Fishwick, The Red House, Moretonhamstead, Devon, England

T. Greaves, Commonwealth Forestry Institute, Oxford, England

K. A. Longman, Institute of Terrestrial Ecology, Penicuik, Midlothian, Scotland

E. D. Morgan, Department of Chemistry, Keele University, Keele, Staffordshire, England

J. E. Newby, Market Rasen, Lincolnshire, England

O. Rackham, Corpus Christi College, Cambridge University, Cambridge, England

C. E. Ranson, Sudbury, Suffolk, England

L. Roche, Department of Forestry and Wood Science, University of North Wales, Bangor, Gwynedd, Wales

G. E. Wickens, Royal Botanic Gardens, Kew, Richmond, Surrey, England

P. J. Wood, Senior Research Officer, Commonwealth Forestry Institute, Oxford, England

United States

Agricultural Research and Education Center, University of Florida, Homestead, Florida

F. Almeda, Director and Curator, Botanical Gardens-Herbarium, University of California, Los Angeles, California

Arizona Agricultural Experiment Station, University of Arizona, Tucson, Arizona

J. H. Arnold, Energy and Transport, Boston, Massachusetts

P. S. Ashton, The Arnold Arboretum, Harvard University, Cambridge, Massachusetts

E. S. Ayensu, Director, Office of Biological Conservation, Smithsonian Institution, Washington, D.C.

M. D. Benge, Agency for International Development, Washington, D.C.

J. S. Bethel, Dean, College of Forest Resources, University of Washington, Seattle, Washington

S. Bosken, Agency for International Development, Washington, D.C.

J. L. Brewbaker, Department of Horticulture, University of Hawaii, Honolulu, Hawaii

C. B. Briscoe, Consulting Forester, Wayne, Oklahoma

W. H. Brooks, Office of Arid Lands Studies, University of Arizona, Tucson, Arizona

N. R. Brouard, East Africa Projects, The World Bank, Washington, D.C.

R. L. Bryant, International Paper Co., Southlands Experiment Forest, Bainbridge, Georgia

Chihuahua Desert Research Institute, Alpine, Texas

L. Conde, School of Forest Resources and Conservation, University of Florida, Gainesville, Florida

H. A. Core, Professor of Forestry, University of Tennessee, Knoxville, Tennessee

L. J. Cummins, Santa Rosa, California

D. S. DeBell, Pacific Northwest Forest and Range Experiment Station, U.S. Forest Service, Olympia, Washington

M. B. Dickerman, Alexandria, Virginia

S. A. Draper, Agriculture and Rural Development, The World Bank, Washington, D.C.
J. W. Duffield, School of Forest Resources, North Carolina State University, Raleigh, North Carolina
E. P. Eckholm, Worldwatch Institute, Washington, D.C.
E. Ernst, Bethesda, Maryland
J. Ewel, Botany Department, University of Florida, Gainesville, Florida
D. Farno, The World Bank, Washington, D.C.
P. J. Felker, University of California, Riverside, California
A. Fege, Energy Research and Development Administration, Washington, D.C.
D. French, Agency for International Development, Washington, D.C.
T. F. Geary, U.S. Forest Service, Leigh Acres, Florida
H. S. Gentry, Research Biology Laboratory and Herbarium, Desert Botanical Garden, Phoenix, Arizona
H. M. Gregersen, College of Forestry, University of Minnesota, Minneapolis, Minnesota
R. A. Howard, The Arnold Arboretum, Harvard University, Jamaica Plain, Massachusetts
H. M. Hull, School of Renewable Natural Resources, University of Arizona, Tucson, Arizona
R. E. Inman, The Mitre Corporation, McLean, Virginia
F. D. Johnson, Wildlife and Range Experiment Station, University of Idaho, Moscow, Idaho
H. C. Johnston, University of Texas, Austin, Texas
E. H. Ketchledge, Department of Botany and Forest Pathology, State University of New York, Syracuse, New York
O. D. Knipe, Rocky Mountain Station, Forest Sciences, Arizona State University, Tempe, Arizona
S. L. Krugman, Principal Research Forest Geneticist, Timber Management Research, U.S. Forest Service, Washington, D.C.
B. F. Kukachka, Forest Products Laboratory, U.S. Forest Service, Madison, Wisconsin
C. H. Lamoureux, University of Hawaii, Honolulu, Hawaii
W. H. Lewis, Department of Biology, Washington, University, St. Louis, Missouri
R. V. Long, Resource Management, Division of Forestry, Carson City, Nevada
A. E. Lugo, Project Leader, Institute of Tropical Forestry, U.S. Forest Service, Rio Piedras, Puerto Rico
W. Magrath, Ann Arbor, Michigan
R. McAlister, Wood Products Research Project, U.S. Forest Service, Athens, Georgia
W. McGinnies, Director Emeritus, Office of Arid Lands Studies, University of Arizona, Tucson, Arizona
K. G. McIndoe, Leesburg, Florida
W. Meijer, School of Biological Sciences, University of Kentucky, Lexington, Kentucky
F. Mergen, School of Forestry and Environmental Sciences, Yale University, New Haven, Connecticut
J. F. Morton, Director, Morton Collectanea, University of Miami, Coral Gables, Florida
G. P. Nabhan, University of Arizona, Tucson, Arizona
W. Parham, Agency for International Development, Washington, D.C.
T. Plowman, Botanical Museum, Harvard University, Cambridge, Massachusetts
D. L. Plucknett, Chief, Agriculture and Rural Development Division, Bureau for Asia, Agency for International Development, Washington, D.C.
J. L. Ruehle, U.S. Forest Service, Southeastern Forest Experiment Station, Athens, Georgia
D. Salo, The Mitre Corporation, McLean, Virginia
J. R. Saucier, U.S. Forest Service, Southeastern Forest Experiment Station, Athens, Georgia
J. Sauer, Department of Geography, University of California, Los Angeles, California
B. G. Schubert, Curator, The Arnold Arboretum of Harvard University, Cambridge, Massachusetts
T. S. Schubert, Institute of Tropical Forestry, U.S. Forest Service, Rio Piedras, Puerto Rico
R. E. Schultes, Director, Botanical Museum, Harvard University, Cambridge, Massachusetts
F. Simerly, County of Los Angeles, Arboreta and Botanic Gardens, Arcadia, California
B. B. Simpson, Department of Botany, University of Texas, Austin, Texas
R. G. Skolmen, Institute of Pacific Islands Forestry, U.S. Forest Service, Honolulu, Hawaii
S. C. Snedaker, School of Marine and Atmospheric Science, Miami, Florida
P. K. Soderholm, U.S. Department of Agriculture, Subtropical Horticulture Research Unit, Miami, Florida
O. T. Solbrig, Gray Herbarium, Harvard University, Cambridge, Massachusetts
J. Spears, Agriculture and Rural Development, The World Bank, Washington, D.C.
G. Stanford, Greenhills Center, Cedar Hill, Texas
Y. N. Tamimi, Agricultural Experiment Station, University of Hawaii, Hilo, Hawaii
H. J. Teas, Department of Biology, University of Miami, Coral Gables, Florida
Texas A&M University, College Station, Texas
United States Forest Tree Seed Center, U.S. Forest Service, Macon, Georgia
F. H. Wadsworth, Institute of Tropical Forestry, U.S. Forest Service, Rio Piedras, Puerto Rico
C. W. Weber, Nutrition and Food Science, University of Arizona, Tucson, Arizona
R. Whistler, Department of Biochemistry, Purdue University, Lafayette, Indiana
J. A. White, Agency for International Development, Washington, D.C.
C. D. Whitesell, Silviculturist, U.S. Forest Service, Honolulu, Hawaii
J. L. Whitmore, Institute of Tropical Forestry, U.S. Forest Service, Rio Piedras, Puerto Rico
F. Wiseman, Tropical Botanist, Arizona-Sonora Desert Museum, Tucson, Arizona

Uruguay

G. M. Caldevilla, Director, Ministerio de Agricultura y Pesca, Dirección Forestal, Parques y Fauna, Montevideo

Venezuela

J. A. Steyermark, Ministry of Agriculture and Animal Breeding, Botany Institute, Caracas

Western Samoa

J. A. Walter, General Manager, Samoa Forest Products, Apia

Yemen Arab Republic

Ministry of Agriculture, Sana

Zaire

Département de Foresterie, Institut Facultaire des Sciences Agronomiques, Yangambi

Zambia

J. B. Brown, Forest Department, Ndola

D. E. Greenwood, Chief Forest Research Officer, Division of Forest Research, Ministry of Lands and Natural Resources, Kitwe

A. S. Hans, Tree Improvement Research Centre, National Council for Scientific Research, Kitwe

Zimbabwe

D. T. Carter, Chief of Research, Forest Research Centre, Highlands, Salisbury

T. Muller, c/o National Herbarium and Botanic Gardens, Causeway, Salisbury

APPENDIX 10

Index of Plants

Species described or referred to in the body of the report are indexed below. Additional plants are listed in the Master List of Firewood Species, Appendix 4, p. 187.

Advisory Committee on Technology Innovation

GEORGE BUGLIARELLO, President, Polytechnic Institute of New York, Brooklyn, New York (Chairman through December 1978)

HUGH POPENOE, Director, International Programs in Agriculture, University of Florida, Gainesville, Florida, *Chairman*

Members

HAROLD DREGNE, Director, International Center for Arid and Semi-Arid Land Studies, Texas Tech University, Lubbock, Texas

ELMER L. GADEN, JR., Department of Chemical Engineering, University of Virginia, Charlottesville, Virginia

ANDREW HAY, President, Calvert-Peat, Inc., New York, New York

CARL N. HODGES, Director, Environmental Research Laboratory, Tucson, Arizona

CYRUS MCKELL, Institute of Land Rehabilitation, Utah State University, Logan, Utah

FRANÇOIS MERGEN, Pinchot Professor of Forestry, School of Forestry and Environmental Studies, Yale University, New Haven, Connecticut

E. RAY PARISER, Senior Research Scientist, Department of Nutrition and Food Science, Massachusetts Institute of Technology, Cambridge, Massachusetts (member through 1977)

CHARLES A. ROSEN, Staff Scientist, Stanford Research Institute, Menlo Park, California (member through 1977)

VIRGINIA WALBOT, Assistant Professor, Department of Biology, Washington University, St. Louis, Missouri (member through 1977)

Board on Science and Technology for International Development

GEORGE BUGLIARELLO, President, Polytechnic Institute of New York, Brooklyn, New York (member 1972–1978), *Chairman*

DAVID PIMENTEL, Professor, Department of Entomology and Section of Ecology and Systematics, Cornell University, Ithaca, New York, *Chairman* 1975–1979

Members

RUTH ADAMS, Editor, *The Bulletin of the Atomic Scientists*, Chicago, Illinois (member through December 1977)

EDWARD S. AYENSU, Director, Office of Biological Conservation, Smithsonian Institution, Washington, D.C.

PEDRO BARBOSA, Department of Entomology, University of Maryland, College Park, Maryland (member through December 1979)

DWIGHT S. BROTHERS, International Economist and Consultant, Fairhaven Hill, Concord, Massachusetts (member through December 1978)

HARRISON BROWN, Director, Resource Systems Institute, The East-West Center, Honolulu, Hawaii

JOHN H. BRYANT, Chairman, Committee on International Health, Institute of Medicine, *ex-officio* (through July 1978)

DORIS CALLOWAY, Department of Nutrition Sciences, University of California, Berkeley, California

MARY E. CARTER, Director, Southern Regional Research Center, U.S. Department of Agriculture, New Orleans, Louisiana

ELIZABETH COLSON, Department of Anthropology, University of California, Berkeley, California

CHARLES DENNISON, Consultant, New York, New York (member through December 1977)

BREWSTER C. DENNY, Dean, Graduate School of Public Affairs, University of Washington, Seattle, Washington

ROLAND I. FUCHS, Chairman, Department of Geography, University of Hawaii at Manoa, Honolulu, Hawaii, *ex-officio*

HERBERT I. FUSFELD, Director, Center for Science and Technology Policy, Graduate School of Public Administration, New York University, New York, New York (member through December 1979)

JOHN H. GIBBONS, Director, Environment Center, University of Tennessee, Knoxville, Tennessee (member January 1979-June 1979)

MARTIN GOLAND, President, Southwest Research Institute, San Antonio, Texas (member through December 1979)

LINCOLN GORDON, Senior Fellow, Resources for the Future, Washington, D.C. (member through September 1980)

JAMES P. GRANT, President, Overseas Development Council, Washington, D.C. (member through December 1979)

GEORGE S. HAMMOND, Foreign Secretary, National Academy of Sciences, *ex-officio* (through June 1978)

N. BRUCE HANNAY, Foreign Secretary, National Academy of Engineering, *ex-officio*

GEORGE R. HERBERT, President, Research Triangle Institute, Research Triangle Park, North Carolina

WILLIAM N. HUBBARD, JR., President, The Upjohn Company, Kalamazoo, Michigan

WILLIAM A. W. KREBS, Vice President, Arthur D. Little, Inc., Cambridge, Massachusetts (member through December 1977)

ROBERT N. KREIDLER, Private Foundation Consultant, New York, New York (member 1969-1974 and 1980-)

RICHARD LASTER, Executive Vice President, General Foods Corporation, White Plains, New York

JOHN LISTON, Director, Institute for Food Science & Technology, College of Fisheries, University of Washington, Seattle, Washington

THOMAS F. MALONE, Foreign Secretary, National Academy of Sciences, *ex-officio*

FRANCOIS MERGEN, Pinchot Professor of Forestry, School of Forestry and Environmental Studies, Yale University, New Haven, Connecticut

FREDERICK T. MOORE, Economic Advisor, International Bank for Reconstruction and Development, Washington, D.C. (member through December 1978)

W. HENRY MOSLEY, Director, Cholera Research Laboratory, Dacca, Bangladesh, and Associate, Johns Hopkins University, Baltimore, Maryland (member through December 1978)

RODNEY W. NICHOLS, Executive Vice President, Rockefeller University, New York, New York

DANIEL A. OKUN, Department of Environmental Sciences and Engineering, School of Public Health, University of North Carolina, Chapel Hill, North Carolina

E. RAY PARISER, Senior Research Scientist, Department of Nutrition and Food Science, Massachusetts Institute of Technology, Cambridge, Massachusetts

JOSEPH PETTIT, President, Georgia Institute of Technology, Atlanta, Georgia (member through December 1977)

JOSEPH B. PLATT, President, Claremont University Center, Claremont, California (member through December 1977)

HUGH POPENOE, Director, International Programs in Agriculture, University of Florida, Gainesville, Florida

JAMES BRIAN QUINN, Amos Tuck School of Business Administration, Dartmouth College, Hanover, New Hampshire (member through December 1979)

PRISCILLA C. REINING, Director, Project on Desertification, International Division, American Association for the Advancement of Science, Washington, D.C.

RALPH W. RICHARDSON, JR., Department of Horticulture, Pennsylvania State University, University Park, Pennsylvania (member through December 1979)

FREDERICK SEITZ, President Emeritus, Rockefeller University, New York, New York

H. GUYFORD STEVER, Consultant, Washington, D.C.

BILL C. WRIGHT, Program Officer, Africa and the Middle East, International Agricultural Development Service, New York, New York

VICTOR RABINOWITCH, Director

MICHAEL G. C. McDONALD DOW, Deputy Director

JOHN G. HURLEY, Deputy Director

Board on Science and Technology for International Development
Commission on International Relations
National Academy of Sciences–National Research Council
2101 Constitution Avenue, Washington, D.C. 20418, USA

Advisory Studies and Special Reports

Reports published by the Board on Science and Technology for International Development are sponsored in most instances by the U.S. Agency for International Development and are intended for free distribution primarily to readers in developing countries. A limited number of copies is available for distribution on a courtesy basis to readers in the United States and other industrialized countries who have institutional affiliation with government, education, or research and who have professional interest in the subject areas treated by the reports.

Single copies of published reports listed below are available free while the supplies last. Requests should be made on your organization's letterhead. Other interested readers may buy the reports listed here from the National Technical Information Service (NTIS) whose address appears below.

5. **The Role of U.S. Engineering Schools in Development Assistance.** 1976. 30 pp. Examines opportunities and constraints facing U.S. engineering schools in mobilizing their resources to aid developing countries. NTIS Accession No. PB 262-055. $6.00.

7. **U.S. International Firms and R, D&E in Developing Countries.** 1973. 92 pp. Discusses aims and interests of international firms and developing-country hosts and suggests that differences could be mitigated by sustained efforts by the firms to strengthen local R, D&E capabilities. NTIS Accession No. PB 222-787. $8.00.

8. **Ferrocement: Applications in Developing Countries.** 1973. 89 pp. Assesses state of the art and cites applications of particular interest to developing countries—boat-building, construction, food and water storage facilities, etc. NTIS Accession No. PB 220-825. $9.00.

16. **Underexploited Tropical Plants with Promising Economic Value.** 1975. 187 pp. Describes 36 little-known tropical plants that, with research, could become important cash and food crops in the future. Includes cereals, roots and tubers, vegetables, fruits, oilseeds, forage plants, and others. NTIS Accession No. PB 251-656. $12.00.

17. **The Winged Bean: A High Protein Crop for the Tropics.** 1975. 43 pp. Describes a neglected tropical legume from Southeast Asia and Papua New Guinea that appears to have promise for combatting malnutrition worldwide. NTIS Accession No. PB 243-442. $6.00.

18. **Energy for Rural Development: Renewable Resources and Alternative Technologies for Developing Countries.** 1976. 305 pp. Examines energy technologies with power capabilities of 10-100 kilowatts at village or rural level in terms of short- and intermediate-term availability. Identifies specific research and development efforts needed to make intermediate-term applications feasible in areas offering realistic promise. NTIS Accession No. PB 260-606. $17.00 (French language edition is available through NTIS, Accession No. PB 286-467, $17.00.)

21. **Making Aquatic Weeds Useful: Some Perspectives for Developing Countries.** 1976. 175 pp. Describes ways to exploit aquatic weeds for grazing, and by harvesting and processing for use as compost, animal feed, pulp, paper, and fuel. Also describes utilization for sewage and industrial wastewater treatment. Examines certain plants with potential for aquaculture. NTIS Accession No. PB 265-161. $12.00.

22. **Guayule: An Alternative Source of Natural Rubber.** 1977. 80 pp. Describes a little-known bush that grows wild in deserts of North America and produces a rubber virtually identical with that from the rubber tree. Recommends funding for guayule development. NTIS Accession No. PB 264-170. $8.00.

23. **Resource Sensing from Space: Prospects for Developing Countries.** 1977. 203 pp. An examination of current and prospective applications of interest to the LDCs, certain implications for long-term governance of a remote sensing system, and desirable technical cooperation initiatives to diffuse user capabilities. NTIS Accession No. PB 264-171. $13.00. (French language edition available from BOSTID or through NTIS, Accession No. PB 80-168-693, $14.00.)

25. **Tropical Legumes: Resources for the Future.** 1979. 331 pp. Describes plants of the family Leguminosae, including root crops, pulses, fruits, forages, timber and wood products, ornamentals, and others. NTIS Accession No. PB 298-423. $18.00.

26. **Leucaena: Promising Forage and Tree Crop for the Tropics.** 1977. 118 pp. Describes *Leucaena leucocephala*, a little known Mexican plant with vigorously growing, bushy types that produce nutritious forage and organic fertilizer as well as tree types that produce timber, firewood, and pulp and paper. The plant is also useful for revegetating hillslopes and providing firebreaks, shade, and city beautification. NTIS Accession No. PB 268-124. $10.00.

27. **Firewood Crops: Shrub and Tree Species for Energy Production.** 1980. 237 pp.

28. **Microbial Processes: Promising Technologies for Developing Countries.** 1979. 198 pp. Discusses the potential importance of microbiology in developing countries in food and feed, plant nutrition, pest control, fuel and energy, waste treatment and utilization, and health. NTIS Accession No. 80-144-686. $13.00.

29. **Postharvest Food Losses in Developing Countries.** 1978. 202 pp. Assesses potential and limitations of food loss reduction efforts; summarizes existing work and information about losses of major food crops and fish; discusses economic and social factors involved; identifies major areas of need; and suggests policy and program options for developing countries and technical assistance agencies. NTIS Accession No. PB 290-421. $13.00.

30. **U.S. Science and Technology for Development: Contributions to the UN Conference.** 1978. 226 pp. Serves the U.S. Department of State as a major background document for the U.S. national paper, 1979 United Nations Conference on Science and Technology for Development. Includes an overview section plus five substantive sections as follows: 1) industrialization; 2) health, nutrition, and population; 3) food, climate, soil, and water; 4) energy, natural resources and environment; and 5) urbanization, transportation, and communication.

Related Publications

Other reports (prepared in cooperation with BOSTID) available from the above address are:

An International Centre for Manatee Research. 1975. 34 pp. Describes the use of the manatee, a large, almost extinct, marine mammal, to clear aquatic weeds from canals. Proposes a research laboratory to develop manatee reproduction and husbandry. Published by the National Science Research Council of Guyana. NTIS Accession No. PB 240-244. $6.00.

Workshop on Solar Energy for the Villages of Tanzania. 1978. 167 pp. Report of a workshop with the Tanzania National Scientific Research Council, Dar es Salaam, Tanzania. Reviews state-of-the-art of small-scale solar energy devices, and suggests short and long-range projects using them in villages. Published by Tanzania National Scientific Research Council. NTIS Accession No. PB 282-941. $12.00.

Out-of-Print Publications

The following out-of-print BOSTID reports are available only from the National Technical Information Service (NTIS) unless otherwise noted. Persons located in the United States, Canada and Mexico wishing out-of-print reports should order directly from NTIS giving report title, NTIS Accession Number, and the amount indicated. Prices are current for January 1980 and are subject to change without notice. Payment must accompany all orders. Send orders to:

National Technical Information Service
Springfield, Virginia 22161 USA

Persons located outside the United States, Canada and Mexico should order through the local NTIS Representative (usually located in the national research council or national documentation center) if there is one in your country. If there is no representative, send your order including title, accession number, and payment of double the price indicated to the address below. Payment must be in one of the following forms: checks drawn on a U.S. bank; international money order; international postal coupons, or UNESCO coupons. Send orders to:

National Technical Information Service (NTIS)
Springfield, Virginia 22161 USA
ATTENTION: Order Control Unit, Mildred Johnson

1. **East Pakistan Land and Water Development as Related to Agriculture.** January 1971. 67 pp. Reviews World Bank proposed action program in land and water management. NTIS Accession No. PB 203-328. $7.00.

2. **The International Development Institute.** July 1971. 57 pp. Endorses concept of new science-based technical assistance agency as successor to AID; examines its character, purposes, and functions. NTIS Accession No. PB 203-331. $7.00.

3. **Solar Energy in Developing Countries: Perspectives and Prospects.** March 1972. 49 pp. Assesses state of art, identifies promising areas for R&D and proposes multipurpose regional energy research institute for developing world. NTIS Accession No. PB 208-550. $7.00.

4. **Scientific and Technical Information for Developing Countries.** April 1972. 80 pp. Examines problem of developing world's access to scientific and technical information sources, provides rationale for assistance in this field, and suggests programs for strengthening information infrastructure and promoting information transfer. NTIS Accession No. PB 210-107. $8.00.

6. **Research Management and Technical Entrepreneurship: A U.S. Role in Improving Skills in Developing Countries.** 1973. 40 pp. Recommends initiation of a systematic program and indicates priority elements. NTIS Accession No. PB 225-129. $6.00.

9. **Mosquito Control: Some Perspectives for Developing Countries.** 1973. 63 pp. Examines biological control alternatives to conventional pesticides; evaluates state of knowledge and research potential of several approaches. NTIS Accession No. PB 224-749. $8.00.

10. **Food Science in Developing Countries: A Selection of Unsolved Problems.** 1974. 81 pp. Describes 42 unsolved technical problems with background information, possible approaches to a solution, and information sources. NTIS Accession No. PB 235-410. $8.00.

11. **Some Prospects for Aquatic Weed Management in Guyana.** 1973. 44 pp. Report on workshop with the National Science Research Council of Guyana describes new methods of aquatic weed control suitable for tropical developing countries. NTIS Accession No. PB 228-660. $7.00.

12. **Roofing in Developing Countries: Research for New Technologies.** 1974. 74 pp. Emphasizes the need for research on low cost roofs, particularly using materials available in developing countries. NTIS Accession No. PB 234-503. $8.00.

13. **Meeting the Challenge of Industrialization: A Feasibility Study for an International Industrialization Institute.** 1973. 133 pp. Advances concept of an independent, interdisciplinary research institute to illuminate new policy options confronting all nations. NTIS Accession No. PB 228-348. $10.00.

14. **More Water for Arid Lands: Promising Technologies and Research Opportunities.** 1974. 153 pp. Outlines little-known but promising technologies to supply and conserve water in arid areas. NTIS Accession No. PB 239-472. $11.00 (French language edition is available from BOSTID or through NTIS, Accession No. PB 274-612, $11.00.)

15. **International Development Programs of the Office of the Foreign Secretary**, by Harrison Brown and Theresa Tellez. 1973. 68 pp. History and analysis, 1963-1973; lists staff/participants and publications. NTIS Accession No. PB 230-543. $7.00.

19. **Methane Generation from Human, Animal and Agricultural Wastes.** 1977. 131 pp. Discusses means by which natural process of anaerobic fermentation can be controlled by man for his benefit, and how the methane generated can be used as a fuel. NTIS Accession No. PB 276-469. $10.00.

20. **Systems Analysis and Operations Research: A Tool for Policy and Program Planning for Developing Countries.** 1976. 98 pp. Examines utility and limitations of SA/OR methodology for developing country application and means for acquiring indigenous capabilities. NTIS Accession No. PB 251-639. $9.00.

24. **Appropriate Technologies for Developing Countries.** 1977. 140 pp. Examines fundamental issues and interrelationships among economic, political and social factors relating to choice of technologies in developing countries. Discusses criteria of appropriateness and suggest policies for improving technical decisions. NTIS Accession No. PB 80-124-118. $11.00.

Other out-of-print reports (prepared in cooperation with BOSTID) available from the National Technical Information service are:

Products from Jojoba: A Promising New Crop for Arid Lands. 1975. 30 pp. Describes the chemistry of the oil obtained from the North American desert shrub, *Simmondsia chinensis.* NTIS Accession No. PB 253-126. $6.00.

Aquatic Weed Management: Some prospects for the Sudan and the Nile Basin. 1975. 57 pp. Report of a 1975 workshop with the Sudanese National Council for Research. Suggests modern and innovative methods for managing the water hyacinth. Published by National Council for Research – Agricultural Research Council of Sudan. NTIS Accession No. PB 259-990. $7.00.

Natural Products for Sri Lanka's Future. 1975. 53 pp. Report of a 1975 workshop with the National Science Council of Sri Lanka. Identifies neglected and unconventional plant products that can significantly contribute to Sri Lanka's economic development. Published by National Science Council of Sri Lanka. NTIS Accession No. PB 251-520. $7.00.

Ferrocement, a Versatile Construction Material: Its Increasing Use in Asia. 1976. 106 pp. Report of a 1974 workshop with the Asian Institute of Technology, Bangkok, Thailand. Surveys applications of ferrocement technology in Asia and the Pacific Islands. Includes construction of grain silos, water tanks, roofs, and boats. Published by Asian Institute of Technology. NTIS Accession No. PB 261-818. $9.00.

International Consultation of Ipil-Ipil Research. 1978. 172 pp. Report of a 1976 conference cosponsored with the Philippine Council for Agriculture and Resources Research, Los Banos, Laguna, Philippines. Contains background papers and workshop session summary reports on ipil-ipil (*Leucaena* spp.). (Companion volume to report no. 26 above.) NTIS Accession No. PB 280-161. $11.00.

Aquatic Weed Management: Integrated Control Techniques for the Gezira Irrigation Scheme. Report of a 1978 workshop with the University of Gezira. Considers biological, herbicidal, and mechanical methods for controlling aquatic vegetation in canals. Published by the University of Gezira. NTIS Accession No. PB 80-122-286. $8.00.

Reports in Preparation (working titles)

BOSTID will fill requests for single copies of reports in preparation upon publication as outlined at the beginning of this section.

31. Food, Fuel, and Fertilizer from Organic Wastes.
32. The Water Buffalo: Its Potential for Development.
33. The Potential for Alcohol Fuels in Developing Countries.
34. Revegetating the Range: Selected Research and Development Opportunities.
35. Sowing Forests from the Air.
36. Producer Gas: a Little-Known Fuel for Motor Transport.
37. The Winged Bean: A High Protein Crop for the Tropics (Second Edition).

Firewood Crops

Shrub and Tree Species for Energy Production Volume 2

Report of an Ad Hoc Panel of the
Advisory Committee on Technology Innovation
Board on Science and Technology for International Development
Office of International Affairs

Funding for this study was provided by
the Agency for International Development

NATIONAL ACADEMY PRESS
Washington, D.C. 1983

NOTICE: The project that is the subject of this report was approved by the Governing Board of the National Research Council, whose members are drawn from the councils of the National Academy of Sciences, the National Academy of Engineering, and the Institute of Medicine. The members of the committee responsible for the report were chosen for their special competences and with regard for appropriate balance.

This report has been reviewed by a group other than the authors according to procedures approved by a Report Review Committee consisting of members of the National Academy of Sciences, the National Academy of Engineering, and the Institute of Medicine.

The National Research Council was established by the National Academy of Sciences in 1916 to associate the broad community of science and technology with the Academy's purposes of furthering knowledge and of advising the federal government. The Council operates in accordance with general policies determined by the Academy under the authority of its congressional charter of 1863, which establishes the Academy as a private, nonprofit, self-governing membership corporation. The Council has become the principal operating agency of both the National Academy of Sciences and the National Academy of Engineering in the conduct of their services to the government, the public, and the scientific and engineering communities. It is administered jointly by both Academies and the Institute of Medicine. The National Academy of Engineering and the Institute of Medicine were established in 1964 and 1970, respectively, under the charter of the National Academy of Sciences.

The Board on Science and Technology for International Development (BOSTID) of the Office of International Affairs addresses a range of issues arising from the ways in which science and technology in developing countries can stimulate and complement the complex processes of social and economic development. It oversees a broad program of bilateral workshops with scientific organizations in developing countries and conducts special studies. BOSTID's Advisory Committee on Technology Innovation publishes topical reviews of technical processes and biological resources of potential importance to developing countries.

This report has been prepared by an ad hoc advisory panel of the Advisory Committee on Technology Innovation, Board on Science and Technology for International Development, Office of International Affairs, National Research Council, for the Office of the Science Advisor, Agency for International Development, Washington, D.C., under Grant No. DAN-5538-G-00-1023-00.

Library of Congress Catalog Card Number: 80-83796

PANEL ON FIREWOOD CROPS

EDWARD S. AYENSU, Director, Office of Biological Conservation, Smithsonian Institution, Washington D.C., USA, *Chairman*

JOHN G. BENE, Senior Advisor to the President, International Development Research Centre, Ottawa, Canada

JAMES S. BETHEL, Former Dean, School of Forest Resources, University of Washington, Seattle, Washington, USA

LOUTFY BOULOS, Professor, National Research Centre, Dokki, Cairo, Egypt

GERARDO BUDOWSKI, Head, Natural Renewable Resources Program, Centro Agronomico Tropical de Investigacion y Ensenanza, Turrialba, Costa Rica

JEFFERY BURLEY, University Lecturer, Forestry Department, Commonwealth Forestry Institute, University of Oxford, Oxford, England

ERICK ECKHOLM, International Institute for Environment and Development, Washington, D.C., USA

HANS M. GREGERSEN, Professor, College of Forestry, University of Minnesota, Minneapolis, Minnesota, USA

ANTHONY V. HALL, Senior Lecturer and Assistant Curator, Bolus Herbarium, University of Cape Town, Rondebosch, South Africa

R. D. JOHNSTON, Principal Research Scientist, Division of Forest Research, Commonwealth Scientific and Industrial Research Organization, Canberra, Australia

STANLEY L. KRUGMAN, Principal Research Forest Geneticist, Timber Management Research, United States Department of Agriculture-Forest Service, Washington, D.C., USA

BOHUMIL F. KUKACHKA, Former Wood Technologist, Pioneer Research Unit, Forest Products Laboratory, Madison, Wisconsin, USA

HANS-JURGEN VON MAYDELL, Professor, Federal Research Centre for Forestry and Forest Products, Hamburg-Reinbek, West Germany

FRANÇOIS MERGEN, Professor, School of Forestry and Environmental Studies, Yale University, New Haven, Connecticut, USA

JULIA MORTON, Director, Morton Collectanea, University of Miami, Coral Gables, Florida, USA

H. Y. MOHAN RAM, Head, Department of Botany, University of Delhi, Delhi, India

JOZEF SWIDERSKI, Chief, Mechanical Wood Products Branch, Forest Industries and Trade Division, Food and Agriculture Organization of the United Nations, Rome, Italy

FRANK H. WADSWORTH, Former Director, Institute of Tropical Forestry, United States Department of Agriculture-Forest Service, Rio Piedras, Puerto Rico

K. FREERK WIERSUM, Staff Member, Department of Silviculture, Agricultural University, Wageningen, The Netherlands.

National Research Council Staff

MARY JANE ENGQUIST, Staff Associate, *Study Director*

NOEL D. VIETMEYER, Professional Associate

F. R. RUSKIN, Editor

Preface

In 1980 the National Academy of Sciences published *Firewood Crops: Shrub and Tree Species for Energy Production*, a report prepared by an ad hoc panel of the Advisory Committee on Technology Innovation (ACTI). It is one of a series of reports that propose unconventional scientific approaches to problems of developing countries.

The original book describes 60 tree species or groups of species with promise for cultivation around the home or village or in firewood plantations. It does not suggest a comprehensive solution to the firewood crisis, but it does examine one part of the solution: the selection of species suitable for cultivation in different climatic zones.

During the final preparation of the first report, information was received on other species that also seemed to have potential as firewood crops but could not be included because of time constraints and lack of information. After publication of the book, still more information on further species was received. It was decided that these additional species should be evaluated and described in a second volume.

The purpose of this report, as of its predecessor, is not to delineate strategies for growing and utilizing firewood in any given region of the world, but to provide some general concepts and methods for planners and technicians to consider.

Primary emphasis is on species suitable for growing firewood for individual family needs. However, species suited to plantation cultivation for fueling small industrial factories, electric generators, and crop driers are also discussed. Most of the plants are little known in traditional forest production. Some are woody shrubs rather than forest trees, but these many-branched, crooked, sometimes short-lived species may better meet requirements for small-scale village use.

We particularly looked for multipurpose plants that have uses in addition to providing fuel—plants that adapt well to different sites, establish easily, and require little care; plants that thrive in problem environments such as steep hillslopes, low-nutrient or toxic soils, arid zones, and tropical highlands; and plants that have characteristics such as nitrogen-fixing ability, rapid growth, ability to coppice, and high calorific value.

In addition to the first report, other ACTI publications in the series *Innovations in Tropical Reforestation* that contain information on some exceptionally promising firewood species and related technologies are:*

- *Sowing Forests from the Air* (1981)
- *Mangium and Other Fast-Growing Acacias of the Humid Tropics* (1983)
- *Calliandra: A Versatile Small Tree for the Humid Tropics* (1983)
- *Casuarinas: Nitrogen-Fixing Trees for Adverse Sites* (1984)
- *Leucaena: Promising Forage and Tree Crop for the Tropics* (Second edition in preparation)

*For information on how to order these and other reports, see page 90.

Information on promising, fast-growing trees is also contained in Report No. 25, *Tropical Legumes: Resources for the Future* (1979).

These publication activities are supported largely by the U.S. Agency for International Development (AID). This study was sponsored by AID's Office of the Science Advisor, which also made possible the free distribution of this report.

A related report was written by E. L. Little, Jr., under a contract funded by AID and administered by the Forest Service, U.S. Department of Agriculture (*Common Fuelwood Crops: A Handbook for Their Identification*. Communi-Tech Associates, Morgantown, West Virginia. Available for purchase from Communi-Tech Associates, Post Office Box 3170, Morgantown, West Virginia 26503, USA. Single copy price: $13.50 [softcover]). The main objective of the handbook is to aid in the identification of the common trees and shrubs grown as fuelwood crops in plantations and forests chiefly in tropical regions. It was prepared as a companion volume to *Firewood Crops: Shrub and Tree Species for Energy Production* (Report No. 27).

WARNING

This book, if misunderstood, is potentially dangerous. Because of the severity of the firewood crisis, the panel has selected trees and shrubs that are aggressive and grow rapidly. These seem appropriate for cultivation in areas of extreme fuel shortage, particularly where climates and soil conditions are harsh. However, in more equable environments, and where no fuelwood shortages exist, such potentially invasive plants should be introduced only with great care and with serious consideration for the threat posed by their weediness. **In any trials of fuelwood plantations, local species should always be given priority.**

Contents

Introduction

The problems of firewood shortages and the promise of firewood plantations were described extensively in the first volume.

No less than one-and-a-half billion people in developing countries derive at least 90 percent of their energy requirements from wood and charcoal. Another billion people meet at least 50 percent of their energy needs this way. Indeed, it has been estimated that at least half the timber cut in the world still serves its original role for mankind: as fuel for cooking and heating.*

This essential resource, however, is seriously threatened. The developing world is facing a critical firewood shortage. The growth in human population is far outpacing the growth of new trees — not surprising when the average user burns as much as a ton of firewood a year. The results are soaring prices for wood, a growing drain on incomes and physical energy expended to satisfy basic fuel needs, the wasteful burning of animal manures to cook food rather than help produce it, and an ecologically disastrous and potentially irreversible spread of treeless landscapes.

There is no single magical solution to the firewood scarcity, but some blend of fuel conservation, tree planting, and new technologies could certainly ease its severity in any country. The failure of many affected countries to meet the firewood challenge does not, in the final analysis, reflect an absence of suitable technologies, but rather a failure of political systems, of social organization.

Many of the species included in this book would not be appropriate under certain conditions for fuelwood; for example, some of them have much higher value for other uses. The advantage of *Albizia* and a number of pine species, for instance, is that they have multiple uses. Should fuel needs diminish, the species can be turned to these alternative uses. *Albizia* is used for pulp; it could at the same time serve as firewood if that were needed.

In the Korean fuelwood plantation program, 30 percent of the species planted were required to have commercial uses in addition to fuelwood. The idea was that if fuelwood demands turned out to be less than initially anticipated, the trees would still serve some useful purpose.

Setting up a dependable national supply of firewood seems an ideal project for combined action by a country's government and its people.

*E. Eckholm 1975. *The Other Energy Crisis: Firewood.* Worldwatch Paper 1. Worldwatch Institute, Washington, D.C., USA.

I Fuelwood Species for Humid Tropics

Of the one billion people in the humid tropics, perhaps a fifth live in or adjacent to forests. To provide fuelwood and charcoal for their daily needs, these 200 million people fell acres of trees each year. In addition, much fuelwood is cut for shipment to urban markets. Together with commercial timber logging and expansion of agriculture, the result is depletion of tropical forests at a rate that exceeds any present attempts to replenish them.

This study, the second of two volumes, is concerned with ways of establishing self-replenishing sources of fuelwood. In some countries trees are already being profitably cultivated in energy plantations, and the idea has become more attractive as the price of other fuels rises.

Conditions in the humid tropics promote the growth of many kinds of trees and shrubs, with the heat and dampness ensuring high rates of photosynthesis and short harvest cycles. The question is, which trees are the best candidates for deliberate cultivation?

The first volume of this report described the following species as worth testing for fuelwood crops in the humid tropics:

- *Acacia auriculiformis*
- *Calliandra calothyrsus*
- *Casuarina equisetifolia*
- *Derris indica*
- *Gliricidia sepium*
- *Gmelina arborea*
- *Guazuma ulmifolia*
- *Leucaena leucocephala*
- Mangroves
- *Mimosa scabrella*
- *Muntingia calabura*
- *Sesbania bispinosa*
- *Sesbania grandiflora*
- *Syzygium cumini*
- *Terminalia catappa*
- *Trema* spp

This section of the present volume describes further shrub and tree species that hold promise for cultivation in humid tropical regions both for profit and for staving off the threat of fuelwood depletion in developing tropical countries.

Species included in other sections of this volume that are worth testing in the humid tropics are *Melaleuca quinquenervia* and *Robinia pseudoacacia*.

Pinus caribaea (see page 22) 8 months old, intercropped with corn 1 month old in Turrialba, Costa Rica. (R. Sal

Albizia falcataria

Botanic Name *Albizia falcataria* (L.) Fosberg

Synonyms *Albizzia falcata* (L.) Backer, *A. moluccana* Miq.

Common Names Batai, Molucca albizzia, Moluccan sau, djeungjing, sengon, falcata, vaivai, puah, white albizzia, kayu macis, tamalini, mara, placata, plakata

Family Leguminosae (Mimosoidae)

Main Attributes Because of its rapid growth, vigorous coppicing, and usefulness when grown in combination with agricultural crops, it should be considered for firewood, notwithstanding its low specific gravity and low caloric value. In some regions like Western Samoa and Java the species is already used as fuelwood.

Description One of the fastest growing of all trees, *Albizia falcataria* reaches (under favorable conditions) 15 m in height in 3 years, 30 m in 10 years, and 44 m in 17 years. When grown in the open, its crown spreads to form a large umbrella-shaped canopy, but in plantations it has a narrow crown. Its flowers are creamy white and have a slight fragrance.

Distribution *Albizia falcataria* is native to Papua New Guinea, West Irian, the Solomon Islands, and the Moluccas. The species is grown in plantations, notably in the Philippines and elsewhere in Southeast Asia and in Fiji. It has also been used in variety trials in India, Sri Lanka, Taiwan, and Western Samoa. It is now naturalized in many urban areas of Sabah. On Java, especially west Java, it is planted in home gardens and gardens mixed with herbaceous and tree crops (for example fruit trees) as a source of timber and firewood. The practice is quite widespread in the area and can yield a substantial income for villagers.

Use as Firewood The wood is light, soft, and lacks strength (specific gravity 0.24-0.49). Tests performed in Java show that it makes good charcoal and that its caloric potential varies from 2,865–3,357 kcal per kg.

Yield On better soils growth can be up to 50 m^3 per ha per year, while average sites yield up to 39 m^3 per ha per year on a 10-year rotation. (Growth tends to culminate at or before 10 years.)

Other Uses

- Wood. The wood is excellent for fiber board and particle board. It is also used as veneer core stock and for pallets and crating, furniture components, matches, and boxes.
- Pulp. The species is most suitable for ground wood pulp because of its low density, pale color, and fiber length and qualities. In the Philippines it is successfully used for newsprint.
- Shade. *Albizia falcataria* is widely used as a shade tree for cattle and for cocoa, coffee, banana, and tea plantations.

Environmental Requirements

- Temperature. *Albizia falcataria* requires a temperature range of 22°-29°C.
- Altitude. Within its natural range, this species occurs up to 1,000 m elevation. It is grown at altitudes up to 1,500 m, but generally does best in plantations below 800 m.
- Rainfall. This tropical species thrives in humid tropical climates with no more than a slight dry season (0–2 months). The best Philippine growth has been achieved on sites with an annual rainfall of 4,500 mm and no dry period.
- Soil. *Albizia falcataria* does best on well-drained, deep soils with reasonably high fertility. Like some other legumes, it is likely to perform better on slightly alkaline than on acid soils.

Establishment The tree regenerates easily by natural seeding on any clearing. It produces seeds prolifically and regularly from an early age, usually from 3 to 4 years on.

- Seed treatment. Best germination occurs after mechanical scarification or treatment with sulfuric acid for 12 minutes followed by water for 15 minutes. Seed may also be placed in boiling water, removed from heat, and wrapped in wet cloth for 24 hours.
- Ability to compete with weeds. Remarkably fast initial growth is typical of *Albizia falcataria*; therefore, weeding normally can be limited to one complete weeding and three spot weedings during the first year after planting. Occasional elimination of vines may be necessary after that.

Pests and Diseases Various white and brown rots are the most serious diseases afflicting *Albizia falcataria* plantations. These diseases normally follow wind damage. Plantings in the Philippines are subject to attack by the fungus *Coiticum salmonicolor*, causing "albizia canker,"

Albizia falcataria growing on Mt. Vaea, near Apia, Western Samoa, where it is widely used for firewood. (B. Cahusac)

which can be devastating unless controlled with Bordeaux mixture or other fungicides. Harmful insects are *Eurema blanda, Eurema hecabe, Semiothise* spp., *Xystocera festiva* (wood borer), and *Penthicodes jarinosa.*

Limitations *Albizia falcataria* is extremely susceptible to uprooting and breakage of branches by high winds. Once exposed, the low-density wood is easily attacked by fungi, which bring about rotting of the trees. The tree has a massive root system that is known to contribute to soil erosion, and it should not be planted on steep hillsides. Large tree roots, rather than holding the soil, have an opposite effect. Grass and herbaceous plants are soil retainers.

Bursera simaruba

Botanic Name *Bursera simaruba* Sarg.

Synonyms *Bursera gummifera* L., *B. ovalifolia* Engler, *Elaphrium simaruba* Rose

Common Names Gumbo limbo, gum-elemi, gumtree, West Indian birch, red birch, almacigo, indio desnudo, palo mulato, galo de incienso, turpentine tree, and many others

Family Burseraceae

Main Attributes This handsome tree is esteemed for fuel and for living fence posts throughout Central America and the Caribbean. So far, however, it is virtually unknown elsewhere. It is easily propagated—green branches thrust into the ground take root quickly and grow vigorously. The trees regenerate swiftly after cutting. In fact, even trees blown down by winds send up shoots, which soon become trunks as big as the original.

Description The gumbo limbo as it grows in forests is an erect, straight, resinous tree reaching 20-30 m, with a stout trunk (to 75 cm) that is often forked about 2 m from the ground, forming thick vertical branches that fork again higher up. The bark, which peels off in thin flakes, is typically copper colored (sometimes silvery) and glossy. When growing in the open, the branches spread and form a broad crown. The tree is bare during the cool, dry season. The leaves are compound and have a turpentine odor when crushed.

Distribution This handsome tree is native to and esteemed in areas from central Florida through the Bahamas and West Indies and from southern Mexico to the northern parts of South America.

Use as Firewood The wood has a high moisture content, but when thoroughly dry it is commonly burned as firewood and charcoal. Because of its flammability, Indians of Yucatan use it for kindling. The wood has a specific gravity, green and oven dried, of 0.30-0.40.

Yield No studies have been made of the potential firewood yield. In Belize, surveys have shown populations of up to 57 trees per ha in the wild.

Other Uses

- Wood. The wood seasons well, with slight shrinkage; is easily worked; saws, planes, and polishes satisfactorily; holds nails firmly; and is used commercially for a veneer that resembles paper birch and for plywood for interior use. It is fairly strong, but is not durable for outside use, being attacked by borers, beetles, and termites. Locally, it is made into boxes and crates, soles for sandals, light furniture, and matchsticks and toothpicks.
- Beautification. The gumbo limbo is much used as an ornamental. In southern Florida it is planted as a landscape tree in new developments and along streets, mainly because of its quick growth, but also because of the newly awakened appreciation of native species. It has long been planted as a living fence throughout the Caribbean area.
- Resin. The aromatic resin, which oozes from an incision as a thick, amber gum, is concentrated and dried and the chips offered as tithes or burned as incense in South American churches.

Environmental Requirements

- Temperature. Gumbo limbo requires a subtropical or tropical climate. Fully grown trees stand occasional brief winter frosts.
- Altitude. It is found from sea level to 1,800 m elevation in Guatemala, but generally occurs below 1,000 m.
- Rainfall. The tree abounds where annual rainfall averages 500-1,400 mm.
- Soil. Gumbo limbo endures extreme soil types ranging from fertile, moist forest habitats to dry, barren limestone, but it grows best in rich lowlands. Under the severely arid conditions of some Caribbean islands it is rather stunted and crooked, but survives. It has a high degree of salt tolerance.

Establishment While the tree reseeds itself in its natural areas, it is seldom deliberately propagated by seed.

- Ability to compete with weeds. Being a forest tree, the gumbo limbo tolerates shade at all stages of growth. It is not retarded by the shade of competing vegetation even in the juvenile period.

Pests and Diseases The resin is a natural insect repellent, and no pests or diseases are reported in available literature.

Limitations Apart from cold sensitivity, the handicaps seem to be only the brittleness of branches, which may be snapped off by strong gusts of wind; the perishability of exposed wood; and the discoloration of the timber by the

Bursera simaruba. (From *500 Plants of South Florida* by J. Morton)

sap-stain fungi to which it is prone because of its high moisture content. (Such staining can be avoided by immediate kiln drying or by immediate spraying of the cut logs with a fungicide.)

Coccoloba uvifera

Botanic Name *Coccoloba uvifera* (L.) L.

Synonyms *Coccolobis uvifera* Jacq., *Polygonum uvifera* L.

Common Names Seagrape, seaside grape, bay grape, baya de praga, mangle de falda, raisin la mer, uva de mar, uva de playa, uva caleta, uvero, horsewood, hopwood

Family Polygonaceae

Main Attributes Seagrape is one of the first woody species to become established on sandy shores, being more hardy in these exposed places and more tolerant of salt than most trees.

Description Fully exposed on windswept seacoasts, the seagrape is dwarfed and bushy (to 2.5 m high) and forms dense colonies. On leeward shores and inland locations it becomes a round-topped, spreading, low-branching tree up to 15 m high, with thick, smooth branches and a stout trunk to 1 m in diameter. The stiff and leathery leaves are large, thick, and almost circular. They turn scarlet or yellow and drop off in the winter dry season, but they are soon replaced by silky new growth with a bronze hue. Small and fragrant white flowers are borne profusely in spring and develop into red-purple grapelike clusters of velvet-skinned edible fruits. Male and female flowers are borne on separate trees.

Distribution The seagrape is native to southern Florida, Bermuda, the Bahamas, and West Indies. It also grows from the northeast coast of Mexico along the Atlantic coast of Central America, reaching down both coasts of South America as far as Peru and Brazil. It is often planted as a seashore windbreak in the Hawaiian Islands. It was introduced into the Philippines and Zanzibar about 40 years ago.

Use as Firewood Throughout its range, the wood is commonly employed as firewood and for making charcoal. In Mexico the charcoal has been particularly prized by blacksmiths and silversmiths. In an ignition trial, small dry branches and twigs were lit with matches. The first flames gave off a little smoke; within 5 minutes there were red coals and only wisps of smoke; and in another 5 minutes there were bright, smokeless, even flames whose intense heat surpassed the maximum 204°C on a thermometer held 40 cm above the fire, proving the seagrape to be ideal for cooking. Its specific gravity is reported to be from 0.7 to 0.96.

Yield The seagrape is fast growing, multi-stemmed if pruned, and profusely branched. Yield should be exceptional, though no statistics are available.

Other Uses

- Wood. The wood is fine grained, compact, heavy, and hard. It takes a high polish and is valued for furniture and cabinetwork, but is subject to dry-wood termites.
- Ornamental. The seagrape is widely grown as an ornamental tree and is a great favorite for Florida landscaping in even the most sophisticated sites, either free form or trimmed to a globular head. It is also close planted and trimmed as a hedge. It flourishes inland as well as in coastal locations and is long lived.
- Fruit. The fruits are popular in the Caribbean, where they are sold by street vendors and in local markets. They make excellent jelly.
- Honey. The flowers yield abundant nectar, and the honey, though high in moisture, is of good quality, light amber in color, and spicy.
- Bark. The bark is rich in tannin. In the past, the red-brown evaporated decoction of the wood and bark, called kino, was regularly exported to Europe for use in tanning and dyeing.

Environmental Requirements

- Temperature. The seagrape requires a subtropical or tropical, climate. It does need full sun.
- Altitude. In the Philippines the tree grows well from sea level to an elevation of 500 m. In Florida it occurs up to the central part of the peninsula, and it has withstood brief freezes that have killed mature mango trees.
- Rainfall. Throughout its range, the seagrape is subject to great variations in annual precipitation, from the 1,400-mm Florida average to the 500-mm average of the near-arid islands of the Caribbean.
- Soil. The seagrape is remarkable in its ability to thrive in nearly pure sand and on rocky coasts, as well as in marl, oolitic limestone, and diabase, dry or wet, but with good drainage. It is extremely tolerant of salt.

Establishment Seeds germinate readily, but a seedling seagrape may not fruit for 6–8 years. Vegetative propagation is preferred because it is

Coccoloba uvifera tree and fruit. (K. and J. Morton)

quicker and is the only way to assure reproduction of female trees or selected cultivars. Air layering, ground layering, and grafting are practiced, but mature wood cuttings are most commonly employed for mass multiplication.

- Seed treatment. None required.
- Ability to compete with weeds. The seagrape cannot tolerate shade; therefore, seedlings should be weeded until they outgrow competition. For this reason, cuttings are more practical because they become established quickly and outdistance the weeds.

Pests and Diseases In Hawaii, seagrapes planted inland are attacked by the rose beetle *Adoretus sinicus*. Generally, the seagrape is virtually free of pests and diseases, but occasionally, under adverse conditions, the leaves are affected by rust or fungal or algal leaf spots, and there have been reports of mushroom root rot.

Limitations The seagrape is not adapted to extremely hot and humid environments. It needs full sun, low altitudes, and mild winter weather where only brief frosts may occur. Leaf fall may be a nuisance in cold dry seasons.

Eucalyptus brassiana

Botanic Name *Eucalyptus brassiana* S. T. Blake

Common Names Cape York red gum, gum-topped peppermint (Queensland); karo (Papua New Guinea)

Family Myrtaceae

Main Attributes *Eucalyptus brassiana* is a hardy, fast-growing tree adapted to a variety of site conditions in the lowland tropics. It tolerates periodic flooding and readily coppices.

Description The tree often grows from 7 to 15 m high on infertile sites but reaches more than 30 m on better sites. The trunk may be moderately straight for half the tree's height, but is often badly shaped. Typically, it divides into several large ascending branches.

Distribution A tropical eucalypt, *Eucalyptus brassiana* occurs in southwestern Papua New Guinea and on Cape York Peninsula in Queensland, northern Australia, between latitudes 8° and 16° S. The best development is on lowland sites in Papua. It has recently been introduced to several subhumid to humid tropical areas.

Use as Firewood *Eucalyptus brassiana* is a fast-growing species with the capacity to produce coppice shoots. The wood is moderately dense, probably similar to that of *Eucalyptus camaldulensis.*

Yield Trees in experimental plots in Malaysia have reached as high as 7.6 m, with a diameter of 6.3 cm at 2.5 years of age. In Bangladesh the tree's growth rate is similar to that of *Eucalyptus camaldulensis*, while it has shown average-to-good growth rates in trial plots in Vietnam, parts of East and West Africa, and Brazil.

Other Uses

- Wood. The brown or pale reddish brown heartwood is hard, heavy, strong, and durable. Because the natural stands occur in areas of low population density, the timber has been used very little. On Daru Island, Papua New Guinea, it is regarded as an excellent building timber for general construction and posts.

Environmental Requirements

- Temperature. In Queensland and Papua New Guinea, the natural habitat of the species, the climate is tropical and humid to subhumid. The trees withstand high temperatures, since the mean maximum temperature of the hottest month is around 32°C. Frosts have not been recorded, and the seedlings are likely to be frost-sensitive.
- Altitude. The natural stands occur at low altitudes from sea level to 650 m.
- Rainfall. In its native habitat the tree tolerates a dry season lasting from 3 to 5 months; the minimum annual rainfall is 1,000 mm.
- Soil. The species grows from well-drained rocky slopes to undulating plains and on seasonally inundated flats and depressions. The soils are typically infertile. It has grown well on the margins of rice fields in Bangladesh.

Establishment *Eucalyptus brassiana* regularly produces good seed crops. Seed germination is high, and the seeds are long lived when kept sealed and dry in cold storage. The plantable stock raised in the nursery is likely to be about 25 percent of the viable seeds sown. The seedlings can be raised in trays and transplanted into containers or sown directly in containers. Under favorable conditions, plantable seedlings can be produced within 6-10 weeks after sowing.

- Seed treatment. None required.
- Ability to compete with weeds. Poor. Rapid initial growth is dependent on complete weeding during the first year.

Limitations There is little experience with planting this species in large areas, and its silvicultural requirements are not well known, but it must be kept free of weeds during the establishment phase. Provenance trials are required to determine the optimum seed sources.

Related Species The tree is related to the well-known fuelwood species *Eucalyptus tereticornis* and *Eucalyptus camaldulensis*, and to *Eucalyptus exserta*, which is widely planted in the People's Republic of China.

Eucalyptus brassiana, Australia. (J. Doran)

Eucalyptus deglupta

Botanic Name *Eucalyptus deglupta* Blume

Synonym *E. naudiniana* F. Muell.

Common Names Kamarere (Papua New Guinea); bagras (Philippines); leda (Indonesia)

Family Myrtaceae

Main Attributes Kamarere is one of the world's fastest growing trees. It is capable of colonizing land eroded by landslides and areas of recent volcanic activity.

Description This is a large tree, generally reaching from 35 to 60 m high but occasionally reaching up to 75 m, with diameters of 0.5–2 m or greater. On unstable soils and river alluvia the tree often develops buttresses 3–4 m high. The trunk is one-half to two-thirds of the tree height and is typically straight.

Distribution A wholly tropical eucalypt, kamarere has a markedly discontinuous distribution that includes Mindanao in the Philippines; Sulawesi, Ceram, and Irian Jaya in Indonesia; and parts of the mainland and New Britain in Papua New Guinea. The best development occurs on riverine banks at altitudes mainly below 150 m in New Britain. The species has been introduced to many humid tropical areas.

Use as Firewood The main use of *Eucalyptus deglupta* is as a pulpwood. It is normally considered too valuable for use as firewood. However, it grows so quickly that in suitable fuel-short areas it could be considered for both uses. Its wood has a specific gravity of 0.40–0.80. Trees greater than 15 years of age yield good charcoal.

Yield Test plots in Papua New Guinea suggest that in 15 years' time it may reach 44 m in height, with a diameter at breast height of 54 cm. Yields of 20–40 m^3 per ha per year are common to that age in several countries.

Other Uses

- Wood. The wood from natural stands has been used for heavy construction, furniture, and flooring. Shrinkage and fibre collapse have been slight.
- Pulp. Pulps of good yield have been obtained using the sulfate process.

Environmental Requirements

- Temperature. Kamarere's native climate is subequatorial in coastal areas and tropical elsewhere. Near sea level the mean maximum temperature of all months may be 30°–32°C, but at higher altitudes the mean maximum of the coolest month may be only 24°C.
- Altitude. It is found from sea level to about 1,800 m.
- Rainfall. The mean annual rainfall is mainly 2,500–3,500 mm in its native habitat, but there is no upper limit for planted stands on well-drained sites.
- Soil. The tree grows best on deep, moderately fertile sandy loams, but also occurs on soils derived from volcanic ash and pumice. Although most stands occur on alluvium along rivers, landslides and volcanic disturbances provide sites for small stands inland.

Establishment Under plantation conditions, kamarere flowers at an early age, usually by 3–4 years. Normally, the seedlings are field planted as tubed stock.

- Seed treatment. None required. The seed germinates quickly and subsequent seedling growth is usually fast.
- Ability to compete with weeds. The planting fields for kamarere must be weeded for the first 6 months and some weeding is necessary until the end of the second year.

Pests and Diseases In both natural stands and plantations, termites are the most serious pests of *Eucalyptus deglupta*. Young trees in Malaysia have been damaged severely by the cossid moth *Zeuzera coffeae* and a ring bark borer (family Hepialidae). Tip dieback of young trees in the Solomon Islands is caused by the coreid bug. Ant control must be continuous in Brazil. Trees of some provenances have been attacked by a stem and bark borer (*Agrilis* spp) in Papua New Guinea and the Philippines.

Limitations The bark is very thin and the tree is sensitive to fire, susceptible to frost damage, and intolerant of drought. Young trees in particular are brittle and easily damaged by strong winds, thereby decreasing the viability of seed. Kamarere usually does not coppice.

Eucalyptus deglupta. Four-year-old tree planted in Costa Rica by a small farmer on land used for agroforestry.
(H. J. von Maydell)

Eucalyptus pellita

Botanic Name *Eucalyptus pellita* F. Muell.

Synonym *E. spectabilis* F. Muell.

Common Name Red mahogany, red stringybark, large-fruited red mahogany, Daintree stringybark

Family Myrtaceae

Main Attributes Red mahogany grows quickly in humid and subhumid tropical lowland regions. Good provenances from the Cape York Peninsula could provide the basis for important new eucalypt trials in tropical countries.

Description Red mahogany is a medium to large tree, usually 20–25 m in height but exceeding 35 m under favorable conditions. At its best its trunk grows straight up to half the tree height, with a large, heavily branched crown. On bare rock above beaches it may be reduced to a bushy shrub.

Distribution This eucalypt occurs on lower coastal slopes of Australia. Northern populations occur from near the top of Cape York Peninsula to between Cairns and Townsville (latitude 12°–18°S) and southern ones from near Brisbane to the Batemans Bay area in southern, coastal New South Wales (27°–36°S).

Use as Firewood The tree has a high density (990 kg per m^3) suitable for charcoal and fuelwood.

Yield In Espirito Santo, Brazil, a northern provenance reached a mean height of 13.7 m and a diameter of 12.6 cm in 3.5 years. In Minas Gerais State, the same provenance was 14.9 m tall with a diameter of 13.3 cm at the same age.

Other Uses

- Wood. The heartwood is moderately heavy, strong, and durable in and out of the ground. Although the grain is somewhat interlocked, the wood is not difficult to work. It has a wide range of uses for building and heavy ornamental work.
- Honey. The flowers are a minor source of thin, strong-flavored honey from January to February in Queensland. They provide large quantities of pollen for the bee colonies.

Environmental Requirements

- Temperature. Red mahogany is found mainly in the humid climatic zone. The mean maximum temperature of the hottest month varies from 24° to 33°C and the mean minimum of the coolest from 6° to 16°C; frosts are absent or occur less frequently than five times a year.
- Altitude. From sea level to about 750 m.
- Rainfall. The mean annual rainfall in the tree's natural Australian habitat is 900–2,300 mm.
- Soil. The tree grows mainly on gentle-to-moderate topography, though it is found to a limited extent on steep, well-drained slopes. It prefers moist sites—the lower slopes of large ridges and even alongside small streams in the drier and hotter parts of its occurrence. Soil types range from shallow sandy podsols derived from sandstone to deep forest loams. However, red mahogany is recommended for planting only in well-drained, sandy soil.

Establishment

- Seed treatment. No pretreatment is required. The seeds remain viable for many years under controlled storage conditions.
- Ability to compete with weeds. Once established, the tree forms a dense crown that shades out weeds at an early stage.

Pests and Diseases *Eucalyptus pellita* is moderately resistant to the stem canker *Cryphonectria cubensis* (Bruner) Hodges.

Limitations In areas other than Australia where it has been grown, *Eucalyptus pellita*'s growth rate has not been good enough to demonstrate an advantage over the better-known *Eucalyptus resinifera*. However, these results appear to be based largely on the southern provenances. Northern provenances have grown well in moist tropical areas, especially in Brazil.

Eucalyptus pellita, Australia. (R. D. Johnston)

Eucalyptus urophylla

Botanic Name *Eucalyptus urophylla* S. T. Blake

Synonyms *E. decaisneana*, also erroneously referred to as *E. alba*

Common Names Ampupu, Timor mountain gum

Family Myrtaceae

Main Attributes *Eucalyptus urophylla* shows promise in tropical or subtropical countries where the rainfall exceeds 1,000 mm annually. It is highly resistant to the stem canker *Cryophonectria cubensis* (Bruner) Hodges, which has caused serious losses to eucalypts in Latin America.

Description On favorable sites this species can exceed 50 m in height, with diameters of up to 2 m and straight clear boles for half or two-thirds of the tree height. There is considerable variation within the species in the degree of retention of rough bark, ranging from only a short stocking to boles fully clothed in rough fibrous bark.

Distribution *Eucalyptus urophylla* occurs in Timor and other Indonesian islands. It was introduced into Brazil in 1919 under the name *Eucalyptus alba*, and the progeny from this introduction were used to establish large areas of plantations. The progeny were mainly hybrid stock, and these trees have become known as "Brazil alba." The pure species has grown extremely well in Brazil, Cameroon, Ivory Coast, and the People's Republic of Congo and is showing considerable promise in areas in Southeast Asia and the Pacific where there is a distinct but not too severe dry period in the cool season.

Use as Firewood Although one of the less dense eucalypts, young trees make a satisfactory fuelwood. It coppices well.

Yield Yields of 20-30 m³ per ha per year have been reported under favorable growing conditions. Low-altitude provenances have usually given the highest yields.

Other Uses

- Wood. The wood of older trees is used in heavy construction, building poles, and fence posts and gives a high pulp yield.

Environmental Requirements

- Temperature. Mean monthly temperatures range from 18° to 28°C in *Eucalyptus urophylla*'s natural habitat.
- Altitude. This is primarily a mountainous species, growing at altitudes from 300 m to almost 3,000 m.
- Rainfall. The rainfall can be as low as 1,300–1,400 mm, but the tree does best with rainfall of 2,000–2,500 mm.
- Soil. This eucalypt grows best on deep, moist, free-draining, medium-to-heavy soils derived from noncalcareous rock.

Establishment The species seems to grow best under nursery conditions. In one trial in Costa Rica, seedlings were transplanted 55 days after sowing to polyethylene bags containing loam-sand mixture. In northeast Brazil the seedlings are grown in polyethylene bags in the nursery. They reach an optimum size of about 25 cm in 75 days. A technique to raise *Eucalyptus urophylla* and a hybrid of *Eucalyptus urophylla* × *Eucalyptus grandis* vegetatively by cuttings has been developed in Brazil and the Congo.

- Seed treatment. None required.
- Ability to compete with weeds. The tree is highly sensitive to competition in the early stages, and the plantation must be kept weed free for 6–12 months after planting. After that time, the dense crown inhibits competing weeds.

Pests and Diseases *Eucalyptus urophylla* is susceptible to termite attack. In Australia defoliation by a range of insects has reduced the growth considerably. The trees are resistant to the canker disease in Brazil. Leaf-cutting ants, *Atta* spp. and *Acromyrmex* spp., can be a problem in parts of South America; control measures are essential where the ants are present.

Limitations Seeds are not yet readily available from plantations or natural forests; however, a seed orchard has been planted in the Congo.

Eucalyptus urophylla, Brazil production stand. (CSIRO, Brisbane, Australia)

Hibiscus tiliaceus

Botanic Name *Hibiscus tiliaceus* L.

Synonyms *Pariti tiliaceum* Britton, *Paritium tiliaceum* St. Hil.

Common Names Mahoe, majagua, majagua de playa, sea hibiscus, seaside mahoe, emajagua, hau, among others

Family Malvaceae

Main Attributes The wood of this fast-growing tree is used chiefly for fuel, although the fibrous bark is much used for cordage. The trees sprout from stumps and, when cut back, produce long, vigorous shoots from which quantities of rope can be made.

Description The mahoe is an evergreen that may grow 12 m tall. Normally, it has a short trunk and a broad, rounded crown. In wetlands it is branched close to the ground, forming a low and sprawling massive mound. The lowest branches bend down, take root, and put forth new growth, thus forming extensive tangled thickets. The handsome leaves are somewhat heart shaped.

Distribution The mahoe is pantropic, growing just inshore of the mangroves on the coastal fringe and extending up estuaries and rivers.

Use as Firewood In Puerto Rico, India, and Malaya, the wood is used principally for fuel; it is also employed as firewood in Ghana. In Malaya, the tree is recommended for planting on poor soils as a firewood resource. In American Samoa and other Pacific islands, it is prized for making fire by friction because it ignites readily when a hardwood stick is twirled rapidly against it. In the past, in Hawaii, oiled sticks of mahoe wood were set afire and thrown from cliffs in quick succession as "fireworks." The wood has a specific gravity of 0.6.

Coppicing is remarkably swift and profuse, with multiple long stems arising from cut stumps.

Yield No systematic study has been made of the productivity of mahoe for firewood. Its rapid growth and spreading habit, however, augur well for high volume per hectare.

Other Uses

- Wood. The tree has been of such inestimable value to Pacific seafarers that it was formerly necessary to obtain permission before cutting a single branch. The heartwood is yellowish or purplish, light but firm, flexible, porous, close grained, easy to work, and takes a high polish. Durable in salt water, it is popular for fishing-net floats, floats for outriggers and for light boats, planking, and pilings for dwellings in wetlands. It is also used for laths for huts and for tool handles and cabinetwork. It has been pulped for papermaking, but the product is of low quality because of the wood's short fibers and is fit only for wrapping purposes.
- Bark. The bark has always been used for making rope and twine; for long, strong cables for hauling logs and ships; for harpoon lines and fish traps; and for mats, household strainers, tapa cloth, and hula skirts. Easily stripped from the branches in long ribbons, its rough outer surface is peeled off, leaving a smooth surface. When reduced to fiber, it is similar to jute. It is stronger wet than dry.
- Erosion control and dune fixation. The mahoe is ofen planted to stablize sand dunes and, on muddy shores, to trap soil to reinforce the coastline. In India it is planted to prevent erosion on the banks of rivers and reservoirs.
- Beautification. The mahoe is widely valued as an ornamental because of its lush foliage and attractive flowers. On high, well-drained land it becomes an elegant landscape plant with a symmetrical, rounded head and clear trunk, providing it is "raised up" by pruning the base when the tree is young. Enough space must be available for development of its form; this is not a tree for small gardens. It is sometimes close planted as a living fence.

Environmental Requirements

- Temperature. The species is adapted to warm, humid areas, from near-tropical (for example southern Florida) to ultratropical.
- Altitude. The mahoe grows from sea level to 500 m elevation.
- Rainfall. It thrives where there is annual precipitation of 1,400 mm. Some scattered specimens are found on the coasts of extremely dry regions.
- Soil. The tree has no particular soil requirements; it seems to grow equally well in mud, marl, sand, and limestone, assuming greater stature on high, dry sites inland than in its native habitat—shallow brackish swamps. It is highly salt tolerant.

Hibiscus tiliaceus. (From *500 Plants of South Florida* by J. Morton)

Establishment The seed capsules may float for months until they reach favorable shores, where the seeds readily germinate and colonize the area. In cultivation the tree is easily raised from seed or cuttings, or by air layering. Growth is rapid; in 2–3 years the tree is large enough to provide shade.

• Ability to compete with weeds. Seedlings should be weeded in the early stages if planted inland. If planted in swampy, saline sites, there will be little vegetative competition and they will be self-sustaining.

Pests and Diseases The mahoe is an extraordinarily healthy tree with few problems. Spring and summer wilting was observed at Lahore, India, and attributed to *Alternaria dianthi*, which is locally prevalent on other plants. The tree recovers by the following season without treatment. In Florida it is sometimes affected by leaf spot caused by *Gloeosporium* spp or *Phyllosticta hibiscina*.

Limitations The mahoe is limited to warm, humid, low elevations. It is intolerant of cold, though mature trees have not been seriously injured by brief frosts in southern Florida.

Maesopsis eminii

Botanic Name *Maesopsis eminii* Engl.

Common Names Musizi, ndunga, muhumula, esenge

Family Rhamnaceae

Main Attributes *Maesopsis eminii* is a quick-growing, short-lived species suggested for enrichment planting in the humid tropics where it grows well if the canopy opening is sufficient and there is no dense shade. It is also recommended for well-drained grassy areas of watercourses.

Description This fast-growing deciduous tree is cylindrical, straight, free of branches for 9–20 m, and has a wide-spreading crown. The size of the tree decreases across the African continent from east to west. In Nigeria it is seldom over 15 m high, while in Uganda it grows to 36 m.

Distribution The species is native to tropical Central Africa from Liberia to Tanzania, covering a range of latitudes from 8°N to 2°S. Plantations have been established or natural forests managed in Fiji, Indonesia, Malaysia, Kenya, Tanzania, Uganda, and Zaire. Trials have been established in other African countries and in Costa Rica, Hawaii, Western Samoa, and the Solomon Islands.

Use as Firewood Plantation wood of *Maesopsis eminii* is light and therefore is not an ideal fuelwood. Its specific gravity is 0.38–0.48.

Yield On good soils in Africa the tree may grow 2-3 m in height per year while it is young. In Uganda, musizi forest plantations are managed on a 30-year rotation, producing a final volume of 20–30 m^3 per ha per year. In many areas, yields of 8–20 m^3 per ha are common; on fertile volcanic soils of Malaysia and Indonesia annual yields even reached 20–30 m^3 per ha with a maximum of 33 m^3 and 40 m^3 per ha, respectively.

Other Uses

- Fruit. The fruits, which are often eaten by animals, contain an edible oil.
- Wood. The wood of *Maesopsis eminii* is open grained and light brown and, although light weight, is tough and easy to work. It tends to warp, split, or collapse during seasoning, but is used for general indoor construction, joinery, and veneers for matchboxes. It is also suitable for plywood and pulpwood. On Java, Indonesia, the species has been proposed as a substitute for *Albizia falcataria* because of its similar wood quality and its resistance to the *Xystrocera* wood-borer, which damages the *Albizia*.
- Miscellaneous. In Zaire the trees are planted to shade coffee and cocoa plantations.

Environmental Requirements

- Temperature. *Maesopsis eminii* grows in an area of mean annual temperatures ranging from 22° to 27°C, while the mean minimal temperatures of the coldest months are 16°–24°C.
- Altitude. The tree is normally found between 100 m and 700 m elevation, but in several countries, including Uganda, it grows well up to 1,200 m.
- Rainfall. The species requires 1,200–3,000 mm of uniform summer rains.
- Soil. *Maesopsis eminii* is considered best suited to moderately fertile, well-drained, light-to medium-textured soils, with a neutral to acid pH.

Establishment Natural regeneration is good; large quantities of seed are eaten and distributed by birds. The seeds number about 700 per kg. Artificial regeneration can be effected by direct seeding, but it is safer to raise seedlings in nurseries. Nursery stock should be planted out as pot plants at an early age, since the plants rapidly form long taproots. Stumps cannot be used.

- Seed treatment. Seeds must soak in cold water for 2-3 days.
- Ability to compete with weeds. The tree cannot tolerate *Imperata* grass competition during its early growth, but if it has been well tended during its early years, the rather dense umbrella crown rapidly suppresses competition. It is a light-demanding species that can be planted in pure plantations, but it can also be used for underplanting or enrichment planting in open woodlands.

Pests and Diseases Young trees are susceptible to cankers caused by *Fusarium solani* and other fungi. Bacterial wilt has been reported from a site in Malaysia with impeded drainage. In Zaire, damage by the *Monochamus scabiosus* borer has been reported.

Limitations The species cannot stand heavy weed competition or waterlogged soils. It is too wide crowned for use where maximum productivity per unit of land is the aim. It does not like

Maesopsis eminii, Sudan. Left: 2-year-old seedlings. Below: 25-year-old tree. (R. L. Plumptre)

grass competition. In the tropics, well-dried seeds can be stored, but not more than 3 months. Trees in poorly drained, infertile soil and on hillsides or in valleys are highly canker prone.

Pinus caribaea

Botanic Name *Pinus caribaea* Morelet*

Synonyms *Pinus bahamensis* Griseb., *Pinus hondurensis* Loock

Common Names Caribbean pine, pitch pine, pino de la costa, pino colorado, ocote blanco, pino caribaea de Honduras, pino macho

Family Pinaceae

Main Attributes Caribbean pine, which has fast initial growth, has become one of the more important pines for commercial timber plantations in tropical areas below 1,000 m. Because of the knowledge about the tree and its adaptability to equatorial lowlands or degraded lowland sites, it also deserves consideration as a fuelwood species.

Description *Pinus caribaea* is a large tree that under the best conditions may grow to 45 m in height and 135 cm in diameter. Generally the stem is straight and the branching is regular. However, there is great variation in tree form, particularly in the branching habits of natural stands.

Distribution Three geographic varieties of *Pinus caribaea* are distinguished. *Pinus caribaea* var *caribaea* (typical), or Caribbean pine, is found in western Cuba and the Isle of Pines. *Pinus caribaea* var *bahamensis*, Bahamas pine, is native to the Bahamas and Caicos Islands. *Pinus caribaea* var *hondurensis*, Honduran pine or pino hondureño, grows in the Atlantic lowland of Central America from northern Belize to northern Nicaragua.

Use as Firewood Young trees often grow rapidly, and both thinnings and branches from sawlog or pulp plantations could yield valuable fuel. The wood is soft, moderately lightweight (specific gravity, 0.4–0.66), and resinous.

Yield A mean annual increment of 21–40 m^3 per ha can be achieved on suitable sites up to the thirteenth year of growth.

Other Uses

- Pulp and papermaking. *Pinus caribaea* from natural forests is used in Nicaragua and Honduras for general-purpose pulpwood.
- Wood. The wood is excellent for particle board and noncompressed fiber board. It also is used for boat building, heavy construction, interior joinery and furniture components, and (form permitting) for veneer.

Environmental Requirements

- Temperature. The tree requires a warm, equable climate with no frost. In its native habitat, mean temperatures range from 22° to 28°C, with a maximum of 37°C and minimum of 5°C sometimes experienced.
- Altitude. In the equatorial zone, the species grows best at altitudes below 1,000 m, although it is planted up to 1,500 m.
- Rainfall. Throughout its natural habitat, *Pinus caribaea* grows in a climate with summer rainfall and a winter dry season. On the islands and on most inland areas, rainfall varies between 1,000 mm and 1,800 mm. In coastal areas rainfall may reach 3,900 mm, with poorly drained sites sometimes becoming waterlogged at the height of the rains. Some stands, however, grow on sites where annual rainfall may be as low as 660 mm.
- Soil. Soils are usually loams or sandy loams, sometimes with high amounts of gravel and generally well drained. The pH is usually between 5.0 and 5.5, although the very thin layer of soil on top of the coral platforms of the Bahamas has a pH of 8.4. In Cuba the soil is a deep, fine-grained clay that allows easy movement of air and water. In mainland coastal plains, *Pinus caribaea* occupies the well-aerated sands and silts and the levee banks where pH is between 4 and 5. Inland it can be found on a wide range of parent materials.

Establishment Easy germination and rapid early growth reduce costs of seedling production. It is essential to ensure that the correct mycorrhizae are available in the nursery soil to ensure root infection.

- Seed treatment. The seeds sometimes respond to moderate stratification, that is, a 24-hour water soak followed by 2–3 days of cold storage at 4°–5°C. They can be stored up to 10 years at 0°–5°C with moisture content below 10 percent.
- Ability to compete with weeds. *Pinus caribaea* can bear severe competition from shallow-rooted grasses, except when rainfall is below 1,000 mm. It is light demanding and cannot tolerate much overhead shade from tall grasses. Fast initial growth is typical of the species, and therefore weeding is often required for only one year after planting; weeding must

*This species was often confused with *Pinus elliottii* Engelm. Before 1950 it included *Pinus elliottii*.

Pinus caribaea, East Kalimantan, Indonesia. Two-year-old plantation growing in ultisol soil (R. L. Voss)

Pinus caribaea var *hondurensis* in Guatemala. (R. H. Kemp)

continue longer when the trees compete with climbing vines because of little shade production.

Pests and Diseases This species is susceptible to soil-borne pathogens that cause "damping off" and to various fungi that cause foliage blights, needle cast, dieback, stem and root damage, heart rot, and sap stain, depending on the location of the plantation. Serious insect pests are *Dendroctonus* (in Central American natural populations); *Ips* beetles; the pine aphid; leaf-cutting ants; termites; the Australian case moth, *Hyalarcta*; and the Nantucket tip moth, *Rhyacionia*. Damage to young plantations has been caused by squirrels, porcupines, and tree shrews feeding on the bark of stems and branches.

Limitations The great genetic variation existing within this species suggests abundant possibilities for genetic improvement in yield, stem form, growth, branching habit, and pest resistance. Indigenous seed of *Pinus caribaea* is available through commercial sources. However, in the short term, difficulties may be encountered in procuring seed for large plantations because the tree is so successful that seed is in high demand. In addition, outside its native habitat *Pinus caribaea* produces cones at low altitudes in the tropics, which reduces seed supplies.

Psidium guajava

Botanic Name *Psidium guajava* L.

Common Names Guava, guayaba

Family Myrtaceae

Main Attributes In high-population areas guava should be a good tree for use as firewood. It coppices readily and its growth rate is rapid at first and slower as the tree ages. Its ability to withstand repeated cutting should make it an exceptionally fine firewood, although the plant can be a problem weed and its wider planting should be encouraged only where enough utilization to keep it in check can be assured.

Description Guava is generally a shrub or low, wide-spreading evergreen tree 3-10 m high. It branches close to the ground and often produces suckers from roots near the base of the trunk. The small, fragrant white flowers are solitary or grow in small clusters on new axiliary shoots.

Distribution Guava is indigenous to the American tropics, where it occurs wild and cultivated. At an early date the Spanish took it to the Philippines and the Portuguese to India. It then spread throughout the tropics. It has been naturalized in many countries, being spread by birds; in some places it has become a troublesome weed in pastures and has been declared a noxious weed in Fiji.

Use as Firewood The hard, strong, heavy wood has a specific gravity of 0.8 and makes excellent firewood and charcoal. The gross calorific value per gram of dry matter (ash free) is 4,792 kcal with 0.85 percent ash.

Yield Firewood cutting causes the guava to spread by suckering, and it may become a pest in high-rainfall areas.

Other Uses

- Fruits. Guava fruits are made into preserves, jam, jelly, paste, juice, and nectar. They contain two to five times the vitamin C content of fresh orange juice, and red guavas are also a good source of vitamin A.
- Leaves. In some countries the tannin-rich leaves and green fruits are used for dyeing and tanning.
- Wood. Because of its strength the wood has been used for tool handles and implements.

Environmental Requirements

- Temperature. The trees are susceptible to frost, but if the freeze is not too severe they grow back quickly by suckering from below the soil. They require light.
- Altitude. Guava grows in the tropics from sea level to 1,500 m. However, it normally grows slowly at altitudes above 800 m.
- Rainfall. *Psidium guajava* does well in areas of 1,000 mm or more of well-distributed rainfall, and it can endure 4–5 months of drought.
- Soil. Guava does well on slightly to strongly acid soils. It flourishes in Florida on limestone and marl with a pH of 7.5–8.0. It grows relatively poorly in heavy clay soils; however, it is reported to tolerate flooding and to grow in areas where drainage is poor.

Establishment The tree can be propagated from seeds, cuttings, and suckers, or by inarching. In the past, most guavas have been grown from seed, but because of the great variability of guava fruits, vegetative propagation (for example, air layering) of superior clones is more common.

- Seed treatment. None required. The seeds remain viable for several months and will germinate in 3–5 weeks in the warm season.
- Ability to compete with weeds. Excellent. Guava is an aggressive plant that can withstand partial shading and can take over a site when it receives full sunlight.

Pests and Diseases Fruit flies, mealy bugs, scale insects, and thrips are reported to cause damage to the guava, mostly to the fruit. Wilt, the most serious enemy of the tree in India, reportedly occurs on soil above pH 7.5 and is usually fatal. Bark canker and dieback are diseases that kill branches. A bark-eating caterpillar and mealy scale may cause severe damage.

Limitations The trees can become serious pests, particularly with the root suckering induced by firewood cutting.

Related Species

- *Psidium littorale* is a native of Brazil that grows to a height of 8 m and can grow in dense shade as well as in the open. It is a good understory firewood in eucalyptus plantations in wet areas (5,000–6,000 mm rainfall). It also tolerates considerably lower temperatures and higher elevations than *Psidium guajava*.

Psidium guajava tree. (H. Y. Nakasone)

Eucalyptus robusta (see page 30) with a *Psidium cattleianum* understory on the island of Hawaii. (R. G. Skolmen)

II Fuelwood Species for Tropical Highlands

In the highlands (those areas above 1,000 m elevation) the problem of deforestation has become critical in many tropical developing countries. In part, the forests have been cut to make way for cultivation, but much of the denuded land results from the local population cutting trees and shrubs for fuelwood. The forests are receding rapidly up the hillsides as villages are forced to go higher in their constant search for fuel.

Only 10 percent of the population of tropical countries lives in the highlands, but the 40 percent living in the adjacent lowlands is also affected by the deforestation of hillsides that can no longer retain rainwater. The result is flash flooding, siltation, and drying up of streams needed for livestock and irrigation, all of which cut farmland productivity.

Reforestation of the highlands is official policy in many developing countries where subsistence farming is traditional, and efforts are under way to identify tree species best suited for these areas.

The first volume of this report described a number of species worth testing for highland fuelwood crops:

Acacia mearnsii
Ailanthus altissima
Alnus acuminata
Alnus nepalensis
Alnus rubra
Eucalyptus globulus
Eucalyptus grandis
Grevillea robusta
Inga vera

This section describes further species of trees and shrubs that merit testing in tropical highland areas.

Another species described in section III of this book that should be tested in the tropical highlands is *Dalbergia sissoo.*

Eucalyptus in Peru. (M. L. Barker)

Acacia decurrens

Botanic Name *Acacia decurrens* (Wendl.) Willd.

Synonym *Acacia decurrens* var *normalis* Benth. The species is closely related to *Acacia mearnsii* (black wattle) and *Acacia dealbata* (see *Firewood Crops* Vol. I), and in the literature these species have often been confused.

Common Names Green wattle, Sydney black wattle, king or queen wattle

Family Leguminosae (Mimosoideae)

Main Attributes *Acacia decurrens* yields excellent fuelwood and charcoal and has proved suitable for growing in many countries. The wood is little used for fuel except in Australia, but the bark is esteemed second only to that of *Acacia mearnsii* for tanning purposes. Like other acacias, green wattle fixes atmospheric nitrogen.

Description This is a beautiful tree with strong, upright growth, usually reaching 6-12 m in height. In the Nilgiris (India) it attains a height of 30 m and a diameter of 37 cm. It has a fern- or feather-like green foliage and in spring has a crowded head of fluffy, golden flowers. The slender seedpods, to 10 cm long, snap open when mature and throw the seeds a good distance.

Distribution This species is native to the coastal strip within 200 km of Sydney, Australia, and to adjacent lower montane valleys. It has been planted fairly widely on moist sites in southeastern Australia. It has also been introduced to Sri Lanka, Fiji, India, Kenya, South Africa, Hawaii, the Philippines, and parts of Central and South America.

Use as Firewood The wood is not suitable for sawtimber because of its small dimensions. It does make a good firewood, with a specific gravity of 0.50–0.70. According to one report, it has a caloric potential of 3,530–3,940 kcal per kg.

Yield A yield of 6–16 m^3 per ha per year has been reported, and in Sri Lanka the yield of firewood on a 15-year rotation averaged 25 m^3 per ha.

Other Uses

• Tanning. The bark of *Acacia decurrens* yields 35–40 percent good-quality tannin. However, it contains undesirable coloring matter, which reduces the value of the leather; planting was therefore stopped in favor of other acacias. Recent research, however, has shown that this problem can be eliminated by changing the tanning process or by adding other suitable tanning materials.

• Wood. The wood is used for building poles, mine props, fence posts, and, in recent years, for hardboard.

• Shelterbelt. The tree is used for shade and windbreaks. It has also been used as hedges on tea estates in Sri Lanka.

Environmental Requirements

• Temperature. *Acacia decurrens* thrives in a climate of 12°–25°C mean annual temperature. It is frost resistant and shade tolerant.

• Altitude. In its Australian home the tree grows in low valleys and on hillslopes (25–1,000 m), but mainly below 700 m. Elsewhere the tree may range up to 2,500 m.

• Rainfall. The mean annual rainfall for this species is 900–2,600 mm.

• Soil. *Acacia decurrens* prefers deep soils that are light to medium and free draining. It occurs naturally on soils of only moderate fertility: acid and neutral yellow earths, acid-bleached red duplex soils, podsols, and some brown friable earths derived principally from shales. The species also occurs on basalt-derived soils.

Establishment The species spreads rapidly by seed and root suckers and regenerates by coppicing. Seeds can be germinated after many years of storage in a cool, dry place. Seeds germinate in 7–14 days, and seedlings can be transplanted in 5–7 months.

• Seed treatment. The seeds are soaked 2 hours in acid or dipped in boiling water and left to cool and soak.

• Ability to compete with weeds. The tree spreads rapidly, forming solid stands too dense to permit grass or other vegetation to intrude.

Pests and Diseases *Acacia decurrens* is susceptible to the defoliator *Acanthopsyche junode*, but less so than *Acacia mearnsii*. Severe attacks by the rust fungus *Uromycladium* in the 1920s caused most plantations in New Zealand to be felled.

Limitations *Acacia decurrens*, introduced to Hawaii about 1890, has been declared noxious for state land leases because it spreads rapidly by seed and root suckers, crowding out other plants.

Acacia decurrens shelterbelt, South Africa. (A. P. G. Schönau)

Related species

• *Acacia dealbata* and *Acacia mearnsii* occur much more widely than *Acacia decurrens*. There are several species of relatively restricted distribution—for example, *Acacia parramattensis*, *Acacia sylvestris*, and *Acacia fulva*—with similar site requirements, but these have not yet been tested in cultivation.

Eucalyptus robusta

Botanic Name *Eucalyptus robusta* Sm.

Synonym *Eucalyptus multiflora* Poir.

Common Names Swamp mahogany, white mahogany, Australian brown mahogany, swamp messmate (Australia); robusta, beakpod eucalyptus (USA); brown gum, red gum (Sri Lanka and India)

Family Myrtaceae

Main Attributes Swamp mahogany is one of the most widely planted eucalypts. It grows well in plantations on good sites, but because of its ability to grow on both poorly drained and droughty locations, it is usually planted on adverse sites.

Description Swamp mahogany is a tree normally attaining heights of 25–30 m and diameters of 1-1.2 m (however, in Hawaii some giant specimens have reached 55 m with a 25-m trunk). The trunk is typically straight and branch free for about half the height of the tree. In open-grown trees the crown has long, spreading, irregular, and brittle branches, forming a dense canopy. In closely spaced plantations the branches are almost erect, so that little crown spread occurs. In plantations in humid climates a portion of the trees form aerial roots on the main trunk as far as 6 m above ground.

Distribution The species occurs naturally in a narrow coastal belt from Queensland to south of Bega, New South Wales, Australia. It grows from sea level to about 100 m. It has been planted in many other countries and adapts to varied conditions from equatorial regions to about latitude 35°, provided frosts are not severe. In Madagascar 150,000 ha have been planted.

Use as Firewood *Eucalyptus robusta* has been used as firewood in Madagascar, Uganda, Cameroon, Papua New Guinea, Sri Lanka, and Malaysia. It is also used for charcoal. The wood has a specific gravity of 0.70–0.80. The species coppices well up to age 25.

Yield Good yields have been reported from Brazil, Chile, Madagascar (10–35 m^3 per ha per year), Mauritius (10 m^3 per ha per year), Malawi, India, Malaysia, Papua New Guinea (21 m^3 per ha per year), Zaire (20-30 m^3 per ha per year), and Florida (16.8 m^3 per ha per year from 10-year-old trees). In Hawaii, 23- to 38-year-old plantations have a mean annual increment of 26 m^3 per ha.

Other Uses

- Wood. The timber is used in the round and sawn for lumber, but there are limited supplies. In Hawaii, three companies were reported in 1961 to be making veneer and plywood of the timber. The wood can also be used for poles, fencing, wharf and bridge work, and general construction.
- Shelterbelt. This species is suitable for planting in coastal areas as shelterbelts, and it makes a good roadside shade tree. It is intolerant of salt spray but is quite wind firm and is used as a windbreak.
- Pulp. The timber is used for pulpwood, but the pulp is dark reddish brown and is not as good for this purpose as some other species of eucalypts. The bark must be removed from the stem before pulping.
- Tannin. The gum contains about 30 percent tannin.

Environmental Requirements

- Temperature. Mean minimum temperature in the coldest winter month is 3°–5°C. Winter frosts are rare, but 5 or 10 frosts may occur as long as they are not severe. Summer temperatures are warm, with a mean maximum of 30°–32°C. Humidity is high in all seasons of the year.
- Altitude. The tree grows on coastal sites from sea level to 1,600 m, depending on latitude.
- Rainfall. *Eucalyptus robusta* does best in areas with evenly distributed rainfall of more than 2,000 mm a year. It will grow reasonably well when there is rainfall of 1,000–1,500 mm with a 4-month dry season.
- Soil. In its natural habitat in Australia the tree occurs mainly on bottom slopes, in swamps, and on the edges of saltwater estuaries and lagoons. When artificially assisted, it grows much more vigorously in better soils if there is no competition from other eucalypts. It does relatively well in stiff clays and leached sandy loams, but has done poorly on droughty sands in northern Brazil.

Establishment *Eucalyptus robusta* is most commonly grown from seeds in containers in nurseries. Vegetative propagation by grafting and rooting of stem cuttings has been done with young trees, but it is not a common method of reproduction. Seedlings are best planted early in the

Eucalyptus robusta near Coffs Harbour, New South Wales, Australia. (Division of Forest Research, CSIRO, Australia)

rainy season. Natural regeneration on bare ground adjacent to plantations is common in many countries where the tree has been introduced.

- Seed treatment. None required.
- Ability to compete with weeds. Removal of vegetative cover before planting is essential. Planted seedlings are susceptible to competition and shading and generally require two weedings in the first 6 months.

Pests and Diseases In São Paulo, Brazil, *Eucalyptus robusta* has been attacked by the bacterium *Phytomonas tumifaciens.* This same organism has been detected in plants originating in the United States and Chile. The most common injuries in swamp plantations in Uganda are windthrow and root rot. The tree is also susceptible to attack by the *Gonipterus* beetle and to termites when it is young.

Limitations *Eucalyptus robusta* hybridizes with many other eucalypt species, which can make the collection of seed of true origin difficult. In California, USA, this tree has been abandoned for street planting because the tops break readily in strong winds.

Eucalyptus tereticornis

Botanic Name *Eucalyptus tereticornis* Sm.

Synonym *Eucalyptus umbellata* (Gaertn.) Domin.

Common Names Forest red gum, blue gum, mountain gum, red iron gum, Queensland blue gum, red gum (Australia); Mysore gum (India)

Family Myrtaceae

Main Attributes *Eucalyptus tereticornis* grows rapidly and withstands periodic flooding for short periods. It coppices vigorously and is one of the principal eucalypts grown as fuelwood.

Description A moderately large tree, *Eucalyptus tereticornis* attains a height of 30–45 m and a diameter of 1–2 m. The trunk is usually straight and at least half the total height. The crown is large and somewhat open. Small clusters of white flowers appear every year, but heavy blooming occurs only every 3 or 4 years in spring and summer.

Distribution This species occurs naturally in the widest latitudinal range of any eucalypt (6°-38°S) along the eastern Australian coast from southern Victoria to Queensland. It grows also in the savanna woodlands of Papua New Guinea's south coast. It is found in open forest or as scattered trees on alluvial flats and along stream banks, including brackish water. It has been introduced to many tropical and subtropical countries in Africa, Asia, and South America. More than 400,000 ha have been planted in India.

Use as Firewood The wood is hard, heavy, and strong (specific gravity 0.75 or higher). Density is lower in plantations in the tropics. The tree produces first-class fuelwood, which also makes good charcoal. Coppice regeneration has been widely used and it can be done three to four times on a 10-year rotation.

Yield The yield is very dependent on moisture. Higher yields are reported along canal banks and under irrigated conditions. In unirrigated plantations under good conditions in Africa the tree will yield 20–25 m^3 per ha per year for the first 15 years; thereafter, the yield drops to 10–15 m^3 per ha, unless the trees are coppiced.

Other Uses

- Wood. The wood is immune to termites and dry rot and is therefore one of the most durable of timbers, valued for construction, especially underground. It has a variety of other uses including poles, posts, fiberboard, and particle board.
- Pulp. This is considered one of the best trees for fiber for paper pulp and rayon-grade pulp in India, Africa, the Pacific basin, and Latin America.
- Intercropping. *Eucalyptus tereticornis* has been used as an intercrop; for example, in Pakistan it is used with maize, especially during the first 6–12 months after planting, and in India with tapioca (cassava) during the first 2 years.
- Sand dune reclamation. In Uruguay and Costa Rica *Eucalyptus tereticornis* has been used for sand dune reclamation.
- Afforestation and reforestation. The species is extensively used in afforestation works in India, from the coastal plains to the mountains in the Himalayas, and in West Africa, notably Zaire.
- Oil. The leaves are among the commercial sources of eucalyptus oil.
- Honey and pollen. This species is an important provider of nectar and pollen for honeybee colonies.

Environmental Requirements

- Temperature. *Eucalyptus tereticornis* occurs within a wide climatic range with mean annual temperatures from 17°C to 38°C. This tree may withstand up to 15 frosts a year in the southern end of its natural range. In South China and Pakistan it is reported to survive temperatures down to −7°C.
- Altitude. In its natural habitat this species ranges from 600 m in Papua New Guinea to 1,000 m in Australia. As an exotic, it is planted from sea level (Zaire) to more than 1,000 m.
- Rainfall. The tree is widely planted in areas of summer rainfall with moderate to severe dry seasons. The optimum precipitation appears to be between 800 mm and 1,500 mm, but trees have been planted in areas with lower rainfall (400 mm in India, 550 mm in Israel, and 580 mm in Zimbabwe) and with considerably higher rainfall (2,180 mm in Colombia and 3,500 mm in Papua New Guinea).
- Soil. This species does best on deep, well-drained, light-textured soils that are neutral or slightly acid. It will tolerate some flooding, but not seasonal inundation. Outside its range it has

Eucalyptus tereticornis, northern Queensland, Australia. (Division of Forest Research, CSIRO, Australia)

Afforestation with *Eucalyptus tereticornis* in Karnataka, India. (K. A. Kushalappa)

been planted on a wide variety of sites that include alluvial soils, silts, and sandy clays.

Establishment For *Eucalyptus tereticornis* the site of the seed collection is important, and advice should be sought in selecting the most appropriate provenance in a new locality.

In most countries, seedlings 15–25 cm high are used as planting stock, and they can be produced in the nursery in 3–4 months.

- Seed treatment. None required; however, seeds from native stands in both Australia and Papua New Guinea may need cold, moist stratification.
- Ability to compete with weeds. Eucalypts are very sensitive to grass competition, and weeds should be removed until the canopy closes. Periodic soil working between rows by tractors has given good yield.

Pests and Diseases Most plantations have been free of pests and diseases. Termites will attack the young plants unless they are protected, and some countries have reported problems with varied fungi like *Cylindrocladium*, the climber *Merremia*, the snout beetle *Gonipterus*, and the mole cricket.

Limitations Because it is closely related to river red gum (*Eucalyptus camaldulensis*) and intermediate forms occur, care needs to be taken in assessing reports of growth.

Related Species

- *Eucalyptus rostricta* Schlecht. has similar wood that is used for the same purpose.

A 12-year-old seed stand of *Eucalyptus terreticornis* in Bangalore, India (K. A. Kushalappa)

• *Eucalyptus camaldulensis* Dehnh. is closely related, and intermediate forms occur commonly in Queensland.

• *Eucalyptus glaucina* (Blakely) L. Johnson, now regarded as a species, was formerly *Eucalyptus tereticornis* Sm. var *glaucina* (Blakely) Cameron. It occurs in northern coastal New South Wales.

• *Eucalyptus amplifolia* Naudin is a somewhat smaller tree that grows on clay soils (sometimes waterlogged) often at slightly higher altitudes than *Eucalyptus tereticornis* in eastern New South Wales.

• A number of land races, some of which were thought to be hybrids, occur. Some of these are:

– "Mysore hybrid," Mysore gum, India
– Eucalyptus "C," Zanzibar
– Eucalyptus "12ABL," Madagascar, Zaire.

Gleditsia triacanthos

Botanic Name *Gleditsia triacanthos* L.

Synonym The genus name has also been spelled *Gleditschia*

Common Names Honeylocust, soetpeul, sweet locust, thorntree, three-thorned acacia

Family Leguminosae (Caesalpinioideae)

Main Attributes Honeylocust is widely planted as an ornamental and for windbreaks, fodder, and hedges. It is a moderately fast-growing tree with hard, strong wood that is quite resistant to decay. It is tolerant of low temperature, drought, and salt. The tree's pods are rich in sugar and are enjoyed by people, livestock, and wildlife.

Description A broad-crowned, flat-topped tree, the honeylocust typically attains a height of 24 m, with trunk diameters of 60-90 cm. Although some bisexual types are known, most specimens are either male (pollen producing) or female (fruit producing). Trunks and limbs of wild trees usually bear branching thorns, but a thornless form (f. *inermis* Schneid.) has been developed for cultivation; it has small, greenish, fragrant flowers in spring and flat, crooked pods in fall.

Distribution This valuable tree legume, well suited to cooler zones, can be found in North America from Ontario to Texas. It has also been naturalized in Europe and introduced into parts of Africa, Australia, New Zealand, and South America.

Use as Firewood Honeylocust has strong and durable wood of specific gravity 0.70-0.80. The trees coppice freely.

Yield Growth is fairly rapid on good sites under average conditions. Trees 18-35 years old in plantations in the Central Plains of the United States had an average increase in diameter of 4.6 cm each 10 years. Height growth in shelterbelts throughout the Central States has averaged 0.5 m per year for 7 years.

Other Uses

- Timber. The timber is hard, heavy, and strong, and the heartwood is durable in contact with the ground. The wood is used principally for posts, but on a smaller scale for furniture, structural work, and general utility purposes.
- Pods. The long pods, borne in profusion, have sweet-tasting pulp that is relished by people and animals alike, and so the tree is often planted for pasture shade. The pods make high-quality livestock feed, especially when crushed to make the high-protein seeds more digestible.
- Erosion control. A strong taproot, many-branched lateral roots, tolerance of alkaline and saline sites, and ease of reproduction make honeylocust a valuable erosion control species for temperate and subtropical areas. It provides light shade that encourages a grass cover.
- Ornamental and shade tree. In temperate areas of many countries such as South Africa, the United States, and Australia, the species is often planted for shade and beautification along roads and in towns. It is also used for hedges (the thorny kind makes impenetrable hedges) and windbreaks.

Environmental Requirements

- Temperature. Honeylocust is tolerant of low temperatures, and northern races are hardy to −34°C. Southern races are subject to frost damage when planted in cooler areas. The tree needs light.
- Altitude. This species can be grown up to altitudes of 1,500 m or more. A plantation in Colorado had good survival at 2,100 m, although the trees averaged only 2.4 m tall. The tree seems particularly worth testing as a new crop for tropical highlands such as those found in Nepal, northern Thailand, Central Africa, and Latin America.
- Rainfall. Honeylocust has deep roots and can make use of the moisture reserves of the subsoil, even if the surface is dry. In this way it is able to survive on all but the driest sites. Normal annual precipitation for good growth varies from 500 mm to more than 1,500 mm.
- Soil. The tree grows in most soil types, but occurs most commonly on rich alluvial floodplains of major rivers and on soils of limestone origin. The species tolerates both alkaline and acid soils, but develops best on those with a pH between 6.0 and 8.0.

Establishment Propagation by seedlings is easy and can be achieved by grafting, budding, or cuttings taken from hardwood, softwood, or the roots. Root cuttings and budding appear to be the best methods of reproducing desirable strains as fruit trees. Large seed crops are produced every 1-2 years, and the seeds will remain viable for up to 2 years when stored at room temperature and

Gleditsia triacanthos L. Honeylocust on Ohio State University campus, Columbus, Ohio, USA. (U.S. Forest Service)

for several years if stored in sealed containers at 0°–7°C.

• Seed treatment. Before germination can occur, the hard seedcoat must be made permeable, either by covering with hot water (88°C) until the seeds swell or by soaking in concentrated sulfuric acid for 1–2 hours and then washing in water. Treated seeds cannot be stored and must be sown promptly.

• Ability to compete with weeds. The tree needs weeding until it is well established.

Pests and Diseases Honeylocust is relatively disease free, but it is subject to a canker that can be fatal, and also to several wood rots. The leaves are eaten by mimosa webworm in some areas of North America.

Limitations Honeylocust may suffer some crown damage during high winds and is easily damaged by fire. Its extensive root system can block sewers and drains.

Melaleuca quinquenervia

Botanic Name *Melaleuca quinquenervia* (Cav.) S. T. Blake

Synonyms *Melaleuca quinquenervia* is one of a group of six closely related species (notably *Melaleuca leucadendron, Melaleuca cajuputi,* and *Melaleuca viridiflora*) that occur in Oceania and have been cultivated and naturalized in other areas. Their identities have been confused, and many references (both past and recent) must be treated with caution.

Common Names Broad-leaved tea tree, tea tree, paperbark tea tree, belbowrie (Australia); melaleuca, cajeput, paperbark, punktree (USA); niaouli (New Caledonia)

Family Myrtaceae

Main Attributes *Melaleuca quinquenervia* is a moderately fast-growing tree suitable for plantation cultivation on wet soils, including those subject to high rainfall and even inundation. It will grow at high planting densities, which quickly shade out competing vegetation. It survives fires, but has only slight resistance to freezes (although rootstocks survive and sprout). The species has shown promise in plantations in northern Nigeria.

Description The tree is usually medium sized, growing in Australia up to 25 m tall; the bole is short, 4-5 m high (and to 15 m in rare cases), and often crooked or twisted. However, if grown in dense stands it develops relatively straight, clear stems. It is easily recognized by its odd, whitish, thick, and spongy bark, which splits and peels in many cork-like layers and becomes rough and shaggy. The tree has narrow, stiff, aromatic leaves, 5-9 cm long, and "bottlebrush-shaped" clusters of fuzzy, white, malodorous flowers.

Distribution This species is native to Australia from Sydney north along the coast to the Cape York Peninsula in Papua New Guinea and to New Caledonia. It typically grows in almost pure stands or with only a few associates such as *Casuarina glauca, Eucalyptus robusta,* or *Eucalyptus tereticornis.* It has been planted and naturalized in many tropical regions. In southern Florida it escaped cultivation on seasonally wet sites and is constantly multiplying as a weed tree.

Use as Firewood The wood is an excellent fuel, but the thick bark of mature trees must be peeled off because only the outer layers will burn. The tree could be used as a woody biomass fuel; the problems of dust from the bark of mature trees and the low density of the bark can be overcome by properly engineered systems. Bark of young seedlings has a greater heating value per unit of oven-dry weight than the wood itself. The wood has a specific gravity of 0.60–0.74.

Yield In the Florida Everglades, unmanaged melaleuca stands that appear mature may have 7,000–20,000 stems per ha with an outside bark basal area up to 133 m^2 per ha and an outside bark volume of 792 m^3 per ha. Forty-year-old trees on good sites in Hawaiian plantations average 50 cm in diameter and 18 m in height at 6 m × 6 m spacing. The largest trees there reach a diameter of 90 cm and grow 24 m tall. Stumps sprout readily.

Other Uses

- Wood. The wood of paperbark is valued for wharf piling, boat knees, railway ties, mine braces, posts, fence rails, flooring, and rafters, and, if carefully seasoned, for gunstocks, cabinetwork, and carving. When the wood is well finished, some people consider it more beautiful than mahogany.
- Bark. The bark of paperbark is an effective insulator and was first used for packing the walls of cold storage rooms in Australia in 1861. It is mildew resistant and is long lasting under water, so that sheets of bark have been used for caulking boats. When chopped, it makes an excellent filler in nursery potting mixes.
- Honey. The tree blooms much of the year, providing abundant pollen and nectar. The honey is strongly flavored and dark and is not used as table honey. Because of its low cost, there is a strong demand for it by the baking industry and by natural health food dealers in Florida.
- Ornamental. Seedlings are transplanted from the wild (especially mature trees) for landscaping new developments in Florida because of their availability and low cost.
- Oil. Leaves, twigs, and seed capsules are crushed and distilled to produce niaouli oil, which together with cajeput oil from *Melaleuca cajeputi* has pharmaceutical and other specialty uses.

Environmental Requirements

- Temperature. Paperbark grows in areas with mean annual temepratures ranging from 18°C to 34°C.

A mature swamp stand of *Melaleuca quinquenervia* in southern Florida during the dry season. This was a *Taxodium distichum* forest that was killed by fire and invaded by paperbark. (T. F. Geary)

• Altitude. In Australia the tree occurs in low altitudes up to 500 m in the coastal belt, most commonly in seasonal swamps and the edges of tidal waters. In Hawaii it grows well in wet conditions up to 1,400 m altitude.

• Rainfall. The trees grow well in rainfall of as little as 1,000 mm at lower elevations and as much as 5,000 mm at higher elevations.

• Soil. In Australia paperbark can grow down to the sea's edge and is also found 40 km inland. It occurs on old and new alluvial soils, on shallow soils, and on degraded soils left after shifting cultivation. Trees grow to great size in oolitic limestone in Florida.

Establishment The seeds need wet soil for germination, and once established, seedlings can survive complete immersion for several weeks.

• Seed treatment. No pretreatment is necessary. The seeds are small (30,000 seeds per g).

• Ability to compete with weeds. The tree successfully competes with and outgrows other vegetation. The deep shade beneath a dense melaleuca canopy, combined with intense root competition and possible allelopathic influences, restricts the need for weed control to about the first 6 months.

Pests and Diseases None of importance has been reported.

Limitations Because of its high incidence of seeding, adaptability to a variety of growth conditions, rapid growth, and resistance to damage by disease, insects, flooding, and fire, the paperbark tree has spread rapidly throughout southern Florida. In wetlands it has crowded out native vegetation and destroyed wildlife habitats. Control by conventional means is difficult.

Volatile substances excreted by the flowers have been implicated as the cause of acute respiratory problems in Florida.

Related Species

In Thailand, *Melaleuca cajuputi* is the species cultivated; it occurs naturally from tropical Australia to Burma and Vietnam. *Melaleuca leucadendon* occurs in northern Australia, New Guinea, and northwestward to Amboina; *Melaleuca viridiflora* occurs in western and northern Australia and New Guinea.

Melia azedarach

Botanic Name *Melia azedarach* L.

Synonym *Melia orientalis* M. Roem.

Common Names Chinaberry, Persian lilac, alelaila, paraiso, pride of India, pride of China, bead tree, umbrella tree, bastard cedar, cape lilac, white cedar, drek

Family Meliaceae

Main Attributes The chinaberry is well known and has already been planted in many countries as an ornamental. Its quick growth and small dimensions make it a good choice for fast fuelwood production for household needs. In the Middle East the species has already been used for this purpose.

Description Chinaberry is a medium-sized deciduous tree, 6-30 m tall and 50-80 cm in diameter. It is popular for its showy clusters of pale purplish spreading flowers and for the shade of its dense, dark green foliage in summer. Its clusters of small, fragant flowers are succeeded by glossy, golden, berrylike fruits that remain long after the leaves fall. The tree is closely related to the neem tree, *Azadirachta indica* (see *Firewood Crops* Vol. I), which differs in having longer leaflets, white flowers, and a more spreading and open crown.

Distribution This native of Asia is probably from Baluchistan and Kashmir, but has long been cultivated throughout the Middle East and India and is now cultivated and naturalized in most tropical and subtropical countries. It is grown throughout the West Indies, southern United States and Mexico, Argentina and Brazil, West and East Africa, Southeast Asia, and Australia.

Use as Firewood The species is often planted for fuel supply in the Middle East as well as in Puerto Rico and Nigeria. It is planted in reforestation projects in Thailand for veneer and fuel, and in Assam (India) it is grown on tea estates for fuel. Its specific gravity is about 0.66 (calorific value 5,043-5,176 kcal per kg).

Yield Under good conditions the chinaberry grows fast. In Uganda it may grow 1.70 m in height each year. But growth slows down before large dimensions are attained, and trees of large girth are often hollow. Thus the trees are grown on short rotations. They regenerate readily from stump sprouts or root suckers.

Other Uses

- Wood. The wood is moderately soft and is weak, brittle, and susceptible to attack by drywood termites. Uses of wood include tool handles, cabinets, furniture, face veneer for plywood, cigar boxes, and the manufacture of writing and printing paper.
- Insecticide. Chinaberry, like neem, has insecticidal properties and its leaves and fruits are used to protect stored clothing and other articles against insects. Leaves, seeds, and fresh fruit contain substances that inhibit the feeding of the desert locust.
- Fodder. The leaves may be used for goat fodder.
- Seeds. Chinaberry seeds are often used for beads and rosaries.
- Ornamental. The tree is widely grown in warmer parts of the world for its scent and for shade. It is also used to shade coffee trees and to shelter cattle.

Environmental Requirements

- Temperature. The species grows in tropical, subtropical, and warm-temperature climates where mean annual temperatures are at least 18°C. Young trees are frost tender, but older trees resist frost (to a minimum of – 15°C).
- Altitude. The tree grows up to 2,000 m in the Himalayas. In Africa the species has been recommended for lowlands and medium elevations.
- Rainfall. The species is drought hardy and can grow in areas with 600-1,000 mm annual precipitation. In drier climates it will perform well on wet soils along rivers or when irrigated, as is done in the Middle East for fuelwood production.
- Soil. The tree grows on a wide range of soils, but best growth is obtained on well-drained, deep, sandy loams.

Establishment Chinaberry is easily propagated from seed or cuttings. Fruits number 1,400-2,500 per kg and may be used directly or macerated to remove the pulp. Each fruit contains 1-5 seeds, which if not extracted may produce several seedlings. Seeds number 4,000-13,000 per kg. Average germination is about 65 percent. As planting stock, 1-year-old seedlings are preferred in temperate climates, while 6-month-old seedlings are used in the tropics.

- Seed treatment. To hasten germination, seeds should be soaked in water for a few days.

Melia azedarach. Fifteen-year-old tree in Pakistan. (M. I. Sheikh)

Seeds retain viability for a year, or for several years if kept in sealed cold storage.

Pests and Diseases The tree—especially if forced into fast growth—is susceptible to wind damage. From Jamaica, attacks of shootborers are reported. Several other pests and diseases have only been observed incidentally and are not particularly significant.

Limitations The tree is short-lived and its brittle limbs are easily broken by the wind. The fruits are bitter and have poisonous or narcotic properties.

Related species

- *Melia azedarach* L. var *australasica* (A. Juss.) C.DC. (white cedar, tulip cedar) from Australia (synonym *Melia dubia*). This species grows to a much larger size than *Melia azedarach* and is widely planted in Australia as a shade tree.
- *Melia a.* forma *umbraculiformis* Berckm. (Texas umbrella, chinaberry).

Robinia pseudoacacia

Botanic Name *Robinia pseudoacacia* L.

Common Names Black locust, false acacia, yellow locust

Family Leguminosae (Papillionaceae)

Main Attributes Black locust is fast growing. It can be planted and regenerated easily, has no serious diseases, is an important nitrogen-fixing species, and tolerates a wide variety of soils. Because of this adaptability, the species is often used for afforestation of gullies and surface-mined areas in the United States. The wood is perhaps the best firewood in North America.

Description A medium-sized, deciduous tree reaching 18–25 m tall, black locust has an open, irregular crown and a straight bole if forest grown. It has short, upright, brittle branches and a wide-spreading, shallow root system with no taproot. There are many varities of the tree, with variable form and growth. Most are spiny, but spineless individuals are known.

Distribution Native to the Appalachian and Ozark mountains of the southeast and central parts of the United States, this species is now grown in most temperate and Mediterranean zones of the world. Hungary, for example, has almost 300,000 ha of black locust plantations. In central and eastern Europe considerable work on selection and improvement of the species has been carried out for more than a century, using a series of cultivars in pure and mixed-wood plantations.

Use as Firewood As firewood, black locust is popular for its high caloric value and its good combustibility, even when wet. It is slow to ignite, but burns like coal, with a bright blue concentrated flame. The specific gravity of the wood is 0.70–0.80, and it is well suited for charcoal production. In India and South Korea the species has been used to establish fuelwood plantations.

Yield Young black locust trees grow fast on good sites, but the species matures early and growth rate decreases rapidly after 30 years. In Korea, on clay loam, the yield of fuel material per year per hectare from the coppice of black locust fuelwood plantations ranges from 10 to 17 tons 3 years after establishment. In Hungary, where the rotation age is 25–35 years, the average standing volume at age 1–10 is 23 m^3 per ha; at age 11–20, 100 m^3 per ha; at age 21–30, 149 m^3 per ha; and, finally, 184 m^3 per ha at age 31–40. Coppicing brings excellent results and may be done twice without reduction of volume increment.

Other Uses

- Wood. *Robinia pseudoacacia* wood is heavy, hard, and strong. It is highly resistant to shock and is extremely durable. It is used extensively for round, hewn, or split mine timbers and for fence posts, poles, railroad ties, stakes, and electric insulator pegs.
- Erosion control. Black locust is commonly planted for erosion control. Because of the nitrifying bacteria in nodules on its roots, it is especially effective in improving the fertility of eroded soils. It has also grown quite well when planted on spoil banks created by mining.
- Shelterbelts and sand dune stabilization. Because of its wide-spreading root system, which suckers vigorously, it is used all over the temperate regions of the world for fixing coastal and continental sand dunes. It has also often been planted in shelterbelts.
- Fodder and wildlife use. In Hungary the leaves are readily eaten by game and grazing animals, with no harmful effects observed. The seed is eaten by quail.
- Ornamental. The tree has attractive flowers and is widely planted as an ornamental on farmsteads and roadsides. The flowers are used intensively by bees for producing honey, which is regarded as one of the finest in the world.

Environmental Requirements

- Temperature. The tree grows in areas with a mean annual temperature of 8°–18°C and a mean minimum temperature in the coldest month of −8.1°C. It is tolerant of temperatures as low as −18°C to −20°C. It is a light-demanding species.
- Altitude. Black locust is found from sea level up to 2,500 m.
- Rainfall. Precipitation in its native habitat is more than 1,000 mm per year, mainly in the summer. The tree has been successfully planted where the rainfall is lower, down to 300–400 mm, and in areas with a winter maximum of rainfall. The species can withstand dry periods of 2–6 months.
- Soil. The species will grow on a variety of soils, including light sand and strongly acid soils and mine spoils, with limestone soils being most

Black locust, Kentucky, USA. (U.S. Forest Service)

Left: Young black locust stumps sprout prolifically after cutting, thus shortening subsequent rotations and making regeneration easy. (S. B. Carpenter)

Above: *Robinia pseudoacacia* stocks, 3 years old, lopped once a year for fuel supply. (Institute of Forest Genetics, Suwon, Korea)

favorable. Compact, plastic soils are unfavorable, since black locust does not grow well in waterlogged areas. It is able to withstand air pollution and is sometimes planted in industrial areas for this reason.

Establishment Good seed crops are produced nearly every year. Seeds number 35,000-70,000 per kg and can be stored for several years under dry, cold conditions. In the past, black locust was established mainly from planted seedlings, but today direct seeding is a common practice, especially on steep slopes on disturbed sites where seeding is easier, safer, and less costly than hand planting.

- Seed treatment. Pretreatment is needed to break seed coat dormancy, either by mechanical scarification, by immersion in concentrated sulphuric acid for 20-60 minutes, or by pouring boiling water (five times as much water as seeds) over seeds. Germination of pretreated seed averages about 70 percent.
- Ability to compete with weeds. Black locust is intolerant of shade but is fairly tolerant of herbaceous competition in humid environments. In the Appalachian area of the United States the trees often coexist with a luxuriant understory of cool-season grasses and forbs.

Pests and Diseases In the United States the saplings are often heavily damaged by the locust borer *Megacyllene robiniae.* Trees in the open and in pure plantations are especially subject to the attack of a woodboring insect (*Cyllene robiniae* Forst.), which frequently renders them unsightly for shade and unfit for timber, but not for fuelwood. However, few pests of any importance have been reported.

Limitations A vigorous growth of suckers with stout spines is produced from the roots, and this can be particularly objectionable in the garden, on the farm, or in public areas.

Although the foliage is used for livestock in some countries, it has been reported that humans, horses, cattle, sheep, and poultry may be poisoned by eating roots, bark, sprouts, seed pods, and trimmings.

Frost damage can decrease growth, and because black locust branches are fragile and the wood splits easily, the tree suffers from wind damage and ice and snow break.

Sapium sebiferum

Botanic Name *Sapium sebiferum* (L.) Roxb.

Synonyms *Stillingia sebifera* Mich., *Croton sebiferums* L., *Excoecaria sebifera* Muell. Arg., *Stillingia sebifera* Boj.

Common Names Chinese tallow tree, Chinese vegetable tallow tree, soap tree, popcorn tree, pau de sebo (Brazil), tarcharbi or pahari shishum (India), arbol de la cera (Cuba)

Family Euphorbiaceae

Main Attributes The Chinese tallow tree holds promise as a highly useful species for short-rotation, intensive silviculture. It is a fast-growing, insect- and disease-resistant tree, and tolerates a wide range of soil conditions, including poorly drained and saline soils. Prolific sprouting occurs from stumps and injured roots.

Description A small, deciduous tree, it seldom grows taller than 12 m. Superficially, it resembles an aspen with glossy, heart-shaped leaves and a rough, gray bark. Form ranges from low, spreading, and multiforked strains to slender and columnar strains with small pendant branches. Seeds are borne in clusters of green capsules, which dry and split to expose white, pea-sized seeds. Their color results from a covering of hard tallow. The trees are notable because they turn brilliant autumn colors even in warm regions where other trees remain green.

Distribution The Chinese tallow tree is native to semitropical areas of central South China. Usually occurring below about 32°N, it may be found at higher latitudes in coastal areas. It has been widely introduced and has been reported in Japan, Taiwan, Vietnam, Burma, northern India, Java, Pakistan, Hawaii, Brazil, Cuba, St. Vincent, Martinique, southern Europe, and the Sudan.* It is very likely present, but simply not recognized, throughout the semitropical and milder temperate zones of the world. In the United States it has become naturalized along the Gulf and southern Atlantic Coasts from North Carolina to Texas, as well as in California.

Use as Firewood Moisture content of tallow tree wood cut in the Texas Gulf Coast ranged from 41 to 45 percent, wet basis. Specific gravity ranged from 0.37–0.48. Energy values ranged from 4,134 to 4,277 kcal per kg. The wood burns well in wood stoves or fireplaces if it has been allowed to dry adequately. It produces a pleasant odor while burning and has been used for Texas barbecues.

Yield In Texas, Chinese tallow plantations with 0.6 m × 0.6 m spacing produced 38.1 tons per ha of oven-dry wood at the end of four seasons of growth (22 m^3 per ha per year). Well-stocked 15-year-old natural stands placed under coppice management produced in excess of 45 tons per ha during 4 years of regrowth (26 m^3 per ha per year). Roots and stumps cut at any time of the year coppice prolifically.

Other Uses

- Wood. In China, the wood reportedly has been used for making implement handles as well as carved-wood products. Its durability and resistance to decay in exterior use are very low. The wood may be suitable for pulp. Fiber lengths in trees from the vicinity of Houston, Texas, averaged 0.8 mm, while similar measurements of trees from Taiwan averaged 1.3 mm.
- Seed. The white, aril-like outer covering of the seed contains a hard, edible tallow (commercially known as Chinese vegetable tallow), which has found use in soap and candle making and other traditional uses of tallow products. The kernel oil (stillingia oil) has been used as an illuminant and is a powerful drying oil that could substitute for similar oils such as tung and linseed. The clusters of white seeds are attractive; in Texas they are used in decorative plant arrangements.
- Flowers. The showy yellow catkins of the male flower first appear during the third growth season and yield a major honey and pollen crop. The honey is moderately dark and exceptionally flavorful.
- Ornamental. The Chinese tallow has been used extensively as an ornamental tree. Because of its rapid growth, it produces quick shade, and some varieties are very colorful in the fall, even in warm climates.

Environmental Requirements

- Temperature. The tree will tolerate frosts and, in the dormant state, will tolerate brief lows of −10°C. There is considerable variability in cold tolerance among strains. Some strains may be sensitive to high temperatures and some may

*D. Hooper. 1904. Chinese or vegetable tallow. Its preparation, uses, and composition. *Agricultural Ledger* 11(2):11–18.

Sapium sebiferum. A single season of growth on the severed stem of a large tree. The cut forks on this tree were 6-8 inches in diameter at the time of severance. (W. Scheld)

require a period of cool weather and dormancy for significant flowering to occur.

- Altitude. In the United States the tree is naturalized mostly in the coastal lowlands but has been observed as a planted ornamental at elevations of several hundred meters. In China it is reported at elevations of around 1,000 m and in India and Pakistan at about 2,000 m, which is probably its upper limit.
- Rainfall. Although the tree tolerates a wide variation in quantities of annual rain, it is generally considered to be a high-moisture plant; trees have survived for more than 2 years in areas flooded by dam construction. But in one instance, the trees grew well in a year when total annual rainfall was a mere 710 mm, and the species' lower limit is probably somewhat below 500 mm. There is no apparent reason why it cannot be grown in areas under irrigation. Preliminary observations suggest, however, that it is not an efficient user of water.
- Soil. The Chinese tallow tree possesses a remarkable capacity for thriving in widely varying soil types. In the United States it has been found growing on barrier islands, competing effectively in pure sand with oak, pine, and palm; in salt marshes and along salt creek borders; in rocky upland soils; and in dense clay soils subject to extended periods of flooding. Seeds have been observed germinating and the seedlings produced have become established in beds of oyster shell. The tree tolerates and competes well with other vegetation on poorly drained, nitrogen- and phosphorus-poor sites, but its response to phosphorus fertilization is dramatic.

Establishment Plantations may be established from seed, seedlings, or cuttings. The most convenient and economical method for establishing stands is direct planting. Seedlings grown in a nursery are ready for planting 6–8 weeks after germination. Special care in handling the seedling is not necessary. Dense plantings are practical.

- Seed treatment. Storage of seeds in the cold in sealed glass or metal containers preserves viability for at least 2 years for most varieties. Pretreatment of the seed to remove the tallow coat is not necessary. Planting in autumn appears to promote more complete germination and allows the earliest possible start for growing seedlings. In some races, a significant percentage of seeds do not germinate until the second season after planting, and seeds planted in the late spring and summer often do not germinate until the following spring.
- Ability to compete with weeds. The large seedling is capable of growing ahead of most annual weeds, and in closely spaced plantings crown closure occurs early in the second season of growth, eliminating significant competition from grass and annual weeds. Grass is the major competitor.

Pests and Diseases In the tree's native range, a number of insect pests appear to have evolved with it. Chinese literature refers to the ailanthus silkworm, poisonous moth, water green moth, and aphid. In India a number of insects have been reported to defoliate the tree. It is not known how serious these might be in firewood plantations. In the United States and probably elsewhere outside the tallow tree's native range, there are no significant diseases or insect pests.

Limitations The Chinese tallow tree can be an aggressive weed. The tallow-covered seeds are spread widely by birds, and once established in agricultural lands, lawns, or gardens, seedlings and sprouts are difficult to eliminate because of their capacity for sprouting from stumps and roots. In general, there is a high resistance to herbicides. The reproductive strategy of the Chinese tallow tree is such that breeding and production of seed of known characteristics will be complex. The sap (latex) is a powerful irritant that blisters the skin.

Sapium sebiferum. Growth form ranges from very low to quite tall even in open growing conditions. Above: This tree possesses a strongly dominant single stem with small, usually pendant, branches. Right: An open-grown form with multiple stems may be the result of stump sprouting. (W. Scheld)

III Fuelwood Species for Arid and Semiarid Regions

The dry regions of the earth—those with less than 500 mm of annual precipitation or that have 6 or more totally rainless months—suffer more serious fuelwood problems than either the humid or highland regions of the tropics.

Four hundred and fifty million people live in these areas. Over the centuries, clearing the forests for fuel, farming, and grazing has destroyed most of the vegetation in these fragile dry habitats. Excessive exploitation of the land has often led to desertification, a process that is being accelerated by the demand for scarce wood fuel.

Most nations with dry zones are seriously in need of reforestation to prevent further loss of land and for wood fuel. The first volume of this report described the following species as worth testing for fuelwood crops in these dry regions.

Acacia brachystachya
Acacia cambagei
Acacia cyclops
Acacia nilotica
Acacia saligna
Acacia senegal
Acacia seyal
Acacia tortilis
Adhatoda vasica
Albizia lebbek
Anogeissus latifolia
Azadirachta indica
Cajanus cajan
Cassia siamea
Colophospermum mopane
Emblica officinalis
Eucalyptus camaldulensis
Eucalyptus citriodora
Eucalyptus gomphocephala
Eucalyptus microtheca
Eucalyptus occidentalis
Haloxylon aphyllum
Haloxylon persicum
Parkinsonia aculeata
Pinus halepensis
Pithecellobium dulce
Prosopis alba
Prosopis chilensis
Prosopis cineraria
Prosopis juliflora
Prosopis pallida
Prosopis tamarugo
Tamarix aphylla
Zizyphus mauritiana
Zizyphus spina-christi

These species have shown the capacity to survive where annual rainfall is 500 mm or less or where rainfall is extremely variable. Their adaptive mechanisms include

Planting seedlings of *Casuarina equisetifolia* (see page 38, Volume I) to prevent sand-dune formation south of Da Nang, Vietnam. (WFP photo by F. Mattioli)

deep root systems that penetrate to subsoil moisture or wide-spreading root systems to gather spare moisture, with some species having both root types; adaptation to the high salinity often found in arid areas; small leaf blades or needle-like leaves to reduce transpiration during drought or other physiological mechanisms to conserve moisture by slowing evaporation through the leaves; and unpalatability or thorniness that discourages grazing animals.

This section outlines further examples of successful adaptations by trees and shrubs to arid and semiarid conditions. Other species described elsewhere in this report that should be included are: *Eucalyptus tereticornis, Gleditsia triacanthos, Melia azedarach*, and *Robinia pseudoacacia*.

Ailanthus excelsa

Botanic Name *Ailanthus excelsa* Roxb.

Common Names Ganges, ardu, aura, maharuka, mothoaraduso, peru, mahanium, coromandel ailanto

Family Simaroubaceae

Main Attributes *Ailanthus excelsa* is a fast-growing tree that regenerates well by coppicing. It provides valuable fodder and shade and makes good shelterbelts. Although it is generally unrecognized as a fuel tree, its pruned branches and cut stems are much used as fuelwood in India, its native region.

Description Ardu is a large tree (18–24 m) with rough, grayish brown bark. The deciduous leaves are variable in shape with coarsely and irregularly toothed leaflets.

Distribution This species is native to central, southern, and western India and is often cultivated in various parts of that country.

Use as Firewood The wood is fairly light (specific gravity 0.45) and therefore is not an ideal fuel; however, as already noted, it is used as firewood.

Yield In Assam, India, the species was planted at a spacing of 3 m × 3 m. The growth rate after 9 years averaged 5–6 m in height.

Other Uses

- Fodder. Twice a year this tree produces highly palatable and nutritive fodder. The leaves are commonly sold in the vegetable markets of Rajasthan, India, especially for nourishing stall-fed goats. The leaves are also dried, ground, and added to concentrate mixtures for livestock feed, notably as a protein supplement to poor-quality roughages.
- Wood. The wood of *Ailanthus excelsa* is yellowish white, very light, soft, and perishable. It is used for making boats, fishing floats, knife and tool handles, toys, spear sheaths, drums, cigar boxes, and packing cases.
- Shelterbelt. The tree is planted in rows along farm boundries and irrigation channels.

Environmental Requirements

- Temperature. Ardu grows in an area where temperatures range from 20°–40°C. It can withstand considerable frost.
- Altitude. It is grown at low altitudes.
- Rainfall. It grows best with annual rainfall of more than 600 mm, although it survives on 400 mm a year.
- Soil. The tree grows in a variety of soils, but sandy loams seem most suitable. Soils should be well drained and not clayish or waterlogged.

Establishment The plant is usually propagated by seed; however, it can be raised by shoot as well as by root cuttings. A sapling 30 cm high and 45–60 days old may be transplanted in pits previously prepared and manured. Monthly watering may be needed during the first year or until the trees are established.

- Seed treatment. The papery fruits (actually winged seeds or samaras) contain one to two seeds each. The seeds, however, cannot be separated from the wings undamaged, and it is necessary to sow the entire fruit. Soaking the pods for 3 days before sowing improves the results.
- Ability to compete with weeds. Good, except under drought conditions when ardu's competitiveness is poor.

Pests and Diseases In the rainy season, leaf growth is attacked by a caterpillar that eats the leaves and discharges a sticky material, making a netting and totally spoiling the leaves. Regular pruning provides some control.

Limitations Survivability is poor if the trees are not sufficiently watered during the first and second years of establishment. The leaves have a disagreeable odor when crushed.

Related Species *Ailanthus altissima* Swingle (see *Firewood Crops* Vol. I, p. 74, *Ailanthus glandulosa* Desf.). A native of northern China, this species has been introduced to Japan, North and Central America, and Mediterranean countries. In Mediterranean regions it is common as an ornamental. It is particularly useful on stony sites. Initially of fast growth up to 25 m high and to a diameter of 60–89 cm, it reproduces by coppice and suckers, and under favorable conditions it may spread and become a weed that invades natural formations (for example, maquis in Corsica and in other parts of southern Italy).

Ailanthus excelsa, India. (N. D. Vietmeyer)

Balanites aegyptiaca

Botanic Name *Balanites aegyptiaca* (L.) Del.

Synonym *Ximenia aegyptiaca* L.

Common Names Desert date, lalob, soapberry tree, thorn tree, Jericho balsam, heglig (Arabic); corona di Jesus, lamunch, shimaron (Curaçao); najlij

Family Balantiaceae (formerly placed in Zygophyllaceae)

Main Attributes This multipurpose, drought-resistant species of the arid zone produces wood that is highly valued for fuel because it produces almost no smoke, making it ideal for use inside dwellings. The trees are slow growing, but they are indifferent to soil conditions and withstand fires.

Description The desert date is a shrub or, more usually, a small tree up to 10 m high with scaly, fissured, gray or brown bark, intricately branched stems, and slender drooping branchlets that bear long green spines. The gray-green leaves with two ovate leaflets are either evergreen or are wholly or partially deciduous during the dry season. The tree has shallow, wide-spreading lateral roots and a narrowly branched taproot that may penetrate several meters to the water table.

Distribution The species is indigenous to African woodlands along the Sahara's southern border from the Atlantic to the Red Sea (Sahel-Sudan region). It is also found on the Arabian peninsula and in Israel and Jordan. Small plantations have been established in Niger and Chad. Individual trees have been planted extensively in African villages far south of its natural range, as well as in India and Puerto Rico. Introduced to the Caribbean island of Curaçao in 1885, it has overrun a large part of the dry east end of the island.

Use as Firewood The pale yellow or yellowish brown wood is hard, heavy, and tough (specific gravity 0.65). It makes excellent firewood and good-quality charcoal. The trees regenerate vigorously after they are cut or lopped. Calorific value is 4,600 kcal per kg.

Yield The tree grows slowly. In Puerto Rico, one trial sowing of 43 trees produced specimens only 2–5 m high after 8 years' growth. In Israel 2- to 3-year-old coppice shoots have reached 1–3 m in height.

Other Uses

- Fences. The thorny branches are massed together to form brushwood fencing; root cuttings strike readily to form living fences.
- Wood. The attractive wood is easily worked, fine grained, durable, and resistant to insects. It saws and planes well and is used for bowls, mortars and pestles, tool handles, gunstocks, and cabinetwork.
- Food. The fruit is oblong, resembling a yellowish date with loose, leathery skin. It has gummy, bittersweet pulp that is edible when fully ripe. The pulp contains about 40 percent sugar and is macerated in water to make a refreshing beverage or is sometimes fermented to make an alcoholic drink. The seed kernel has been made into bread and soup and contains 30–58 percent of an edible oil (Zachon oil). The residue, which is 50 percent protein, is used in cooking and soap making.
- Forage. The young leaves, fruits, and even the thorns are eaten by goats, camels, and wildlife.
- Pest control. Extracts of the fruit and bark have been found to kill the freshwater snails that are intermediary hosts for schistosomiasis (bilharzia), and they also kill the free-swimming life forms of this parasite, making it a suitable tree for planting along banks of irrigation canals. A strong emulsion of the fruits is toxic to fish. It is also lethal to the waterflea that harbors dracunculiasis (guinea worm).
- Sapogenin source. All parts of the plant, including the seed kernel, yield the sapogenins diosgenin and yamogenin, both of which are used in the partial synthesis of steroid drugs.
- Soap substitute. Because of the saponin content of the roots, bark, fruits, and wood chips, all of these parts have been used as "soap" for washing clothes.

Environmental Requirements

- Temperature. The desert date easily withstands high temperatures ranging up to 40.5°C.
- Altitude. The altitudinal range of the tree is from 380 m below sea level to 1,500 m above sea level.
- Rainfall. The tree is commonly found where rainfall is between 250 mm and 800 mm per year. In some drier areas it is confined to sites such as riverbanks with available groundwater.
- Soil. Desert date can be found on a great

Balanites aegyptiaca. (Top: H. A. Musnad; bottom: G. E. Wickens)

variety of soils; however, on sandy soils its growth is poor and only scattered trees occur. It appears sensitive to salinity and does not tolerate prolonged waterlogging.

Establishment Seeds casually distributed by humans or animals germinate readily. The trees can also be reproduced by seed and root suckers.

- Seed treatment. None is needed when fruits or seeds are sown in summer or autumn. In winter, seeds are soaked for 24 hours at room temperature.
- Ability to compete with weeds. The tree's slow growth would make it a poor competitor with weeds in fertile soils. It must be protected from fire and cattle for at least 3 years to ensure the survival of plantations.

Pests and Diseases Seeds are often attacked by a borer.

Limitations The desert date is thorny and, as noted, very slow growing. Browsing by animals can inhibit its growth, but it has run wild in abundance in Curaçao despite free-ranging goats. Because the tree is widely distributed in Africa and other continents, it must have several forms, if not subspecies. It is therefore quite possible to find a form with such requisite characteristics as fast growth and small thorns or no thorns.

Combretum micranthum

Botanic Name *Combretum micranthum* G. Don

Synonyms *Combretum altum* Perr., *C. floribundum* Engl. & Diels, *C. parviflorum* Reichb.

Common Names Kinkeliba, randga (Upper Volta); paramerga, lardaga (Ghana); kolobe, koubou, gugumi, gieze, landaga (West African Sahel)

Family Combretaceae

Main Attributes A tough drought- and fire-resistant shrub or tree, *Combretum micranthum* will survive in barren, rocky wastes where little else will grow.

Description *Combretum micranthum* is normally a spreading shrub or spreading small tree 2-5 m high and 5-8 cm in stem diameter, although it can grow to 15 m with a 1.5 m trunk. Typically, it is reduced to a shrubby habit by repeated burning of the savanna. It has a shaggy, peeling reddish bark, with greenish maroon leaves, white flowers, and four-winged, glossy red fruits.

Distribution This species is common in Savanna woodlands and in some places near the coast from Senegal to northern Nigeria. It has been successfully introduced for cultivation in South Vietnam; plantings around Saigon, made in the 1950s, have flourished.

Use as Firewood The branches and stems of *Combretum micranthum* are slim and can be used for firewood, particularly around the towns where it is abundant. In Sokoto, Nigeria, the wood is much used for charcoal.

Yield No data available.

Other Uses

- Craftwork. Young stems are tough and flexible and are used in basketry, while older ones are used as walking sticks and rafters. Unsplit stems are used for the interior center part of a thatch roof.
- Beverage. *Combretum micranthum*'s main use in Senegal and Mali is for the Kinkeliba beverage. Young leaves are collected, dried, and processed into bundles to be sold in cities and along roads. These leaves are boiled, and the liquid is a popular beverage, believed to be effective against various tropical fevers and alimentary disorders.

Environmental Requirements

- Temperature. *Combretum micranthum* grows in areas with tropical temperatures such as the Sahel.
- Altitude. The shrubs and trees grow from sea level to 1,000 m.
- Rainfall. The species is drought resistant; in Senegal it grows in areas with rainfall from 300 mm to 1,500 mm.
- Soil. The species grows well on arid ironstone plateaus. It is regarded as an indicator of barren or nearly barren soil and is often found on abandoned farmland.

Establishment The tree spreads by root suckers. The germination of seeds in a nursery is difficult, and establishment is not easy.

- Seed treatment. None required.
- Ability to compete with weeds. High.

Pests and Diseases The roots of *Combretum micranthum* are very susceptible to termite attack.

Combretum micranthum. (R. T. Wilson)

Conocarpus lancifolius

Botanic Name *Conocarpus lancifolius* Engler

Common Names Damas, ghalab

Family Combretaceae

Main Attributes A drought-resistant species, *Conocarpus lancifolius* is one of the more promising trees for trials in arid areas. It is recommended for a variety of soil types, including saline soils, and yields excellent charcoal and valuable wood.

Description Damas is an evergreen tree that grows up to 20 m in height and 60-250 cm or more in diameter. However, it is believed that the larger trees have now been almost entirely felled. Whereas it is usually a multibranched tree in its natural habitat, trees planted in the Sudan formed a single, straight stem.

Distribution Natural stands of damas are found beside intermittent watercourses of northern Somalia and in the southwest part of the Arabian Peninsula. Some of these streams are salty and some sulphurous. The tree is also cultivated in Somalia, as it is in Djibouti, Sudan, Kenya, north and south Yemen, and Pakistan. A small plantation has been established in Sudan's Khashm El Birb Arboretum. About 10,000 trees have been planted successfully in limestone near Mombasa, Kenya.

Use as Firewood The wood is light colored and of medium to heavy density (specific gravity 0.81). It makes good firewood (although some reports suggest that the wood smolders rather than burns) and excellent charcoal.

Yield The annual yield of one irrigated plantation was approximately 21 m^3 per ha. Early growth of more than 2.5 m per year has been achieved in irrigated plantations, though 1.0-1.75 m per year is more typical in early years.

Other Uses

- Wood. Damas wood is strong and is used for poles in house construction and in carpentry. At one time it was exported to Aden, where its chief use was for building dhows, and it is still favored for use as ship knees.
- Fodder. The tree is evergreen and its foliage makes a good fodder.
- Other. It is a good shade and roadside tree. In South Yemen it is used for windbreaks around irrigated agricultural areas and for avenue plantings.

Environmental Requirements

- Temperature. Damas grows best in areas where the mean annual temperature ranges from 20°-30°C, but where the maximum summer temperature has reached 50°C. During summer, the dry, hot air of the tree's natural habitat is laden with sand carried by the southwesterly monsoons. Winter temperatures are moderate and free of frost.
- Altitude. The tree grows from sea level up to about 1,000 m.
- Rainfall. The rainfall in its natural habitat is generally between 50 mm and 400 mm, but the tree grows mainly along seasonal watercourses. It can be grown in plantations in areas with less than about 400 mm but grows well only if irrigated or within reach of groundwater. It withstands drought conditions for several months when irrigation fails.
- Soil. Damas does well on deep soils ranging from pure sand to clays and loams, but has difficulty on shallow soils. It will tolerate moderately saline soils. Experiments carried out in Yemen showed that damas can grow on highly saline-alkaline soils of pH 9.5, provided water is in reach.

Establishment The tree seeds prolifically at an early age and the seed, which is small and light, has a germination capacity of about 25 percent. When sown, it should preferably be left uncovered or covered only thinly. Seeds should be sown in rows 5 cm apart, broadcast in moist seedbeds, or planted in sandy soil in containers 30 cm × 15 cm with a number of holes drilled in the bottom to facilitate irrigation. Irrigation should be provided daily for the first 6-8 months, after which the plants are ready for planting out. Plants grown in a semishaded area in polyethelene bags can be transplanted when the seedlings gain 4-5 leaves. Damas can also be raised by cuttings.

- Seed treatment. None required.
- Ability to compete with weeds. Weeding and watering are essential after planting to ensure establishment.

Pests and Diseases Newly germinated seedlings are highly susceptible to damping-off. Damas is attractive to browsing animals, and complete protection is necessary if plantations are to survive.

Limitations Without irrigation or ready access to soil moisture, the tree grows more slowly and on some sites may not reach usable size in a reasonable time.

Conocarpus lancifolius. (Left: Mohsen A. R. Bazara'a; below: R. Melville)

Dalbergia sissoo

Botanic Name *Dalbergia sissoo* Roxb.

Common Names Sissoo, sisu, sarsou, shisham, nelkar, karra, tanach, shewa, tali, yette

Family Leguminosae (Papilionoideae)

Main Attributes *Dalbergia sissoo* is a moderately fast growing tree that adapts well to semiarid conditions and produces first-class firewood. Its timber is highly valued for construction and general utility purposes. Under savanna conditions the trunk is usually crooked, but it is much sought for furniture, shipbuilding, and fuel.

Description A large, fast-growing tree with a spreading crown, sissoo grows to a height of 10–15 m in arid regions and up to 30 m high in the irrigated plains of Pakistan, in the northern part of India with its high rainfall, and along riverbanks. The tree has long, superficial roots, which send up suckers when injured, and small yellowish flowers with a pervasive fragrance.

Distribution Sissoo occurs throughout the Indian subcontinent from the Indus to Assam. It descends the river valleys for some distance into the plains and is planted or self-sown in many parts of India and Pakistan. It has been extensively grown in irrigated plantations, along roads and canals, and around farms and orchards as a windbreak. The tree shows promising results in the Khartoum greenbelt (Sudan) under irrigation. It has been less successful in Ghana, northern Nigeria, northern Cameroon, and Togo; however, it is being increasingly planted as a street tree in southern Florida, where it is running wild. There are experimental forestry plantings in Puerto Rico, the West African Sahel, South America, and the Middle East.

Use as Firewood Sissoo wood is classed as an excellent fuel; calorific values of the sapwood and heartwood are 4,900 and 5,200 kcal per kg, respectively. Specific gravity is between 0.64 and 0.7, depending on the locality. The wood is also suitable for making charcoal. As fuelwood, it is grown on a 10- to 15-year rotation. The tree produces profuse root suckers and coppices well enough to be managed on a short rotation.

Yield Yield on a rotation of 10 years will amount to 61–99 tons per ha per year, or 9–15 m^3. Growth is strongly dependent on soil conditions. Under favorable conditions in Ghana, trees grew an exceptional 3.0–3.7 m in the first year and 9–11 m in 5 years. Compared with other species (for example, eucalyptus) the yield is relatively low with the same input of labor and sissoo should be used in sites not favorable for other species.

Other Uses

- Wood. The wood is hard, elastic, and close grained, with a little white sapwood and dark-brown heartwood that soon becomes dull. The heartwood is durable and valued for furniture, including bentwood items, and for veneers, carved articles, structural work, handles, cart wheels, railway carriages, boat building, ploughs, toys, frames for tennis rackets, hockey sticks, skis, and minute parts of musical instruments.
- Fodder. Young branches and leaves are used for fodder.
- Erosion control. The strong development of root suckers and runners favor sissoo's use as a living barrier against soil movement, and it is planted for this purpose across the bottom of eroding gulleys.
- Ornamental. In many Mediterranean countries the tree is valued as an ornamental.

Environmental Requirements

- Temperature. Sissoo grows in an area where the temperatures range from just below freezing to nearly 50°C. It is considered frost hardy.
- Altitude. Although mainly a species of river bottoms, the tree also ascends to 1,500 m in the Himalaya foothills.
- Rainfall. Sissoo thrives in rainfall ranging from 500 mm to 2,000 mm and can tolerate arid and semiarid conditions.
- Soil. It is most typically found on alluvial ground in and along the beds of streams and rivers or on sand or gravel along the banks of rivers or on islands. It also springs up on landslips and other places where mineral soil is exposed. Porous soil with adequate moisture seems best; stiff clay should be avoided.

Establishment Sissoo is normally propagated by direct seeding. However, in arid regions it is established by transplanting seedlings or by stump plantings. It may also be raised by shoot cuttings. Shade or partial shade is necessary for germination and seedling development.

- Seed treatment. It is difficult to extract seed from the sissoo pod, so the pods are usually

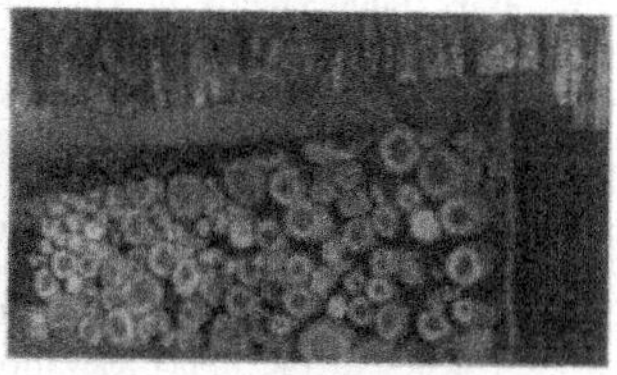

Dalbergia sissoo, Sagarnath, Nepal, on a 9.5-year-old plantation. This tree was 20 cm in diameter and 22 m in height. (D. B. Amatya)

broken and pieces containing one or two seeds are sown. The broken pods should be soaked in water for at least 24 hours before sowing. Sissoo seeds germinate best at 30°C.

• Ability to compete with weeds. Regular weeding is necessary in the plantation during the first 2–3 years.

Pests and Diseases Insects have been noted to damage the root system. Irrigated plantations may be severely damaged by root fungus. In India and Pakistan a leaf fungus (*Phyllactinia*) and a leaf wilt (*Fusarium solani sensa*) have damaged plantations. Sissoo is also susceptible to attack by the pinhole borer. The tree is a host of *Loranthus* spp., *Tapinenthus dodoneifolius*, and the defoliator *Plecoptra reflexa*. Young plants are browsed by deer and cattle.

Limitations There is evidence in Nigeria and Zaire that after two or three coppice rotations the stumps lose their vigor. A change of species or replanting with fresh *Dalbergia sissoo* is probably necessary in these areas. Sissoo needs protection against fire and animal browsing.

*Populus euphratica**

Botanic Name *Populus euphratica* Oliv.

Synonyms *Populus diversifolia* Schrenk, *P. ariana* Dode, *P. mauritanica* Dode, *P. bonnetiana* Dode, *P. litwinowiana* Dode, *P. glaucicomans* Dode, *P. illicitana* Dode, *P. denhardtiorum* Dode, *P. euphratica* f. pruinosa (Schrenk) Nevski

Common Names Euphrates poplar, bahan, bhan, hotung, hodung, gharab, palk, saf-saf, Indian poplar

Family Salicaceae

Main Attributes *Populus euphratica* grows fast. Annual diameter increments of 4.0–5.3 cm have been measured. It can tolerate a high degree of salinity and extremely arid and "continental type" climatic conditions, provided the subsoil is moist. It also grows on land that is seasonally flooded and on which no other form of cultivation appears possible.

Description *Populus euphratica* is a small- to medium-sized tree (7.5–15 m in height) capable of attaining a diameter of 30–70 cm under good conditions. It is often seen in Syria with a bent and nearly always forked stem.

Distribution The tree extends from China to Spain and western Morocco and as far south as Kenya. It is found chiefly in Turkestan, Iran, Iraq, and Syria, where natural stands are economically important, but it occurs also in Turkey, Pakistan, Israel, Egypt, Libya, Algeria, and the Chinese provinces of Hopei, Sinkiang, Shansi, and Manchuria. In India the tree is typically gregarious, occurring in pure stands or associated with other species. Remarkably adaptable, it reaches to the upper elevation of tree growth and is found even in the hottest parts in the plains, where in fact it thrives under river inundations.

Use as Firewood Formerly, *Populus euphratica* constituted dense forests (mixed with willow, tamarisk, and mulberry) along watercourses and their tributaries throughout its vast range, but these forests have been almost completely destroyed to supply firewood. A few remnant forests are still exploited for firewood, but now mainly by coppice on short rotations of one or two years. The wood is only moderately dense (specific gravity up to 0.48). The calorific value is reported to be 5,019 kcal for the sapwood and 5,008 kcal for the heartwood.

Yield Unreported.

Other Uses

- Fodder. The leaves afford a good fodder for sheep, goats, and camels.
- Dental use. The twigs are chewed and used for cleaning teeth.
- Wood. The wood is easy to saw and works to a good finish. It is a good turnery wood and can be peeled off on a rotary cutter. It is used for planking, lacquer work, artificial limbs, match boxes, and splints. It is also suitable for plywood, cricket bats, shoe heels, and bobbins, and can be used for pulp.

Environmental Requirements

- Temperature. *Populus euphratica* grows in areas with a minimum temperatue of −5°C to a maximum of 49°–52°C. It is frost hardy. During the growing season it requires much light and heat.
- Altitude. It is found from below sea level in the Middle East to an altitude of 4,000 m in Ladakh (India) and western Tibet.
- Rainfall. Rainfall range is 75–200 mm, but it mostly grows as a riverine species whose growth does not depend on rainfall.
- Soil. *Populus euphratica* occurs naturally in soils with a high salt content, as much as three parts per thousand.

Establishment Natural reproduction is through root suckers or seeds. The seedlings spring up on fresh alluvial soil after the floods recede. Three sizes of cuttings are used for rooting in the nursery for subsequent planting in India. Planting is done at a spacing of 2 m × 3 m, although wider spacing is sometimes used to provide space for sowing alfalfa for improving soil and to provide fodder. The tree coppices well. It is well adapted for treatment under coppice or coppice-with-standards, reproduction being obtained from root suckers and coppice shoots.

- Seed treatment. None required.
- Ability to compete with weeds. High.

Pests and Diseases In the Near East the tree is

**Populus euphratica* Oliv. is a complex species with many varieties and ecotypes. Formerly classified as *P. inphratica* (now a separate species): (1) pro syn *P. euphratica* Oliv.; (2) pro syn *P. euphratica* Oliv.; (3) pro syn. *P. euphratica* Oliv.; (4) formerly pro syn *P. pruinosa* Schrenk.

subject to attack by various beetles of the genus *Capnodis* and by *Cuscuta monogyna*. It is also attacked by a number of other insect defoliators, borers, and gall-forming pests.

Three-year-old *Populus euphratica* coppice stand in India. (A. L. Griffith)

Limitations The finest specimens of *Populus euphratica* were felled long ago, and the trees that now remain (for example, along the banks of the Euphrates) are inferior in form.

Sesbania sesban

Botanic Name *Sesbania sesban* (Linn.) Merrill

Synonym *Sesbania aegyptiaca* (Poir.) Pers

Common Names Sesban, Egyptian rattle pod, suriminta, soriminta

Family Leguminosae (Papillionaceae)

Main Attributes Sesban is a fast-growing, short-lived tree that regenerates rapidly after pruning. It has a tendency to form root nodules by symbiotic association with soil bacteria capable of fixing atmospheric nitrogen.

Description A shrub or small tree, sesban grows 4.5-6 m high. It is copiously branched, with pinnate leaves, pale yellow flowers, and slender, slightly twisted seedpods up to 25 cm long and containing many seeds.

Distribution Sesban is said to be one of the first garden plants grown in Egypt. It is now widespread in tropical Africa and throughout tropical Asia. It is cultivated and naturalized to some extent in Hawaii.

Use as Firewood In Africa and in India, where fuel is scarce, the tree is sometimes planted to provide firewood, which is used especially for boiling the sap of the sugar palm. The wood yields an excellent gunpowder charcoal.

Yield A stem growth of 5 m in 12 months has been reported, and the yield recorded in India was 30 tons per acre (10 percent moisture) in 1 year.

Other Uses

- Food. In Bihar the flowers are eaten as a vegetable. The leaves are eaten in Thailand. The seeds, high in protein (33.7 percent), are eaten as a famine food in India; they are first soaked for 3 days and then cooked for half an hour to remove the toxic constituent caravanine.
- Fodder. The leaves and young branches are cut for fodder for cattle and sheep.
- Wood. The wood weighs 432 kg per m^3. In Senegal the stems are used for arrows and pipes. In Assam the wood is split and plaited into mats. The very soft wood is made into toys in Burma. In India and Pakistan, stems are used as roofing for huts; in fact, the plant is cultivated as a substitute for bamboo.
- Fiber. The bark fiber is used for making ropes, while the pith serves as fishing floats.
- Other. In India, sesban is extensively planted as a windbreak and shade for vegetable gardens; as support for grape, black pepper, cucurbits, and betel vines; and as a shade for coffee, turmeric, sweet oranges, mandarin oranges, and cotton. In Pakistan it is planted as an intercrop for soil improvement because it is unusually rich in nitrogen. In India it is often grown as green manure in both dry and wet rice fields and plowed in before the crop is planted. It is also grown as a support for sugarcane, each plant bracing six canes. Lopped leaves are composted for use as fertilizer.

Environmental Requirements

- Temperature. In Pakistan the tree requires from 10°C to 45°C.
- Altitude. Sesban grows at 300-500 m in Pakistan. In India it is grown throughout the plains and up to an altitude of 1,200 m.
- Rainfall. It requires 350-1,000 mm rainfall.
- Soil. It tolerates a wide range of soil conditions, withstanding acid soil, periodic flooding, and waterlogging. It can endure 0.4-1.0 percent salt concentration in the seedling stage and 0.9-1.4 percent near maturity.

Establishment

- Seed treatment. None required.
- Ability to compete with weeds. Good; two or three weedings will be required during the first 2 months.

Pests and Diseases The seed is destroyed by insects. Several species of fungi attack the plant. The baterium *Xanthomonas sesbaniae* affects the stems and foliage. The tobacco caterpillar eats the leaves; a weevil, *Alcidodes buho*, damages the plant; and the larvae of *Azygophleps scalaris* tunnels through the main stem. Infested plants must be uprooted and burned.

Limitations The area under cultivation requires protection from cattle; it bears very palatable foliage and is subject to browsing.

Sesbania sesban, 1 year old, Central Soil Salinity Research Institute, Karnal, India. (I. P. Abrol)

Tarchonanthus camphoratus

Botanic Name *Tarchonanthus camphoratus* L.

Synonyms *Tarchonanthus litakunensis* DC., *Tarchonanthus minor* Less., *Tarchonanthus camphoratus* var *litakunensis* (DC) Harv.

Common names Camphor bush, camphor wood, kanferhout, vaalbos (South Africa); leleshua (East Africa)

Family Compositae

Main Attributes This is a many-branched shrub or small tree that is common in open sites in many parts of southern and eastern Africa. It is easy to establish, grows fairly fast, and coppices well. The wood burns with a pleasant-smelling smoke and can be used as fuel even when not completely dry.

Description Under favorable conditions large specimens may reach 9 m, forming a medium-sized tree. More usually, in semiarid areas it remains a densely leaved and bushy shrub some 2-3 m in height. The leaves are oblong to elliptic, 1.3-15 cm long, gray-green above and wooly white below. When crushed they give off a pleasant scent of camphor. The tree is dioecious and has small flowers on different plants, the female forming decorative, many-headed sprays. The fruiting heads that follow are also densely coated with glossy, fluffy white hairs.

Distribution The camphor bush grows from the coast to semiarid inland areas and from the southern tip of Africa to Ethiopia and Somalia. In some areas it is common, forming an extensive shrub savannah.

Use as Firewood The usual multistemmed growth makes the camphor bush easy to harvest for fuel. The bole from which the branches grow adds a substantial piece of firewood. The wood is dense and burns slowly; it ignites well even when only partly dried, making it valuable when shortages of fuel require the harvesting of live wood. People like the pleasantly scented smoke. Good-quality charcoal can be made from the wood.

Yield The camphor bush coppices well under heavy utilization, sprouting even when the bole is burnt or cut almost to the ground. Although precise yield data are not available, its adaptation to harsh environments and fairly rapid growth suggest it as a valuable fuelwood species.

Other Uses

- Windbreak. With its tolerance of wind and extremes of temperature from hot days to freezing nights, the plant makes a valuable windbreak in semiarid areas. It can grow at the shoreline fully exposed to salt spray, making it useful in fixing drifting sands.
- Fodder. The dense, leafy canopy provides good fodder for livestock, especially during droughts and in midwinter when there is a shortage of grazing. In some semiarid areas of Africa, it is the principal source of fodder for wild and domesticated animals.
- Wood. The wood is heavy, tough, close grained, and termite proof. It is good for fence posts, boatbuilding, spear shafts, musical instruments, and decorative work.

Environmental Requirements

- Temperature. The camphor bush thrives in temperate to semidesert climates with daily extremes of temperature.
- Altitude. The plant grows from sea level to about 2,000 m.
- Rainfall. The rainfall ranges from 150 mm to more than 800 mm per year. The plant can tolerate long periods of drought.
- Soil. The camphor bush grows well on open, semiarid plains and rocky places. It is not found in arid areas with deep sands.

Establishment The camphor bush grows easily from seed and may be propagated from cuttings. Seedlings transplant well.

- Seed treatment. Probably none is needed since it seeds and germinates profusely.
- Ability to compete with weeds. It is invasive in overutilized vegetation (bush encroachment), but not in vegetation in good condition.

Pests and Diseases The camphor bush is remarkably free of pests and diseases, being apparently repellant to insects.

Limitations The bush encroaches on grazing lands in East Africa, and because of its tenacious root system it is highly resistant to burning or cutting. Repeated chemical control is successful after the first regeneration. Although it is thornless, a splinter from the wood produces a troublesome sore that heals with difficulty.

Related Species The camphor bush is often seen together with other woody members of the Com-

Tarchonanthus camphoratus (camphor bush) growing near the sea, south of Cape Town, South Africa. (A. V. Hall)

positae family, particularly of the genus *Brachylaena*. In KwaZulu in South Africa, it often grows in association with *Brachylaena ilicifolia*, which is as much used locally for firewood as the camphor bush.

APPENDIX 1

Summary of Characteristics by Climatic Zones

The table following reviews the attributes of the various species duscussed in each section of Volumes I and II of *Firewood Crops: Shrub and Tree Species for Energy Production*. It only includes data contained in the two reports: environmental requirements, yields, calorific values, other uses (uses other than for firewood), coppicing abilities, and nitrogen-fixing abilities.

Our table is based on one prepared by Alan Grainger for his review of Volume I of *Firewood Crops* in *International Tree Crops Journal* Volume 2 No. 1, 1982.

Abbreviations used in table:

Temperature	F Frost tolerant LF Tolerates light frost NF Not frost tolerant
Altitude	Extremes are given in parentheses.
Rainfall	Extremes are given in parentheses. <100 + indicates that trees tolerate less than 100 mm but also a much higher rainfall.
Yields	Estimates are in m^3/ha/year except when stated otherwise. Typical rotations are given in parentheses when available.

Other Uses

B	Bee forage	P	Perfume
DY	Dye	PC	Pest control
F	Food	PW	Pulpwood
Fd	Fodder	R	Resin
Fi	Fiber	SB	Shelterbelts
G	Gum	SC	Soil conservation
GM	Green manure	Sh	Shade
H	Hedges	Si	Silkworm forage
M	Medicine	T	Timber, of whatever quality
O	Oil	Ta	Tannin
Or	Ornamental		

Tropical Highlands

Species	Temperature °C	Altitude m	Rainfall mm	Drought Resistance	Soil Needs	Coppicing Ability	Yields m³/ha/yr	Calorific Value kcal/kg	Other Uses	Nitrogen Fixing
Acacia decurrens	12–25 F	25–1,000 (2,500)	900–1,600	–	Adaptable	Yes	6–16 25 (Sri Lanka) (15)	3,530–3,940	Ta, T, Sh, SB, H	–
Acacia mearnsii	F	300–1,100	500–1,000	–	Adaptable, not calcareous soils	Poor	10–25 (Indonesia) (7–10)	3,500–4,000	Ta, SC, GM, PW	Yes
Ailanthus altissima	F	to 2,000 +	350–600	Up to 8 months	Very adaptable	Yes	–	–	T, SC, Sh, SB	–
Alnus acuminata	4–27 F	1,200–3,200	1,000–3,000 +	–	Adaptable, likes deep well-drained loams or loamy sands	Naturally	10–15 (Costa Rica) (20)	–	T, SC	Yes
Alnus nepalensis	–	1,000–3,000 (300)	over 500	–	Adaptable as above, but needs moist soil	Yes	–	–	T	Probably
Alnus rubra	– 20–45	below 750	600–3,000	–	Adaptable, likes moist soils	Yes	17–21 (USA)	4,600	T, PW, SC	Yes
Eucalyptus globulus	Moderate temperature	to 3,000	800–1,500	No	Adaptable, but needs well-drained soils	Yes	10–30 (Spain) (5–15)	4,800	T, SC, PW, O, B	–
Eucalyptus grandis	– 3–40 LF	to 2,700	1,000–2,500	Moderate	Prefers moist, well-drained soils	Yes	17–45 (Uganda) (6)	–	T, PW	–
Eucalyptus robusta	3–32 LF	to 1,600	1,000–2,000	4 months	Bottom slopes, swamps, and saltwater estuaries	Yes	10–35	–	T, SB, Sh, PW, Ta	–
Eucalyptus tereticornis	17–38 F	600 –1,000	800–1,500 (400–3,500)	Yes	Adaptable	Yes	20–25 (15)	–	T, PW, SC, O, B	–
Gleditsia triacanthos	Varies	1,500 +	500–1,500 +	Yes	Adaptable	Yes	–	–	T, F, Sh, SC, H, SB, Or	–
Grevillea robusta	Mean 20 NF	to 2,300	750–1,500 (400–2,500)	Yes	Adaptable, likes deep, not waterlogged soils	Poorly, but pollards	–	–	T, Sh, B, Or	–
Inga vera	–	Lowland	–	Some	Adaptable	Yes	–	–	T, Sh, B, F	–
Melaleuca quinquenervia	18–34 LF	to 500 (1,400)	1,000–5,000	–	Wet soils	Yes	–	–	T, B, O, Or	–
Melia azedarach	– 15–18	to, 2,000	600–1,000	Yes	Adaptable	Yes	–	5,043–5,176	T, PC, Fd, Sh, Or	–
Robinia pseudoacacia	– 20–18	to 2,500	1,000 (300–400)	2–6 months	Adaptable	Yes	10–17 tons from 3 years (Korea)	High	T, SC, SB, Fd, Or	Yes
Sapium sebiferum	Varies	to 1,000 (2,000)	710 + (500)	–	Adaptable	Yes	26 (4)	4,134–4,277	T, O, B, Or	–

Humid Tropics

Species	Tempera-ture °C	Altitude m	Rainfall mm	Drought Resistance	Soil Needs	Coppicing Ability	Yields m^3/ha/yr	Calorific Value kcal/kg	Other Uses	Nitro-gen Fixing
Acacia auriculiformis	26–30 +	to 600	1,500–1,800	Up to 6 months	Adaptable to even the poorest soils	Poor, but regenerates well	17–20 (Indonesia) (10–12)	4,800–4,900	PW, Sh, Or, SC, Ta	Yes
Albizia falcataria	22–29	to 1,000 (1,500)	4,500	0–2 months	Well-drained, deep with high fertility	Yes	39–50 (10)	2,865–3,357	PW, Sh, Ta, T	–
Bursera simaruba	LF	to 1,000 (1,800)	500–1,400	–	Very adaptable, best in rich lowlands, saline tolerant	Yes	–	–	T, H, R, Or	–
Calliandra calothyrsus	–	150–1,500	1,000	Moderate	Adaptable	Yes	35–65 (Indonesia) (1)	4,500–4,750	SC, Fd, B, Or	Yes
Casuarina equisetifolia	10–33 NF	to 1,500	200–5,000	6–8 months	Tolerates calcareous or slightly saline soils, not heavy clays	Not readily	7.5–20 (Malaysia) tons/ha (7–10)	4,950	T, SC, DY, PW, SB, Ta	Yes
Coccoloba uvifera	LF	to 500	500–1,400	–	Pure sand to inland, good drainage, saline tolerant	Yes	–	–	T, H, F, B, Ta, Or, DY	–
Derris indica	0–50	to 1,200	500–2,500	Yes	Adaptable, highly saline tolerant	Yes	–	4,600	T, Fd, O, PC, Fi, SC	–
Eucalyptus brassiana	to 32 NF	to 650	1,000 minimum	3–5 months	Adaptable, typically infertile	Yes	–	–	T	–
Eucalyptus deglupta	24–32	to 1,800	2,500–3,500	–	Adaptable, colonizes volcanic ash and pumice	Usually not	20–40 (Papua New Guinea) (15)	–	PW, T	–
Eucalyptus pellita	6–33 LF	to 750	900–2,300	–	Adaptable	–	–	–	T, B	–
Eucalyptus urophylla	18–28	300–3,000	1,300–2,500	–	Medium to heavy soil derived from noncalcareous rock	Yes	20–30	–	T, PW	–
Gliricidia sepium	22–30	500 (to 1,600)	1,500–2,300 +	–	Adaptable to both moist and dry soils	Yes	–	4,900	T, Sh, H, Fd, B, GM, Or	Yes
Gmelina arborea	to 52 NF	to 1,000	750–4,500	Some provenances	Adaptable to acid, calcareous and lateritic soils but not water-logged or leached soils or dry sand	Yes	20–35 (5–8)	4,800	T, PW, B	–

Guazuma ulmifolia	Tropical	to 1,200	700–1,500	4–7 months	Adaptable	–	–	Moderately hard	T, Fd, B, Sh, F	Yes
Hibiscus tiliaceus	NF	to 500	1,400	–	Adaptable, highly salt tolerant	Yes	–	–	Fi, T, SC, H, Or	–
Leucaena leucocephala	NF	to 500	600–1,700	Yes	Adaptable, not acid soils	Yes	30–40 (Philippines)	4,200–4,600	F, T, SC, Fd, GM	Yes
Maesopsis eminii	16–27	100–700 (to 1,200)	1,200–3,000	–	Moderate fertile, well-drained soils	–	20–30 (Uganda) (30) 8–20 Common in many areas	–	F, T, Sh, PW	–
Mimosa scabrella	–	to 2,400	–	–	Adaptable, not wet soils	Yes	–	–	PW, GM, Or	Yes
Muntingia calabura	Humid tropical	to 1,300	1,000–2,000	–	Tolerates poor soils, saline tolerant	–	–	–	Sh, PW, Fi, F, Or	–
Pinus caribaea	22–28 (5–37)	1,000 (1,500)	1,000–1,800 (3,900) (660)	–	Variable, generally well drained	–	21–40	–	PW, T	–
Psidium guajava	NF	to 800 (1,500)	1,000	4–5 months	Acid soils, poor in heavy clays	Yes	–	4,792	F, Ta, T, DY	–
Rhizophora species *Avicennia* species	NF	Estuarine/ coastal zones	1,000	–	Tidal areas	Yes	–	4,000–4,300	T, SC, PW, Ta	–
Sesbania bispinosa	Versatile	to 1,200	550–1,100	Yes	Adaptable, tolerates salines, alkaline, and wet soils	–	15 tons/yr (Italy) (2 harvests/yr in the tropics)	–	G, PW, Fi, GM, Fd	Yes
Sesbania grandiflora	NF	to 800	1,000	Few months	Adaptable to dry and moist soils	–	20–25 (Indonesia)	–	F, Fd, SC, PW, G, Ta, Or, SB, H, GM	Yes
Syzygium cumini	–	to 600 (1,800)	1,500–10,000	–	Adaptable	Yes	–	4,800	T, F, H, SB, Ta, B, Or	–
Terminalia catappa	Tropical, humid	to 300	1,000	Yes	Adaptable, saline tolerant	–	2–3.6 tons/ha/yr (10)	–	T, F, Ta, SC, Sh, Si, Or	–
Trema species	–	to 2,000	Varies	Varies	Adaptable to poor soils	Yes	–	4,500	SC, Sh, PW, T, Fd	–

Arid and Semiarid Regions

Species	Tempera-ture °C	Altitude m	Rainfall mm	Drought Resistance	Soil Needs	Coppicing Ability	Yields m^3/ha/yr	Calorific Value kcal/kg	Other Uses	Nitro-gen Fixing
Acacia brachystacha	4–58	to 600	200–500	Yes	Adaptable	–	–	–	T, Fd	–
Acacia cambagei	to 34 +	75–500	125–500	5 months	Adaptable	–	–	–	T	–
Acacia cyclops	5–31 LF	300	200–800	Yes	Adaptable, tolerates salinity and dune crests	Rarely	– (7–10)	–	Fd	–
Acacia nilotica	– LF	to 500	Varies	High	Adaptable	Only occasionally	– (20–30)	4,800–4,950	T, Fd, Ta, G, B	–
Acacia saligna	13–30 NF	to 300	350–600 (1,000)	High	Adaptable to calcareous, acid, alkaline, saline soils	Yes	1.5–10 (5–10)	–	SC, SB, Fd, G	–
Acacia senegal	14–43	100–1,700	300–450 (200–800)	High, 8–11 months	Sand and clay soils, but not waterlogged sites	Yes	–	–	T, G, Fd, F, SG, Fi	Yes
Acacia seyal	Hot	Lowland (to 2,100)	350	Yes	Adaptable, including moist sites	–	–	–	T, Fd, G	Probably
Acacia tortilis	to 50 NF	Lowland	100–1,000	High	Prefers alkaline soils	Yes	54 tons/ha (12)	4,400	T, Fd, SC	–
Adhatoda vasica	NF	to 1,300	500–1,650	–	Adaptable	Yes	–	–	M, GM, PC, DY, H	–
Ailanthus excelsa	20–40	Low	600 (400)	–	Adaptable but not clay or waterlogged	Yes	–	–	Fd, T, SB, Sh	–
Albizia lebbek	– LF	0–1,600	500–2,000	Yes	Likes moist conditions, saline tolerant	Yes	5 (India) (10–15)	5,200	T, Sh, B, Fd, SC, GM	Yes
Anogeissus latifolia	–	to 1,300	600	Yes	Dry, sandy, or rocky soils	Yes	–	4,900	T, G, Ta, DY, Fd, Si, PW	–
Azadirachta indica	0–44	50–1,500	450–1,150 (130 min)	High	Adaptable, not waterlogged or saline soils	Yes	13–17 (Ghana) (8)	High	T, O, SB, Sh, SC, Ta, PC	–
Balanites aegyptiaca	to 40.5	380–1,500	250–800	Yes	Adaptable, sensitive to salinity and waterlogging	Yes	–	4,600	H, T, F, Fd, M, PC	–
Cajanus cajan	18–29 NF	to 3,000	600–1,000 (400–2,500)	Yes	Adaptable, not waterlogged soils	Pollards	2 + woody stalks/ha/growing season	–	F, Fd, Si, Sh, SB, SC, H	Yes
Cassia siamea	NF	Lowland	1,000 +	–	Deep, well-drained soils	Yes	to 15 (5–10)	–	T, SB, SC	–
Colophospermum mopane	36 + LF	to 900	200–450 (125–800)	–	Adaptable, tolerates dry alkaline or saline soils	Yes	–	High	T, Fd, SC	No

Combretum micranthum	tropical	to 1,000	300–1,500	Yes	Arid ironstone plateaus	–	–	–	Fi, F	–
Conocarpus lancifolius	20–30 (50)	to 1,000	50–400	Yes	Adaptable, moderate saline soils	–	21 (irrigated)	–	T, Fd, Sh, SB	–
Dalbergia sissoo	to 50 F	to 1,500	500–2,000	Yes	Porous soil with moisture	Yes	9–15 (10)	4,900–5,200	T, Fd, SC, Or	–
Emblica officinalis	to 46 LF	to 1,800	–	–	Adaptable, likes moist soils, tolerates poor alkaline conditions	Yes	–	5,200	T, F, Fd, GM	–
Eucalyptus camaldulensis	varies	1,200 +	400 – 1,250	–	Various with provenance	Yes	20–25 (Argentina) 30 (Israel) (7–10)	4,800	T, SB, Sh, B, PW	–
E. citriodora	to 29–35 LF	to 900 (2,000 poss.)	900 (600 min)	Yes	Adaptable to poor soils	Yes	15 (Tanzania) (8)	–	T, P, B, Or	–
E. gomphocephala	Mild	to 2,000	700–1,000 (300)	Yes	Adaptable, tolerates saline, but not waterlogged soils	Yes	6–7 (difficult sites) 21–44 (irrigated) (7–10)	–	T, SC, Sh, SB	–
E. microtheca	5–38 F	80–340 (to 700)	200–1,000	Yes	Prefers clay and/or alkaline soils	Yes	– (6)	–	T, SB, Sh, SC	–
E. occidentalis	2–38 F	50–300	300–760	Yes	Prefers clays, saline tolerant	–	–	–	T, Sh	–
Haloxylon aphyllum	– 35–50	Lowland	100 +	Yes	Adaptable, sandy soils, saline tolerant	Yes	1.4 tons/ha/yr (5)	–	SC, Fd, SB	–
Haloxylon persicum	Less resistant to cold than above	–	100 +	High	As above, but less saline tolerant	Yes	–	–	SC, Fd	–
Parkinsonia aculeata	to 36 LF	to 1,300	200–1,000	High	Adaptable to poor soils, saline tolerant, not waterlogged soils	Yes	Fast growth	–	SC, Fd, Or, H	–
Pinus halepensis	18 to high F	to 1,000 (2,000)	250–800	Yes	Adaptable, does not tolerate saline or waterlogged soils	–	3–12 5–6 (Italy)	–	T, R, Or, SC, SB	–
P. eidarica	to 45 F	–	200 +	Yes	As above	–	–	–	–	–
Pithecellobium dulce	–	to 1,800	450–600 (to 1,650)	Yes	Very adaptable	Yes	–	5,200–5,600	T, Sh, SB, F, Fd, O, Ta, H, Or, B	–
Populus euphratica	– 5–52 F	to 4,000	75–200	–	Tolerates high salinity and arid conditions provided subsoil is moist	Yes	–	5,008–5,019	Fd, M, T, PW	–
Prosopis alba	15 + LF	Lowlands (to 1,000)	100–500	Yes	Likes sand with clay, saline tolerant	–	7 (Argentina) (10)	–	T, Fd, F, SB	Yes

Arid and Semiarid Regions (*continued*)

Species	Tempera-ture °C	Altitude m	Rainfall mm	Drought Resistance	Soil Needs	Coppicing Ability	Yields m^3/ha/yr	Calorific Value kcal/kg	Other Uses	Nitro-gen Fixing
P. chilensis	Hot (27) LF	340–1,230 (to 2,900)	200–400	High	Adapted to light soils	–	–	–	Fd, T, Or	Yes
P. cineraria	−6–50 LF	Lowland	75–850	Yes	Adaptable, tolerates salinity or high alkalinity	Yes	21	5,000	Fd, SC, T	Yes
P. juliflora	Hot	to 1,500	150–750	High	Adaptable	Yes	5–6 tons/ha/yr (10)	High	T, B, F, Fd	Yes
P. pallida	NF	0–300	250–1,250	Yes	Adaptable	Yes	–	High	Fd, F, SC	Yes
P. tamarugo	12–36	1,000–1,500	0–10	High	Highly saline tolerant	Yes	Slow growth	–	T, Fd	–
Sesbania sesban	10–45	300–500 (1,200)	350–1,000	–	Adaptable	Yes	–	–	F, Fd, T, Fi, Sh, SB, GM	Yes
Tamarix aphylla	−10–50	–	350–500 (100 +)	Yes	Very adaptable, high saline tolerance	Yes	–	–	T, SC, SB	–
Tarchonanthus camphoratus	–	to 2,000	150–800	Yes	Adaptable but not deep sands	Yes	–	–	SB, SC, Fd, T	–
Zizyphus mauritania	Hot F	< 600	350–500 (to 2,000)	Yes	Adaptable	Yes	Fast growth	4,900	T, F, H, Ta, Si, Fd	–
Zizyphus spina-christi	Hot	to 1,500	100 +	High	Adaptable, likes deep alluvial soils	Yes	–	High	T, Fd, SC, H, SB	–

APPENDIX 2

Selected Readings

Acacia decurrens

Krishnaswamy, V. S. 1957. The wattles or the exotic acacias of Australia introduced in Madras state. *Proceedings of the Seventh British Commonwealth Forestry Conference, Australia and New Zealand.* Madras Forest Department, Madras, India.

Raghavan, M. S. 1939. *Note on the Cultivation of the Green Wattle* Acacia decurrens *in South Africa and South India.* Superintendent, Government Press, Madras, India.

Streets, R. J. 1962. *Exotic Forest Trees in the British Commonwealth.* Clarendon Press, Oxford, England.

Webb, D. B., P. J. Wood, and J. Smith. 1980. *A Guide to Species Selection for Tropical and Subtropical Plantations.* Tropical Forestry Paper No. 16. Commonwealth Forestry Institute, Oxford, England.

Ailanthus excelsa

Bhandari, D. S., and M. L. Gupta. 1972. Studies on the digestibility and nutritive value of ardu (*Ailanthus excelsa* Roxb.). *Indian Veterinary Journal* 49(5): 512-516.

Bhandari, D. S., and M. L. Gupta. 1973. Utilization of ardu leaves (*Ailanthus excelsa*) as a supplement for poor quality roughage. *Indian Veterinary Journal* 50(8):809-812.

Kirtikau, K. R., and D. Basu. 1975. *Indian Medicinal Plants,* Vol. 1, p. 505. Bishshem Singh Mahendra Pal Singh, 23A New Connought Place, Dehra Dun, India.

Singh, N. P., and B. C. Patnayak. 1977. Nutritive value of *Ailanthus excelsa* Roxb. (ardu) leaves for sheep. *Indian Veterinary Journal* 54(3):198-201.

Singh, N. P., and R. D. Mudgal. 1979. Ardu—An ideal fodder tree. *Farmer and Parliament* 14(5):25-26.

Troup, R. S. 1921. *The Silviculture of Indian Trees,* Vol. 1, pp. 171-173. Clarendon Press, Oxford, England.

Albizia falcataria

Burgess, P. F. 1966 (Reprinted 1980). *Timbers of Sabah,* pp. 343-347. Sabah, Malaysia.

Domingo, I. L. 1980. *Bibliography on* Albizia falcataria *(L.) Fosb.* Forest Research Department, Paper Industries Corporation of the Philippines, Bislig, Surigao del Sur, Philippines. (Contains 164 references.)

Fenton, R., R. E. Roper, and G. R. Watt. 1977. *Lowland Tropical Hardwoods.* (An annotated bibliography on selected species with plantation potential.) External Aid Division, Ministry of Foreign Affairs, Wellington, New Zealand.

Fontanilla, C. 1980. FORI has identified the cause of Albizia canker paving the way to an effective control of this disease that is infesting 5 industrial tree plantations of corporate and small tree farmers. *CANOPY* 6(5)1, 8-9; and Editorial, p. 2.

Gerhards, C. C. 1976. *Physical and Mechanical Properties of* Molucca albizzia *Grown in Hawaii.* Research Paper FPL-55. USDA Forest Service, Forest Products Laboratory, Madison, Wisconsin, USA.

Ng, F. S. P. 1980. Legumes in forestry. *Proceedings of Legumes in the Tropics,* pp. 449-456. Universiti Pertanian, Serdang, Selangor, Malaysia.

Skolmen, R. G. 1974. *Some Woods of Hawaii—Properties and Uses of 16 Commercial Species.* General Technical Report PSW-8. USDA Forest Service, Pacific Southwest Forest and Range Experiment Station, Berkeley, California, USA.

Balanites aegyptiaca

Abu El Gasim M. Suliman and J. K. Jackson. 1959. The Heglig tree. *Sudan Silva* 9(1):63-66 (67-70 in Arabic).

Busson, F. 1965. *Plantes alimentaires de l'Ouest Africain.* Leconte, Marseille, France. 568 pp.

Dale, I. R., and P. J. Greenway. 1961. *Kenya Trees and Shrubs.* Buchanan's Kenya Estates, Ltd., Nairobi, and Hatchards, London, England. 654 pp.

Dalziel, J. M. 1948. *Useful Plants of West Tropical Africa* (appendix to the *Flora of West Tropical Africa*). Crown Agents for the Colonies, London, England. 612 pp.

Eggeling, W. J. 1951. *The Indigenous Trees of the Uganda Protectorate,* second revised edition. Government of the Uganda Protectorate, Entebbe, Uganda. 491 pp.

Giffard, P. L. 1974. *L'Arbre dans le Paysage Sénégalais.* Centre Technique Forestier Tropical, Dakar, Senegal. 431 pp.

Hardman, R., and E. A. Sofowora. 1972. A reinvestigation of *Balanites aegyptiaca* as a source of steroidal sapogenins. *Economic Botany* 26(2):169-173.

Hussain, S. A., F. G. Dollear, and R. T. O'Connor. 1949. Oil from the kernels of lalob fruit, *Balanites aegyptiaca. Journal of the American Oil Chemists' Society* 26(12):730-732.

Irvine, F. R. 1961. *Woody Plants of Ghana with Special Reference to Their Uses.* Oxford University Press, London, England. 868 pp.

Karschon, R. 1973. Contributions to the arboreal flora of Israel *Balanites aegyptiaca* (L.) Del. *La-Yaaran* 23(3-4).

Bursera simaruba

Little, E. L., Jr., and F. H. Wadsworth. 1964. *Common Trees of Puerto Rico and the Virgin Islands.* Agricultural Handbook 249. USDA Forest Service, Washington, D.C., USA. 558 pp.

Longwood, F. R. 1962. *Present and Potential Commercial Timbers of the Caribbean.* Agricultural Hand-

book 207. USDA Forest Service, Washington, D.C., USA. 167 pp.
Morton, J. F. 1981. *Atlas of Medicinal Plants of Middle America.* Charles C Thomas, Springfield, Illinois, USA. 1,420 pp.
Pennington, T. D., and J. Sarukhan. 1968. *Arboles Tropicales de Mexico.* FAO and Instituto Nacional de Investigaciones Forestales, Mexico, D.F. 413 pp.
Standley, P. C., and J. A. Steyermark. 1946. *Flora of Guatemala: Fieldiana, Botany,* Vol. 24, Pt. 5. Chicago Natural History Museum Publication 578. Chicago, Illinois, USA. 502 pp.

Coccoloba uvifera

Degener, O. 1946. *Flora Hawaiiensis,* Books 1-4. Author, Riverdale, New York.
Harrar, E. S., and J. G. Harrar. 1946. *Guide to Southern Trees.* McGraw-Hill, New York. 712 pp.
Howard, R. A. 1958. A history of the genus *Coccoloba* in cultivation. *Baileya* 6(4):204-212.
Little, E. L., Jr., and F. H. Wadsworth. 1964. *Common Trees of Puerto Rico and the Virgin Islands.* Agriculture Handbook 249. USDA Forest Service, Washington, D.C., USA. 558 pp.
Morton, J. F. 1981. *Atlas of Medicinal Plants of Middle America.* Charles C Thomas, Springfield, Illinois, USA. 1,420 pp.
Perkin, A. G., and A. E. Everist. 1918. *The Natural Organic Colouring Matters.* Longmans, Green and Co., Bombay, India. 655 pp.
Standley, P. S. 1922. *Trees and Shrubs of Mexico,* Contribution of the U.S. National Herbarium, Vol. 23, Pt. 2, pp. 171-515. Smithsonian Institution, Washington, D.C., USA.
Standley, P. C., and J. A. Steyermark. 1946. *Flora of Guatemala: Fieldiana, Botany,* Vol. 24, Pt. 4. Chicago Natural History Museum Publication 577. Chicago, Illinois, USA. 493 pp.
West, E., and L. E. Arnold. 1946. *Native Trees of Florida.* University of Florida Press, Gainesville, Florida, USA. 212 pp.

Combretum micranthum

Aubreville, A. 1950. *Flore Forestière Soudane-Guinéenne.* Société d'Editions Géographiques Maritimes et Coloniales, Paris, France. 5,223 pp.
Berhaut, J. 1974. *Flore Illustrée du Sénégal,* pp. 347-350. Ministère du Développement Rural, Direction des Eaux et Forêts, Dakar, Senegal.
Bognounou, O. 1975. Note sur une plante médicinale: Le randga ou kinkeliba (*Combretum micranthum* G. Don—Combretaceae). *Notes et Documents Voltaiques* 8(21)36-44. Ouagadougou, Upper Volta.

Conocarpus lancifolius

Boaler, S. B. 1959. *Conocarpus lancifolius* in Somaliland. *Empire Forest Review* 38(4):371-379.
Howes, F. N. 1951. *Conocarpus lancifolius* Engl.—An interesting Somaliland Tree. *Kew Bulletin* 6:323-324.

Dalbergia sissoo

Council of Scientific and Industrial Research. 1952. *Wealth of India, Raw Materials,* Vol. 3, pp. 7-11. CSIR, New Delhi, India.
Food and Agricultural Organization of the United Nations. 1974. *Tree Planting Practices in African Savannas.* FAO Forestry Development Paper No. 19. Rome, Italy.
Gamble, J. S. 1902. *A Manual of Indian Timbers,* pp. 246-250. Sampson Low, Marston and Co., London, England.
Ghosh, R. C. 1977. *Handbook on Afforestation Techniques.* Controller of Publications, Delhi, India.
Goor, A. Y., and C. W. Barney. 1978. *Forest Tree Planting in Arid Zones.* Ronald Press, New York, USA.
Kaul, R. N. 1970. *Afforestation in Arid Zones.* Dr. W. Junk N. V. Publishers, The Hague, Netherlands.
Kumar, A., and H. P. Bhatnagar. 1976. Effect of temperature and substratum on germination of *Dalbergia sissoo* Roxb. seed. *Indian Forestry* 102(9): 608-613.
Street, R. J. 1962. *Exotic Forest Trees in the British Commonwealth,* pp. 259-262. Clarendon Press, Oxford, England.
Troup, R. S. 1921. *The Silviculture of Indian Trees,* Vol. 1, pp. 294-318. Clarendon Press, Oxford, England.

Eucalyptus brassiana

Food and Agriculture Organization of the United Nations. 1979. *Eucalypts for Planting.* FAO Forestry Series No. 11. Rome, Italy. 677 pp.
Turnbull, J., and I. Brooker. 1978. *Cape York Red Gum,* Eucalyptus brassiana *S. T. Blake.* Forest Tree Series No. 213. CSIRO, Melbourne, Australia.

Eucalyptus deglupta

Davidson, J. 1977a. Problems of vegetative propagation of *Eucalyptus. Proceedings of the Third World Consultation on Forest Tree Breeding.* Canberra, Australia.
Davidson, J. 1977b. Breeding *Eucalyptus deglupta*—A case study. *Proceedings of the Third World Consultation on Forest Tree Breeding.* Canberra, Australia.
Fenton, R., R. E. Roper, and G. R. Watt. 1977. *Lowland Tropical Hardwoods.* (An annotated bibliography of selected species with plantation potential.) External Aid Division, Ministry of Foreign Affairs, Wellington, New Zealand.
Food and Agriculture Organization of the United Nations. 1979. *Eucalypts for Planting.* FAO Forestry Series No. 11. Rome, Italy. 677 pp.
Turnbull, J. 1974. *Kamarere.* Eucalyptus deglupta *Blume.* Forest Tree Series No. 175. Department of Primary Industry, Forestry and Timber Bureau, Canberra, Australia.
Hillis, W. E., and A. G. Brown, eds. 1978. *Eucalypts for Wood Production.* CSIRO, Melbourne, Australia. 434 pp.
Webb, D. B., P. J. Wood, and J. Smith. 1980. *A Guide to Species Selection for Tropical and Subtropical Plantations.* Tropical Forestry Paper No. 16. Commonwealth Forestry Institute, Oxford, England.

Eucalyptus pellita

Food and Agriculture Organization of the United Nations. 1979. *Eucalypts for Planting.* FAO Forestry Series No. 11. Rome, Italy. 677 pp.
Golfari, L., R. L. Caser, and V. P. G. Moura. 1978. Zoneamento ecologico esquematico para reflorestamento no Brasil. Instituto Brasileiro de Desenvolvimento e Pesquisa Florestal, Brasilia. FAO order number PNUD/FAO/IBDF/BRA-45 Serie Technica. FAO, Rome, Italy.
Hall, N., and I. Brooker. 1974. *Large-Fruited Red Mahogany.* Eucalyptus pellita *F. Muell.* Forest Tree

Series No. 146. Department of Primary Industry, Forestry and Timber Bureau, Canberra, Australia.
Hillis, W. E., and A. G. Brown, eds. 1978. *Eucalypts for Wood Production*. CSIRO, Melbourne, Australia. 434 pp.
Moura, V. P. G., et al. 1980. Avaliacão de especies a procedencias de *Eucalyptus* em Minas Gerais Espirito Santo. Boletim de Pesquisa No. 1. EMBRAPA Centro de Pesquisa Agropecuaria dos Cerrados, Brasilia.

Eucalyptus robusta

Boland, D. J., M. I. H. Brooker, and J. W. Turnbull. 1980. *Eucalyptus Seed*. CSIRO, Canberra, Australia. 191 pp.
Dabral, S. N. 1957. Notes on some of the exotic forest species established in India. *Proceedings of the Seventh British Commonwealth Forestry Conference, Australia and New Zealand*. Madras Forest Department, Madras, India.
Hall, N., R. D. Johnston, and G. M. Chippendale. 1975. *Forest Trees of Australia*. Australian Government Publishing Service, Canberra, Australia. 334 pp.
Hillis, W. E., and A. G. Brown, eds. 1978. *Eucalypts for Wood Production*. CSIRO, Melbourne, Australia. 434 pp.
Jacobs, M. R. 1979. *Eucalypts for Planting*. FAO Forestry Series No. 11. Rome, Italy. 677 pp.
Krugman, S. L. 1974. *Eucalyptus* L'Herit Eucalyptus. *Seeds of Woody Plants in the United States*, pp. 384–392. Agricultural Handbook 450. USDA Forest Service, Washington, D.C., USA.
Pickford, G. D., and R. K. LeBarron. 1960. *A Study of Forest Plantations for Timber Production on the Island of Hawaii*. Technical Paper 52. USDA Forest Service, Pacific Southwest Forest and Range Experiment Station, Berkeley, California, USA. 17 pp.

Eucalyptus tereticornis

Bell, T. I. W., and T. Evo. 1982. *Energy Plantations in the Fiji Dry Zone*. Fiji Pine Research Paper No. 10. Fiji Pine Commission, Box 521, Lautoka, Fiji.
Boland, D. J., M. I. H. Brooker, and J. W. Turnbull. 1980. *Eucalyptus Seed*. CSIRO, Canberra, Australia. 191 pp.
Hall, N., R. D. Johnston, and G. M. Chippendale. 1975. *Forest Trees of Australia*. Australian Government Publishing Service, Canberra, Australia. 334 pp.
Hillis, W. E., and A. G. Brown, eds. 1978. *Eucalypts for Wood Production*. CSIRO, Melbourne, Australia. 434 pp.
Jacobs, M. R. 1979. *Eucalypts for Planting*. FAO Forestry Series No. 11. Rome, Italy. 677 pp.
Vivekanandan, K. 1979. Performance of provenances of eucalyptus in the dry zone. *The Sri Lanka Forester* 14(1&2):49–57.

Eucalyptus urophylla

Campinohos, E. 1980. More wood of better quality through intensive silviculture and rapid growth improved Brazilian eucalyptus. *TAPPI* 63(11):145–147.
Fenton, R., R. E. Roper, and G. R. Watt. 1977. *Lowland Tropical Hardwoods*. (An annotated bibliography of selected species with plantation potential.) External Aid Division, Ministry of Foreign Affairs, Wellington, New Zealand.
Food and Agriculture Organization of the United Nations. 1979. *Eucalypts for Planting*. FAO Forestry Series No. 11. Rome, Italy. 677 pp.
Hillis, W. E., and A. G. Brown, eds. 1978. *Eucalypts for Wood Production*. CSIRO, Melbourne, Australia. 434 pp.
Martin, C., and C. Cossalter. 1975-76. Les *Eucalyptus* des iles de la Sonde. *Revue des Bois et Forêts des Tropiques*, pp. 163–167.
Moura, U. P. G. 1978. Altitudinal Variation in *Eucalyptus urophylla* S. T. Blake. M. S. thesis, University of Melbourne, Melbourne, Australia.
Turnbull, J., and I. Brooker. 1978. *Timor Mountain Gum* Eucalyptus urophylla, *S. T. Blake*. Forest Tree Series No. 214. CSIRO, Melbourne, Australia.
Webb, D. B., P. J. Wood, and J. Smith. 1980. *A Guide to Species Selection for Tropical and Subtropical Plantations*. Tropical Forestry Paper No. 16. Commonwealth Forestry Institute, Oxford, England.

Gleditsia triacanthos

Bonner, F. T., J. D. Burton, and H. C. Grigsby. 1974. *Gleditsia* L. honeylocust. *Seeds of Woody Plants in the United States*, pp. 431–433. Agricultural Handbook 450. USDA Forest Service, Washington, D.C., USA.
Funk, D. T. 1965. Honeylocust (*Gleditsia triacanthos* L.). *Silvics of Forest Trees in the United States*, pp. 198–201. Agricultural Handbook 271. USDA Forest Service, Washington, D.C., USA.
Goor, A. Y., and C. W. Barney. 1968. *Forest Tree Planting in Arid Zones*. Ronald Press, New York, USA.
Grainger, A., and H. W. Esbenshade. 1978. The development of tree crops for agroforestry systems. Paper presented at the Eighth World Forestry Congress, Jakarta, Indonesia. Food and Agriculture Organization of the United Nations, Rome, Italy. 11 pp.
Sherman, P. J. 1977. *Tropical Forage Legumes*. FAO Plant Production and Protection Series, No. 2. Food and Agriculture Organization of the United Nations, Rome, Italy.
Streets, R. J. 1962. *Exotic Forest Trees in the British Commonwealth*. Clarendon Press, Oxford, England.

Hibiscus tiliaceus

Benthall, A. P. 1946. *Trees of Calcutta and Its Neighbourhood*. Thacker, Spink and Co., Calcutta, India. 513 pp.
Brown, W. H. 1954. *Useful Plants of the Philippines*, Vol. 2. Technical Bulletin 10. Philippine Department of Agriculture and Natural Resources, Manila, Philippines. 513 pp.
Burkill, I. H. 1935. *Dictionary of the Economic Products of the Malay Peninsula*, 2 vols. Crown Agents for the Colonies, London, England. 2,402 pp.
Christophersen, E. 1935. *Flowering Plants of Samoa*. Bulletin 128. Bernice P. Bishop Museum, Honolulu, Hawaii. 221 pp.
Degener, O. 1945. *Plants of Hawaii National Park, Illustrative of the Plants and Customs of the South Seas*. Author, New York Botanical Garden, Bronx Park, New York. 314 pp.
Degener, O. 1946–57, 1963. *Flora Hawaiiensis*, Book 5. Author, Mokuleia Beach, Waialua, Oahu, Hawaii. 481 pp.

Dodge, C. R. 1897. *Descriptive Catalogue of Useful Fiber Plants of the World.* Report No. 9. USDA Office of Fiber Investigations, Washington, D.C., USA. 361 pp.

Irvine, F. R. 1961. *Woody Plants of Ghana, with Special Reference to their Uses.* Oxford University Press, London, England. 868 pp.

Little, E. L. Jr., and F. H. Wadsworth. 1964. *Common Trees of Puerto Rico and the Virgin Islands.* Agriculture Handbook 249. USDA Forest Service, Washington, D.C., USA. 558 pp.

Maiden, J. H. 1889. *Useful Native Plants of Australia (including Tasmania) Technologica.* Museum of New South Wales, Sydney, Australia. 696 pp.

Maesopsis eminii

Food and Agriculture Organization of the United Nations. 1956. *Tree Planting Practices in Tropical Africa.* FAO Forestry Development Paper No. 8. Rome, Italy.

Fenton, R., R. E. Roper, and G. R. Watt. 1977. *Lowland Tropical Hardwoods.* (An annotated bibliography of selected species with plantation potential.) External Aid Division, Ministry of Foreign Affairs, Wellington, New Zealand.

Mitchell, B. A. 1963. Possibilities for forest plantations. *Malayan Forester* 26(4):259–286.

Mugasha, A. G. 1981. *The Silviculture of Tanzanian Indigenous Tree Species II.* Maesopsis eminii. Tanzania Technical Note (New Series) No. 52. Ministry of Cultural Resources and Tourism, Lushoto, Tanzania.

Ofung, A. U. 1974. The incidence of cankers on *Maesopsis eminii. East African Agriculture and Forestry Journal* 39(3): 311–320.

Streets, R. J. 1962. *Exotic Forest Trees in the British Commonwealth.* Clarendon Press, Oxford, England.

Swabey, C. 1954. *Silviculture of Musizi* (Maesopsis eminii). Uganda Forest Department Technical Note No. 10/54. Forest Research Centre, Kampala, Uganda.

Warsopranoto, R. R., R. Sverjono, and R. I. Arkidusuma. 1966. Results of an investigation of *Maesopsis eminii* plantations in South Bandung forest management unit (Java, Indonesia). *Rimba Indonesia* 11(1/6):1–12.

Webb, D. B., P. J. Wood, and J. Smith. 1980. *A Guide to Species Selection for Tropical and Subtropical Plantations.* Tropical Forestry Paper No. 16. Commonwealth Forestry Institute, Oxford, England.

Melaleuca quinquenervia

Blake, S. T. 1968. A Revision of *Melaleuca leucadendron* and Its Allies (Myrtaceae). Contributions from the Queensland Herbarium, No. 1. Department of Primary Industries, Brisbane, Australia. 114 pp.

Geiger, R. K., ed. 1981. *Proceedings of Melaleuca Symposium,* Fort Myers, Florida, September 23–24, 1980. Florida Division of Forestry, Tallahassee, Florida, USA. 140 pp.

Morton, J. F. 1966. The cajeput tree: A boon and an affliction. *Economic Botany* 20(1):31–39.

Morton, J. F. 1969. Some ornamental plants excreting respiratory irritants. *Florida State Horticultural Society* 1969:415–421.

Wade, D., J. Ewel, and R. Hofstetter. 1980. *Fire in South Florida Ecosystems.* Technical Report SE-17. USDA Forest Service, Southeastern Forest Experiment Station, Oskenilles, North Carolina, USA.

Woodell, S. L. 1982. Seed dispersed in *Melaleuca quinquenervia. Florida Scientist* 45(2):81–93.

Melia azedarach

Anonymous. 1963. *Azadirachta indica* and *Melia azedarach*: Silvicultural characteristics and plantation techniques. *Bois et Forêts des Tropiques* 88:23–30 (in French).

Council of Scientific and Industrial Research. 1962. *The Wealth of India, Raw Materials,* Vol. 6, pp. 323–325. New Delhi, India.

Goor, A. Y., and C. W. Barney. 1968. *Forest Tree Planting in Arid Zones.* Ronald Press, New York, USA.

Gupta, J. N. 1958. *Fuelwood Plantations in India.* FAO Occasional Paper No. 5. Food and Agriculture Organization of the United Nations, Rome, Italy. 73 pp.

Streets, R. J. 1962. *Exotic Forest Trees in the British Commonwealth.* Clarendon Press, Oxford, England.

Pinus caribaea

Burley, J., and D. G. Nikles. 1973. *Tropical Provenance and Progeny Research and International Cooperation,* pp. 1–75. Commonwealth Forestry Institute, Oxford, England.

Burley, J., and B. T. Styles, eds. 1976. *Tropical Trees.* Academic Press, New York, USA.

Chatterden, A. E., and E. R. Palmer. 1959. Pulping bails on *Pinus caribaea* wood from British Honduras. *Tropical Science* 1(1):22–40.

Fiji Pine Research Paper series (covers a wide range of topics related to the establishment of pine plantations in Fiji). Fiji Pine Commission, Box 521, Lautoka, Fiji.

Greaves, A. 1980. Review of the *Pinus caribaea* Mor. and *Pinus oocarpa* Schiede international provenance trials, 1978. Commonwealth Forestry Institute Occasional Papers No. 12. Commonwealth Forestry Institute, Oxford, England. 89 pp.

Lamb, A. F. A. 1973. *Pinus caribaea,* Vol. 1. *Fast Growing Timber Trees of the Lowland Tropics* No. 6. Commonwealth Forestry Institute, Oxford, England. 245 pp.

Little, E. L., Jr., R. O. Woodbury, and F. H. Wadsworth. 1974. *Trees of Puerto Rico and the Virgin Islands.* Agricultural Handbook 49. USDA Forest Service, Washington, D.C., USA.

Longwood, F. R. 1962. *Present and Potential Timbers of the Caribbean.* Agricultural Handbook 207. USDA Forest Service, Washington, D.C., USA. 167 pp.

Luckhoff, H. A. 1965. *The Natural Distribution, Growth, and Botanical Variation of* Pinus caribaea *and Its Cultivation in South Africa.* Annual of the University of Stellenbosch, Vol. 39, Series A, No. 1. 160 pp.

Packman, D. F. 1959. The production of hardboard from tropical timbers, IV. Caribbean pitch pine (*Pinus caribaea*). *Tropical Science* 1(1):5–12.

Populus euphratica

Council of Scientific and Industrial Research. 1969. *Wealth of India: Raw Materials,* Poplars, Vol. 8, p. 214. CSIR, New Delhi, India.

Food and Agriculture Organization of the United Nations. 1958. *Poplars in Forestry and Land Use.* FAO Forestry and Forest Products Studies No. 12. Rome, Italy.

Food and Agriculture Organization of the United Nations. 1979. *Poplars and Willows.* FAO Forestry Series No. 10. Rome, Italy.

Qadri, S. M. A. 1969. Artificial regeneration of *Populus euphratica* Olivier. *The Pakistan Journal of Forestry*, April:175–184.

Quraishi, M. A. 1967. *Populus euphratica* Olivier and its fungal enemies. *The Pakistan Journal of Forestry*, January:119–129.

Streets, R. J. 1962. *Exotic Forest Trees in the British Commonwealth*, pp. 627–628. Clarendon Press, Oxford, England.

Psidium guajava

Bakhshi, J. C., and N. S. Randhawa. 1967. Bright prospects for guava cultivation in the Punjab. *Indian Journal of Horticulture* 11(3):3–4.

Bourke, D. O. 1976. *Psidium guajava*—guava. *The Propagation of Tropical Fruit Trees*, eds. R. J. Garner and S. A. Chaudri. Commonwealth Agricultural Bureau, Slough, England. pp. 530–553.

Council of Scientific and Industrial Research. 1969. *Wealth of India: Raw Materials*, Vol. 8, pp. 286–293. CSIR, New Delhi, India.

El Baradi, T. A. 1975. Guava. *Abstracts on Tropical Agriculture* 1(3):9–16.

Leu, L. S., C. W. Kao, C. C. Wang, W. J. Liang, and S. P. Y. Hsieh. 1979. Myxosporium wilt of guava and its control. *Plant Disease Reporter* 63(12):1075–1077.

Nakasone, H. Y. 1973. Guava propagation in Hawaii. Miscellaneous Publication No. 111, pp. 16–19. Cooperative Extension Service, University of Hawaii, Honolulu, Hawaii, USA.

Oakes, A. J. 1970. Herbicidal control of guava (*Psidium guajava* L.). *Turrialba* 20(1):30–36.

Partridge, I. 1979. The guava threat in Fiji. *South Pacific Bulletin* 2d Quarter:28–30.

Prasad, N., R. L. Mathur, and L. S. Chattri. 1966. Bright prospects for guava cultivation in the Punjab. *Indian Journal of Agricultural Science* 36(4):201–209.

Shigeura, G. T. 1973. Culture and management of guava. Miscellaneous Publication No. 111, pp. 10–16. Cooperative Extension Service, University of Hawaii, Honolulu, Hawaii, USA.

Suhag, L. S. 1976. Observations on guava decline in Haryana and its control. *Pesticides* 10(12):42–44.

Robinia pseudoacacia

Carpenter, S. B., D. H. Graves, and R. A. Eigel. 1979. Producing black locust biomass for fuel on southern appalachian surface mines. *Energy Communications* 5(2):101–108.

Funk, D. T., and B. A. Roach. 1961. *Black Locust: A Bibliography.* USDA Forest Service, North Central Forest Experiment Station, St. Paul, Minnesota, USA. 40 pp.

Goor, A. Y., and C. W. Barney. 1968. *Forest Tree Planting in Arid Zones.* Ronald Press, New York, USA.

Hall, N., R. W. Boden, C. S. Christian, R. W. Condon, F. A. Dale, A. J. Hart, J. H. Leigh, J. K. Marshall, A. G. McArthur, V. Russell, and J. W. Turnbull. 1972. *The Use of Trees and Shrubs in the Dry Country of Australia.* Forest and Timber Bureau, Canberra, Australia.

Keresztesi, B. 1980. The black locust. *Unasylva* 32(127): 23–33.

Keresztesi, B. 1979. Mini-monograph on *Robinia pseudoacacia. Technical Consultation on Fast-Growing Plantation Broadleaved Trees for Mediterranean and Temperate Zones*, pp. 301–330. Food and Agriculture Organization of the United Nations, Rome, Italy.

Kim, Chung Suk. 1975. *Studies on the Colchitetraploids of* Robinia pseudoacacia *L.* Research Report of the Institute of Forest Genetics No. 12. Institute of Forest Genetics, Office of Forestry, Suwon, Korea. 108 pp.

Limstrom, G. A. 1960. *Forestation of Strip-Mined Land in the Central States.* Agricultural Handbook 166. USDA Forest Service, Washington, D.C., USA. 74 pp.

McAlister, R. H. 1971. Black locust. *American Woods.* Forest Service Report No. 244. USDA Forest Service, Washington, D.C., USA.

Olson, D. F., Jr. 1974. *Robinia* L. Locust. *Seeds of Woody Plants in the United States*, pp. 728–731. Agricultural Handbook 450. USDA Forest Service, Washington, D.C., USA.

Roach, B. A. 1965. Black locust (*Robinia pseudoacacia* L.). *Silvics of Forest Trees in the United States*, pp. 642–648. Agricultural Handbook 271. USDA Forest Service, Washington, D.C., USA.

Royce, R. D. 1965. *Robinia*—The false acacia. *Journal of Agriculture of Western Australia* 6(9):550–552.

Shah, F. A. 1972. False acacia—A promising bee plant of Kashmir. *Indian Bee Journal* 34(1/2):34–35.

Streets, R. J. 1962. *Exotic Forest Trees in the British Commonwealth.* Clarendon Press, Oxford, England.

Sapium sebiferum

Bolley, D. S., and R. H. McCormack. 1950. Utilization of the seed of the Chinese tallow tree. *Journal of the American Oil Chemists' Society* 27:84–87.

Browne, F. G. 1968. *Pests and Diseases of Forest Plantation Trees.* Oxford University Press, Oxford, England.

Cameron, G. N., and T. W. LaPoint. 1978. Effects of tannins on the decomposition of Chinese tallow leaves by terrestrial and aquatic invertebrates. *Oecologia* 32:349–366.

Chao, S. C., Y. C. Ku, S. J. Lin, and T. T. Pan. 1971. Measurements of fiber dimensions and analysis of chemical composition on Taiwan hardwoods. *Bulletin of Taiwan Forestry Research Institute* 14:1–26.

Christie, W. 1969. The glyceride structure of *Sapium sebiferum* seed oil. *Biochimica et Biophysica Acta* 187:1–5.

Council of Scientific and Industrial Research. *The Wealth of India: Raw Materials*, Vol. 9, pp. 230–231. CSIR, New Delhi, India.

Göetze, H., and G. Schultze-Dewitz. 1972. Structure and physiochemical and strength properties of the wood of tropical Southeast Asian tree species. *Holztechnologie* 13:167–172.

Hayes, B. 1979. The Chinese tallow tree (*Sapium sebiferum*)—Artificial bee pasturage success story. *American Bee Journal* 119:848–849.

Holland, B. R., and W. W. Meinke. 1948. Chinese tallow nut protein I. Isolation, amino acid, and vitamin

analysis. *Journal of the American Oil Chemists' Society* 25:418–419.
Howes, F. N. 1949. The Chinese tallow tree (*Sapium sebiferum* Roxb.)–A source of drying oil. *Kew Bulletin* 4:573–580.
Hsu, B. H. 1928. A systematical examination of Chinese tallow seeds and oil. *The China Journal* 9:244–251.
Lin, W.-C., A.-C. Chen, C.-J. Tseng, and S.-G. Huang. 1958. An investigation and study of Chinese tallow trees in Taiwan (*Sapium sebiferum* Roxb). *Bulletin of Taiwan Forestry Research Institute* 57:1–37.
MacGowan, J. D. 1851. Uses of *Stillingia sebifera* or tallow tree with a notice of the Pe-la, or insect-wax of China. *American Journal of Science* 12:17–22.
Potts, W. M. 1946. The Chinese tallow tree as a chemurgic crop. *The Chemurgic Digest* 5:373–375.
Scheld, H. W., and J. R. Cowles. 1981. Woody biomass potential of the Chinese tallow tree. *Economic Botany* 35:391–397.
Scheld, H. W., N. B. Bell, G. N. Cameron, J. R. Cowles, C. R. Engler, A. D. Krikorian, and E. B. Schultz. 1981. The Chinese tallow tree as a cash and petroleum-substitute crop. *Tree Crops for Energy Co-production on Farms*, pp. 97–111. SERI Bulletin CP-622-1086.

Sesbania sesban

Allen, O. N., and E. K. Allen. 1981. *The Leguminosae.* Macmillan, London, England.
Anonymous. 1970. Cultivation of *Sesbania aegyptiaca. Aranya Vikas (Quarterly Bulletin of A. P. Forest Officers Association)* 17:26–34.

Tarchonanthus camphoratus

Palgrave, K. C., et al. 1977. *Trees of Southern Africa*, p. 910. Struik, Cape Town, South Africa.
Palmer, E., and N. Pitman. 1972. *Trees of Southern Africa*, pp. 2,155–2,157. Balkema, Cape Town, South Africa.
Watt, J. M., and M. G. Breyer-Brandwijk. 1962. *The Medicinal and Poisonous Plants of Southern and Eastern Africa*, pp. 294–295. Livingstone, Edinburgh, Scotland.

APPENDIX 3

Research Contacts

Acacia decurrens

Conservator of Forests (Research), Tamil Nadu, Madurai, Tamil Nadu 625 020 India

R. N. Kaul, Director of Forestry Research and Central Silviculturist, Forest Research Institute and Colleges, P.O. New Forest, Dehra Dun 248 006, U.P., India

Seed Section, CSIRO Division of Forest Research, P.O. Box 4008, Canberra, A.C.T. 2600, Australia

V. R. Nanayakkara, Conservator of Forests, Forest Department, P.O. Box 509, Colombo 2, Sri Lanka

M. D. Tindale, Royal Botanic Gardens, Mrs. Macquarie's Road, Sydney, N.S.W. 2000, Australia.

Ailanthus excelsa

P. K. Khosla, Department of Forestry, Agricultural Complex, Himachal Pradesh Agricultural University, Nauni, Solan 173 230, H.P., India

M. Kolar, Associate Director, Forest Department, Land Development Authority, P.O. Box 45, Kiriat Hayim, Israel (*Ailanthus glandulosa* [*A. altissima*] seed only)

Memendra Kumar Nath, Silviculturist, Forest Department–Government of Assam, South Sarania, Gauhati 7, Assam, India

H. Y. Mohan Ram, Department of Botany, University of Delhi, Delhi 110 007, India

K. M. Siddiqui, Director, Forest Products Research Division, Pakistan Forest Institute, Peshawar, Pakistan

N. P. Singh, Central Sheep and Wood Research Institute, Avikanager, Rajasthan, India

Albizia falcataria

Harun Alrasyid, Silviculturist, Balai Penelitian Hutan (Forest Research Institute), P.O. Box 66, Jalan Gunung Batu, Bogor, Indonesia

Robert M. Bennett, Principal Forest Officer, Department of Agriculture, P.O. Box 129, Vila, Vanuatu, S.W. Pacific

Tan Kee Chong, Sabah Softwoods Sdn. Bhd., P.O. Box 137, Tawau, Sabah, Malaysia

Director, Forest Research Institute, Kepong, Selangor, Malaysia

Ireneo L. Domingo, Forest Research Director, Paper Industries Corporation of the Philippines, Bislig, Surigao del Sur, Philippines

Ta-Wei Hu, Division of Silviculture, Taiwan Forestry Research Institute, 53 Nan-Hai Road, Taipei, Taiwan

Kuswata Kurtawinata, Head, Herbarium Bogoriense, LBN–LIPI, Jalan Raya Juanda 22–24, Bogor, Indonesia

Hua Seng Lee, Silviculturist, Forest Research Section, Forestry Department, Badruddin Road, Kuching, Sarawak, Malaysia

Eduardo B. Principe, Research Officer, Silviculture Division, Forest Research Institute, College, Laguna 3720, Philippines

Jose O. Sargento, Makiling Experimental and Demonstration Forests, College of Forestry, University of the Philippines at Los Baños, College, Laguna 3720, Philippines

Neptali Q. Zabala, Associate Professor of Forest Tree Improvement, Silviculture and Forest Influences Department, University of the Philippines at Los Baños, College, Laguna 3720, Philippines

Balanites aegyptiaca

R. Karschon, Director, Forestry Division, Agricultural Research Organization, 42805 Ilanot, M. P. Lev Hasharon, Israel

Hassan A. R. Musnad, Silviculturist, Arid Zone Forestry Research, Forestry Research Institute, P.O. Box 658, Khartoum, Sudan

SEPASAT, Royal Botanic Gardens, Kew, Richmond, Surrey, TW9 3AE, England, United Kingdom

Ezekiel Abayomi Sofowora, Department of Pharmacognosy, Faculty of Pharmacy, University of Ife, Ile-Ife, Nigeria

H. J. von Maydell, Professor, Federal Research Center for Forestry and Forest Products, Leuschnerstrasse, 91, 2050 Hamburg 80, West Germany

Bursera simaruba

K. S. Bawa, Professor, Department of Biology, University of Massachusetts, Boston, Massachusetts 02125, USA

Director, Direccion de Bosques, INDERENA, Carrera 14, No. 25A66 Piso 2, Bogota, D.E., Colombia

Director, Instituto Forestal Latino-Americano, Apartado 36, Merida, Venezuela

Institute of Tropical Forestry, P.O. Box AQ, Rio Piedras, Puerto Rico 00928, USA

Coccoloba uvifera

Centro Agronomico Tropical de Investigacion y Ensenanza (CATIE), Turrialba, Costa Rica

Director, Direccion de Bosques, INDERENA, Carrera 14, No. 25A66 Piso 2, Bogota, D.E., Colombia

Director, Instituto Forestal Latino-Americano, Apartado 36, Merida, Venezuela

Everglades Sod and Landscape, Inc., 19100 Krome Avenue, Miami, Florida 33187, USA (commercial production)

Institute of Tropical Forestry, P.O. Box AQ, Rio Piedras, Puerto Rico 00928, USA

George N. McHugh, 28590 S.W. 170th Ave., Homestead, Florida 33030, USA (flammability)

Combretum micranthum

Directeur, Institut Sénégalais de Recherches Agricoles,

Centre National de Recherches Forestières, Parc Forestier de Hann, B.P. 2312, Dakar, Sénégal

H. J. von Maydell, Professor, Federal Research Center for Forestry and Forest Products, Leuschnerstrasse, 91, 2050 Hamburg, Germany

J. C. Okafor, Forestry Investigation Unit, No. 2 Works Road, Enugu, Nigeria

SEPASAT, Royal Botanic Gardens, Kew, Richmond, Surrey, TW9 3AE, England, United Kingdom

Conocarpus lancifolius

Mohsen A. R. Bazara'a, Forestry Section, Research and Extension Department, El-Kod Research Station, Ministry of Agriculture, Aden, People's Democratic Republic of Yemen

R. D. Haller, Bamburi Portland Cement Co., Ltd., P.O. Box 90202, Mombasa, Kenya

Hassan A. R. Musnad, Silviculturist, Arid Zone Forestry Research, Forestry Research Institute, P.O. Box 658, Khartoum, Sudan

SEPASAT, Royal Botanic Gardens, Kew, Richmond, Surrey, TW9 3AE, England, United Kingdom

Dalbergia sissoo

Devendra B. Amatya, Forestry Services, Ravi Bhawan Road, Kalimati, Kathmandu, Nepal

Chief Forest Officer, Community Forestry and Afforestation Division, Hattisar, Naxal, Kathmandu, Nepal

P. K. Khosla, Department of Forestry, Agricultural Complex, Himachal Pradesh Agricultural University, Nauni, Solan 173 230, H.P., India

M. Kolar, Associate Director, Forest Department, Land Development Authority, P.O. Box 45, Kiriat Hayim, Israel

Adarsh Kumar, Research Officer, Plant Physiology Branch, Forest Research Institute, P.O. New Forest, Dehra Dun 248 006, U.P., India

Hassan A. R. Musnad, Silviculturist, Arid Zone Forestry Research, Forestry Research Institute, P.O. Box 658, Khartoum, Sudan

K. M. Siddiqui, Director, Forest Products Research Division, Pakistan Forest Institute, Peshawar, Pakistan.

Paul Soderholm, USDA Subtropical Horticulture Research Unit, 13601 Old Cutler Road, Miami, Florida 33158, USA

Eucalyptus brassiana

J. W. Turnbull, CSIRO Division of Forest Research, P.O. Box 4008, Canberra, A.C.T. 2600, Australia

N. H. S. Howcroft, Forest Research Station, P.O. Box 134, Bulolo, Monrobe Province, Papua New Guinea

J. C. Doran, CSIRO Division of Forest Research, P.O. Box 4008, Canberra, A.C.T. 2600, Australia

Eucalyptus deglupta

Tan Kee Chong, Sabah Softwoods Sdn. Bhd., P.O. Box 137, Tawau, Sabah, Malaysia

J. Davidson, Eucalyptus and Forestry Service, P.O. Box 419, Armidale, N.S.W. 2350, Australia

Director, Forest Research Institute, P.O. Box 66, Jalan Gunung Batu, Bogor, Indonesia

Director, Forest Research Institute, College, Laguna 3720, Philippines

Director, OPFLOR, Companhia Florestal de Monte Dourado, Caixa 263, Belém, Pará, Brazil

N. H. S. Howcroft, Forest Research Station, P.O. Box 134, Bulolo, Monrobe Province, Papua New Guinea

C. P. Munoz, 24B Jasmin Street Ext. Carmen, Cagayan de Oro, Philippines

J. W. Turnbull, CSIRO Division of Forest Research, P.O. Box 4008, Canberra, A.C.T. 2600, Australia

Eucalyptus pellita

J. W. Turnbull, CSIRO Division of Forest Research, P.O. Box 4008, Canberra, A.C.T. 2600, Australia

E. Campinhos, Jr., Rua Prof. Lobo, 1.128, Aracruz, Esp. Santo, Brazil

Eucalyptus robusta

T. F. Geary, USDA Forest Service, Southeastern Forest Experiment Station, P.O. Box 938, Lehigh Acres, Florida 33936, USA

Forestry Research Station, Charia Omar Ibn El Khattab, B.P. 763, Rabat (Agdal), Morocco

R. N. Kaul, Director of Forestry Research and Central Silviculturist, Forestry Research Institute and Colleges, P.O. New Forest, Dehra Dun 248 006, U.P., India

M. Kolar, Associate Director, Forest Department, Land Development Authority, P.O. Box 45, Kiriat Hayim, Israel (seed)

George Meskimen, USDA Forest Service, Southeastern Forest Experiment Station, P.O. Box 938, Lehigh Acres, Florida 33936, USA

L. D. Pryor, Professor, Department of Forestry, The Australian National University, P.O. Box 4, Canberra, A.C.T., 2600, Australia

D. R. Quaile, Eucalypt Tree Breeder, Forest Research Centre, P.O. Box HG 595, Highlands, Harare, Zimbabwe

Roger G. Skolmen, USDA Forest Service, Pacific Southwest Forest and Range Experiment Station, Institute of Pacific Islands Forestry, 1151 Punchbowl Street, Room 323, Honolulu, Hawaii 96813, USA

J. W. Turnbull, CSIRO Division of Forest Research, P.O. Box 4008, Canberra, A.C.T. 2600, Australia

Eucalyptus tereticornis

J. C. Delwaulle, Directeur, Centre Technique Forestier Tropical, B.P. 764, Pointe Noire, République Populaire du Congo

Forestry Research Station, Charia Omar Ibn El Khattab, B.P. 763, Rabat (Agdal), Morocco

T. F. Geary, USDA Forest Service, Southeastern Forest Experiment Station, P.O. Box 938, Lehigh Acres, Florida 33936, USA

INDERENA, Apartado Aereo 51912, Bogota, D.E. 2, Colombia

P. K. Karani, Senior Conservator Research, Forest Department, Ministry of Agriculture and Animal Resources, P.O. Box 31, Entebbe, Uganda

M. Kolar, Associate Director, Forest Department, Land Development Authority, P.O. Box 45, Kiriat Hayim, Israel (seed)

Kodira A. Kushalapa, Deputy Conservator of Forests, Yellapur 581 359, U.K. Dist., Karnataka, India

L. D. Pryor, Professor, Department of Forestry, The Australian National University, P.O. Box 4, Canberra, A.C.T., 2600, Australia

D. R. Quaile, Eucalypt Tree Breeder, Forest Research Centre, P.O. Box HG 595, Highlands, Harare, Zimbabwe

K. M. Siddiqui, Diector, Forest Products Research Division, Pakistan Forest Institute, Peshawar, Pakistan

J. W. Turnbull, CSIRO Division of Forest Research, P.O. Box 4008, Canberra, A.C.T. 2600, Australia

K. Vivekanandan, Chief Research Officer, Forest Department, P.O. Box 509, Colombo 2, Sri Lanka
P. J. Wood, Unit of Tropical Silviculture, Commonwealth Forestry Institute, Oxford OX1 3RB, England

Eucalyptus urophylla

E. Campinhos, Jr., Rua Prof. Lobo, 1.128, Aracruz, Esp. Santo, Brazil
J. C. Delwaulle, Directeur, Centre Technique Forestier Tropical, B.P. 764, Pointe Noire, République Populaire du Congo
J. C. Doran, CSIRO Division of Forest Research, P.O. Box 4008, Canberra, A.C.T. 2600, Australia
Director, OPFLOR, Companhia Florestal de Monte Dourado, Caixa Postal 263, Belém, Pará, Brazil
V. P. G. Moura, CPAC/EMBRAPA, Km 18-BR 020, BSB-For., 73300-Planaltina-D.F., Brazil
Senior Forest Officer (Research), Forestry Division, Munda, W/Province, Solomon Islands
J. W. Turnbull, CSIRO Division of Forest Research, P.O. Box 4008, Canberra, A.C.T. 2600, Australia

Gleditsia triacanthos

Robert M. Blair, USDA Forest Service, P.O. Box 7600, SFA Station, Nacogdoches, Texas 75962, USA
James Hanover, Department of Forestry, Michigan State University, East Lansing, Michigan 48824, USA
I. Nahal, Professor of Forestry and Ecology, Faculty of Agriculture, University of Aleppo, Aleppo, Syria
David Scanlon III, Tennessee Valley Authority, Division of Land and Forest Resources, Norris, Tennessee 37828, USA
Gregory Williams, Appalachian Regional Office, International Tree Crops Institute, U.S.A., Inc., Route 1, Gravel Switch, Kentucky 40328, USA

Hibiscus tiliaceus

Central Silviculturist, Forest Research Institute and Colleges, P.O. New Forest, Dehra Dun 248 006, U.P., India
Department of Natural Resources, Forest Research Institute, College, Laguna 3720, Philippines
Forest Products Research Institute, University P.O. Box 63, Kumasi, Ghana
Institute of Tropical Forestry, P.O. Box AQ, Rio Piedras, Puerto Rico 00928, USA

Maesopsis eminii

Commonwealth Forestry Institute, South Parks Road, Oxford OX1 3RB, England, United Kingdom (J. Burley, H. C. Dawkins, R. A. Plumptre)
Director, Forestry Research Institute, Kepong, Selangor, Malaysia
Division of Forestry, Faculty of Agriculture, Forestry, and Veterinary Sciences, University of Dar es Salaam, P.O. Box 643, Morogoro, Tanzania
Hua Seng Lee, Silviculturist, Forest Research Section, Forestry Department, Badruddin Road, Kuching, Sarawak, Malaysia
I. A. Odero, Deputy Director, Kenya Agricultural Research Institute, P.O. Box 74, Kikuyu, Kenya
Senior Silviculturist, Forest Research Centre, P.O. Box 1752, Kampala, Uganda
Silviculturist, Forest Division, Silviculture Research Station, P.O. Box 95, Lushoto, Tanzania

Melaleuca quinquenervia

C. B. Briscoe, Agroforester, P.O. Box 18, Las Cumbres 15, Panama
T. F. Geary, USDA Forest Service, Southeastern Forest Experiment Station, P.O. Box 938, Lehigh Acres, Florida 33936, USA (silviculture and utilization)
Directeur, Institut Sénégalais de Recherches Agricoles, Centre National de Recherches Forestières, Parc Forestier de Hann, B.P. 2312, Dakar, Sénégal
J. B. Huffman, School of Forest Resources and Conservation, University of Florida, Gainesville, Florida 32611, USA
Kuswata Kurtawinata, Head, Herbarium Borgoriense, LBN-LIPI, Jalan Raya Juanda 22-24, Bogor, Indonesia
J. Morton, Morton Collectanea, University of Miami, Box 248204, Coral Gables, Florida 33124, USA
Seed Section, CSIRO Division of Forest Research, P.O. Box 4008, Canberra, A.C.T. 2600, Australia
S. L. Woodall, USDA Forest Service, Southeastern Forest Experiment Station, Lehigh Acres, Florida 33936, USA

Melia azedarach

Sa-ard Boonkird, Emeritus Instructor, College of Forestry, Kasetsart University, Bangkok 9, Thailand
C. B. Briscoe, Agroforester, P.O. Box 18, Las Cumbres 15, Panama
P. K. Khosla, Department of Forestry, Agriculture Complex, Himachal Pradesh Agricultural University, Nauni, Solan 173 230, H.P., India
Sompherm Kittinanda, Royal Forest Department, Ministry of Agriculture and Cooperatives, Phaholyothin Road, Bangkok 9, Thailand
M. Kolar, Associate Director, Forest Department, Land Development Authority, P.O. Box 45, Kiriat Hayim, Israel (seed)
I. Nahal, Professor of Forestry and Ecology, Faculty of Agriculture, University of Aleppo, Syria
M. I. Sheikh, Director General, Pakistan Forest Institute, Peshawar, Pakistan

Pinus caribaea

T. I. W. Bell, Principal Research Officer, Fiji Pine Commission, P.O. Box 521, Lautoka, Fiji
Robert M. Bennett, Principal Forest Officer, Department of Agriculture, P.O. Box 129, Vila, Vanuatu, S.W. Pacific
C. B. Briscoe, Agroforester, P.O. Box 18, Las Cumbres 15, Panama
Chief Conservator of Forests, Forestry Division, M.A.F., Tower Hill, Freetown, Sierra Leone
Chief Forest Officer, Ministry of Natural Resources, Belmopan, Belize
Commonwealth Forestry Institute, South Parks Road, Oxford OX1 3RD, England, United Kingdom (R. D. Barnes, J. Burley, R. A. Plumptre, P. J. Wood)
William Dvorak, Director, CAMCORE Cooperative, North Carolina State University, P.O. Box 5988, Raleigh, North Carolina 27650, USA
Carl M. Gallegos, Senior Research Forester, Central and South American Species, International Paper Company, Mobile, Alabama 36601, USA
INDERENA, Apartado Aereo 51912, Bogota, D.E. 2, Colombia
Hua Seng Lee, Silviculturist, Forest Research Section, Forestry Department, Badruddin Road, Kuching, Sarawak, Malaysia
José A. Lewald, Silviculturist, Seed Export, P.O. Box 543, Guatemala City, Guatemala
Harry E. Murphy, RF/ACF, Resource Management Service, Inc., P.O. Box 43388, Birmingham, Ala-

bama 35243, USA

D. R. Quaile, Eucalpyt Tree Breeder, Forest Research Centre, P.O. Box HG 595, Highlands, Harare, Zimbabwe

Senior Forestry Research Officer, Forest Research Institute of Malawi, P.O. Box 270, Zomba, Malawi

José Rodolfo Salazar, Celulosa de Turrialba, S.A., Ap 17, Turrialba, Costa Rica

Roy Voss, c/o Tropical Reforestation Enterprises, Rt. 3, Box 434, Alachua, Florida 32615, USA

Populus euphratica

Forestry Research Station, Charia Omar Ibn El Khattab, B.P. 763, Rabat (Agdal), Morocco

R. N. Kaul, Director of Forestry Research and Central Silviculturist, Forest Research Institute and Colleges, P.O. New Forest, Dehra Dun 248 006, U.P., India

P. S. Kaushal, Asst. Professor of Forestry, Punjab Agricultural University, Ludhiana, Punjab 141004, India

P. K. Khosla, Department of Forestry, Agricultural Complex, Himachal Pradesh Agricultural University, Nauni, Solan 173 230, H.P., India

I. Nahal, Professor of Forestry and Ecology, Faculty of Agriculture, University of Aleppo, Aleppo, Syria

I. A. Odero, Deputy Director, Kenya Agricultural Research Institute, P.O. Box 74, Kikuyu, Kenya

Salahadin University, College of Agriculture, Aski-Kalak, Arbil, Iraq (Latif Al-Najjar)

K. M. Siddiqui, Director, Forest Products Research Division, Pakistan Forest Institute, Peshawar, Pakistan

Poplar and Fast Growing Exotic Forest Trees Research Institute, Izmit, Turkey

Psidium guajava

C. B. Briscoe, Agroforester, P.O. Box 18, Las Cumbres 15, Panama

Carl W. Campbell, University of Florida, IFAS, AREC, 18905 S.W. 280th St., Homestead, Florida 33031, USA (fruit production)

Centro Agronomico Tropical de Investigacion y Ensenanze (CATIE), Turrialba, Costa Rica

C. Leng Chia, Extension Specialist, Department of Horticulture, University of Hawaii, 3190 Maile Way, Honolulu, Hawaii 96822, USA

Director of Forestry, Federal Department of Forestry, PMB #5011, Ibadan, Nigeria

M. Kolar, Associate Director, Forest Department, Land Development Authority, P.O. Box 45, Kiriat Hayim, Israel (seed)

J. C. Okafor, Forestry Investigation Unit, No. 2 Works Road, Enugu, Nigeria

Robinia pseudoacacia

Stanley B. Carpenter, Department of Forestry, Oklahoma State University, Stillwater, Oklahoma 74078, USA

Valeriu Enescu, Institutul de Cercetàri si Amenajari Silvice (ICAS), R-72904, Bucuresti, Sector 2, SOS, Stefanesti, Romania

D. H. Graves, Department of Forestry, University of Kentucky, Lexington, Kentucky 40546, USA

James Hanover, Department of Forestry, Michigan State University, East Lansing, Michigan 48824, USA

Sin Kyu Hyun, Institute of Forest Genetics, Suwon, P.O. Box 24, Kyunggido, South Korea

Peter Jaciw, Ministry of Natural Resources, Maple, Ontario, Canada L0J 1E0

B. Keresztesi, Forest Research Institute, H-1024 Budapest, Frankel Leo U. 42–44, Hungary

P. K. Khosla, Department of Forestry, Agricultural Complex, Himachal Pradesh Agricultural University, Nauni, Solan 173 230, H.P., India

Chung Suk Kim, Department of Biology, College of Natural Sciences, Gyeongsang National University, Jinju, Gyeongnam 620, South Korea

Jorge A. Lopez, Director, Forest Seeds Center, National Forestry Corporation, Constitucion 291 Casilla 5, Chillan, Chile

I. Nahal, Professor of Forestry and Ecology, Faculty of Agriculture, University of Aleppo, Aleppo, Syria

David H. Scanlon III, Tennessee Valley Authority, Division of Land and Forest Resources, Norris, Tennessee 37828, USA

K. M. Siddiqui, Director, Forest Products Research Division, Pakistan Forest Institute, Peshawar, Pakistan

Willis G. Vogel, Range Scientist, USDA Forest Service, Northeastern Forest Experiment Station, Berea, Kentucky 40403, USA

Gregory Williams, Appalachian Regional Office, International Tree Crops Institute U.S.A., Inc., Route 1, Gravel Switch, Kentucky 40328, USA

Sapium sebiferum

Center for Development Technology, Washington University, Campus Box 1106, St. Louis, Missouri 63130, USA (H. M. Draper, E. B. Shultz, Jr.)

Department of Biology, University of Houston, Houston, Texas 77004, USA (G. N. Cameron, J. R. Cowles, H. W. Scheld)

Department of Biology, University of Houston at Clear Lake City, Houston, Texas 77058, USA (N. B. Bell, M. Vrecenar)

C. R. Engler, Food Protein Research and Development Center, Texas A & M University, College Station, Texas 77843, USA

Ta-Wei Hu, Division of Silviculture, Taiwan Forestry Research Institute, 53 Nan-Hai Road, Taipei, Taiwan

P. K. Khosla, Department of Forestry, Agricultural Complex, Himachal Pradesh Agricultural University, Nauni, Solan 173 230, H.P., India

A. D. Krikorian, Department of Biochemistry, State University of New York at Stony Brook, Stony Brook, New York 11794, USA

S. L. Mathieu, Center for Development Technology, Washington University, Campus Box 1106, St. Louis, Missouri 63130, USA

K. M. Siddiqui, Director, Forest Products Research Division, Pakistan Forest Institute, Peshawar, Pakistan

Sesbania sesban

Central Soil Salinity Research Institute, Karnal 132 001, India (I. P. A. Abrol, R. Chhabra)

Department of Botany, University of the Punjab, Lahore, Pakistan (I. Ahmad, Rab Nawaz, Khan Niazi)

Director, National Agricultural Research Centre, P.O. National Institute of Health, Islamabad, Pakistan

D. O. Evans, College of Tropical Agriculture and Human Resources, Department of Agronomy and Soil Science, 3190 Maile Way, Honolulu, Hawaii 96822, USA

K. M. Siddiqui, Director, Forest Products Research

Division, Pakistan Forest Institute, Peshawar, Pakistan

Tarchonanthus camphoratus

A. Gubb, McGregor Museum, P.O. Box 316, Kimberley, 8300 South Africa

Mark Gandar, Institute of Natural Resources, University of Natal, P.O. Box 375, Pietermaritzburg, 3200 South Africa.

Anthony V. Hall, Associate Professor, Bolus Herbarium, University of Cape Town, Rondebosch, 7700 South Africa

Index of Plants

Species described or referred to in the body of the report are indexed below. An asterisk indicates that the species was described in full in the text of Volume I.

Advisory Committee on Technology Innovation

HUGH POPENOE, Director, International Programs in Agriculture, University of Florida, Gainesville, Florida, *Chairman*

Members

WILLIAM BRADLEY, Consultant, New Hope, Pennsylvania

HAROLD DREGNE, Director, International Center for Arid and Semi-Arid Land Studies, Texas Tech University, Lubbock, Texas (member through 1981)

ELMER L. GADEN, JR., Department of Chemical Engineering, University of Virginia, Charlottesville, Virginia

CARL N. HODGES, Director, Environmental Research Laboratory, Tucson, Arizona

CYRUS MCKELL, Native Plants, Inc., Salt Lake City, Utah

FRANÇOIS MERGEN, Pinchot Professor of Forestry, School of Forestry and Environmental Studies, Yale University, New Haven, Connecticut (member through 1982)

DONALD L. PLUCKNETT, Consultative Group on International Agricultural Research, Washington, D.C.

THEODORE SUDIA, Deputy Science Advisor to the Secretary of Interior, Department of Interior, Washington, D.C.

Board on Science and Technology for International Development

GEORGE BUGLIARELLO, President, Polytechnic Institute of New York, Brooklyn, New York, *Chairman*

Members

SAMUEL P. ASPER, Educational Commission for Foreign Medical Graduates, Philadelphia, Pennsylvania

DAVID BELL, Department of Population Sciences, Harvard School of Public Health, Boston, Massachusetts

LEONARD BERRY, Professor, Graduate School of Geography, Clark University, Worcester, Massachusetts

ERNEST J. BRISKEY, Dean, School of Agriculture, Oregon State University, Corvallis, Oregon

HARRISON S. BROWN, Director, Resources Systems Institute, East-West Center, Honolulu, Hawaii

ROBERT H. BURRIS, Department of Biochemistry, University of Wisconsin, Madison, Wisconsin

CLAUDIA JEAN CARR, Associate Professor, Conservation and Resource Studies, University of California at Berkeley, Berkeley, California

NATE FIELDS, Director, Developing Markets, Control Data Corporation, Minneapolis, Minnesota

ROLAND J. FUCHS, Chairman, Department of Geography, University of Hawaii at Manoa, Honolulu, Hawaii

ELMER L. GADEN, JR., Department of Chemical Engineering, University of Virginia, Charlottesville, Virginia

JOHN HOWARD GIBBONS, Director, U.S. Congress, Office of Technology Assessment, Washington, D.C.

N. BRUCE HANNAY, Foreign Secretary, National Academy of Engineering, Washington, D.C.

Board on Science and Technology for International Development (JH-217D)
Office of International Affairs
National Research Council
2101 Constitution Avenue
Washington, D.C. 20418, USA

How to Order BOSTID Reports

Reports published by the Board on Science and Technology for International Development are sponsored in most instances by the U.S. Agency for International Development and are intended for free distribution primarily to readers in developing countries. A limited number of copies are available without charge to readers in the United States and other industrialized countries who are affiliated with governmental, educational, or research institutions, and who have a professional interest in the subjects treated by the report. Requests should be made on the institution's stationary.

Single copies of published reports listed below are available free from BOSTID at the above address while the supplies last.

Energy

19. **Methane Generation from Human, Animal, and Agricultural Wastes.** 1977. 131 pp. Discusses means by which natural process of anaerobic fermentation can be controlled by man for his benefit and how the methane generated can be used as a fuel.

33. **Alcohol Fuels: Options for Developing Countries.** 1983. 128 pp. Examines the potential for the production and utilization of alcohol fuels in developing countries. Includes information on various tropical crops and their conversion to alcohols through both traditional and novel processes.

36. **Producer Gas: Another Fuel for Motor Transport.** 1983. 112 pp. During World War II Europe and Asia used wood, charcoal, and coal to fuel over a million gasoline and diesel vehicles. However, the technology has since been virtually forgotten. This report reviews producer gas technology in motor vehicles and its modern potential.

38. **Supplement to Energy for Rural Development: Renewable Resources and Alternative Technologies for Developing Countries.** 1981. 240 pp. Updates the 1976 BOSTID publication and offers new material on direct and indirect uses of solar energy. Provides index to both volumes.

39. **Proceedings, International Workshop on Energy Survey Methodologies for Developing Countries.** 1980. 220 pp. Examines past and continuing energy survey efforts in developing countries. Includes rural, urban, industry, and transportation energy survey methodologies, excerpts from 12 background papers, and a directory of energy surveys for developing countries.

48. **Workshop on Energy and Agriculture in Developing Countries.** 1981. Discusses the energy used in agricultural production as well as the energy used to transport, process, and store agricultural products and to prepare food. It reviews current issues and advises USAID of promising opportunities to help solve energy-related problems in agriculture in developing areas.

Technology Options for Developing Countries

8. **Ferrocement: Applications in Developing Countries.** 1973. 89 pp. Assesses state of the art and cites applications of particular interest to developing countries such as boat building, construction, and food and water storage facilities.

14. **More Water for Arid Lands: Promising Technologies and Research Opportunities.** 1974. 153 pp. Outlines little-known but promising small-scale technologies to supply and conserve water in arid areas. (French language edition is also available from BOSTID.)

21. **Making Aquatic Weeds Useful: Some Perspectives for Developing Countries.** 1976. 175 pp. Describes ways to exploit aquatic weeds for grazing, and by harvesting and processing for use as compost, animal feed, pulp, paper, and fuel. Also describes utilization for sewage and industrial wastewater treatment. Examines certain plants with potential for aquaculture.

28. **Microbial Processes: Promising Technologies for Developing Countries.** 1979. 198 pp. Discusses the potential importance of microbiology in developing countries in food and feed, plant nutrition, pest control, fuel and energy, waste treatment and utilization, and health.

31. **Food, Fuel, and Fertilizer for Organic Wastes.** 1981. 150 pp. Examines some of the opportunities for the productive utilization of organic wastes and residues commonly found in the poorer rural areas of the world.

34. **Priorities in Biotechnology Research for International Development: Proceedings of a Workshop.** 1982. 261 pp. Examines opportunities for biotechnology research in developing countries. Includes background papers and specific recommendations in vaccines, animal production, monoclonal antibodies, energy, biological nitrogen fixation, and plant cell and tissue culture.

Plants

16. **Underexploited Tropical Plants with Promising Economic Value.** 1975. 187 pp. Describes 36 little-known tropical plants that, with research, could become important cash and food crops in the future. Includes cereals, roots and tubers, vegetables, fruits, oilseeds, forage plants, and others.

22. **Guayule: An Alternative Source of Natural Rubber.** 1977. 80 pp. Describes a little-known bush that grows wild in deserts of North America and produces a rubber virtually identical with that of the rubber tree.

25. **Tropical Legumes: Resources for the Future.** 1979. 331 pp. Describes plants of the family Leguminosae, including root crops, pulses, fruits, forages, timber and wood products, ornamentals, and others.

37. **The Winged Bean: A High Protein Crop for the Tropics.** (Second Edition) 1981. 59 pp. An update of BOSTID's 1975 report of this neglected tropical legume from Southeast Asia and Papua New Guinea that appears to hold promise for combatting malnutrition worldwide.

47. **Amaranth: Modern Prospects for an Ancient Crop.** 1984. Before the time of Cortez grain amaranths were staple foods of the Aztec and Inca. Today this nutritious food has a bright future. The report also discusses vegetable amaranths.

Innovations in Tropical Reforestation

27. **Firewood Crops: Shrub and Tree Species for Energy Production.** 1980. 237 pp. Examines the selection of species suitable for deliberate cultivation as firewood crops in developing countries.

35. **Sowing Forests from the Air.** 1981. 64 pp. Describes experiences with establishing forests by sowing tree seed from aircraft. Suggests testing and development of the techniques for possible use where forest destruction now outpaces reforestation.

40. **Firewood Crops: Shrub and Tree Species for Energy Production.** Volume 2. 1983. 100 pp. A continuation of BOSTID report number 27. Describes 27 additional species of woody plants that seem suitable candidates for fuelwood plantations in developing countries.

41. **Mangium and Other Fast-Growing Acacias for the Humid Tropics.** 1983. 63 pp. Highlights 10 acacias species native to the tropical rain forest of Australasia. That they could become valuable forestry resources elsewhere is suggested by the exceptional performance of *Acacia mangium* in Malaysia.

42. **Calliandra: A Versatile Small Tree for the Humid Tropics.** 1983. 56 pp. This Latin American shrub is being widely planted by villagers and government agencies in Indonesia to provide firewood, prevent erosion, yield honey, and feed livestock.

43. **Casuarinas: Nitrogen-Fixing Trees for Adverse Sites.** 1984. These robust nitrogen-fixing Australasian trees could become valuable resources for planting on harsh, eroding land to provide fuel and other products. Eighteen species for tropical lowlands and highlands, temperate zones, and semiarid regions are highlighted.

52. **Leucaena: Promising Forage and Tree Crop for the Tropics.** (Second edition in preparation) Describes *Leucaena leucocephala*, a little-known Mexican plant with vigorously growing bushy types that produce nutritious forage as well as tree types that produce timber, firewood, and pulp and paper. The plant is also useful for revegetating hillslopes and for other purposes.

Managing Tropical Animal Resources

32. **The Water Buffalo: New Prospects for an Underutilized Animal.** 1981. 118 pp. Water buffalo are performing well in recent trials in new locations in the United States, Australia, and Brazil. Report discusses the animal's promise, particularly emphasizing its potential for use outside Asia.

44. **Butterfly Farming in Papua New Guinea.** 1983. 36 pp. Indigenous butterflies are being reared in Papua New Guinea villages in a formal government program that both provides a cash income in remote rural areas and contributes to the conservation of wildlife and tropical forests.

45. **Crocodiles as a Resource for the Tropics.** 1983. 60 pp. In most parts of the tropics crocodilian populations are being decimated, but programs in Papua New Guinea and a few other countries demonstrate that, with care, the animals can be raised for profit while the wild populations are being protected.

46. **Little-Known Asian Animals with a Promising Economic Future.** 1983. 124 pp. Describes banteng, madura, mithan, yak, kouprey, babirusa, Javan warty pig, and other obscure, but possibly useful, wild and domesticated indigenous Asian animals.

General

29. **Postharvest Food Losses in Developing Countries.** 1978. 202 pp. Summarizes existing work and information about losses of major food crops and fish; discusses economic and social factors involved; identifies major areas of need; and suggests policy and program options for developing countries and technical assistance agencies.

30. **U.S. Science and Technology for Development: Contributions to the UN Conference.** 1978. 226 pp. Prepared for the U.S. Department of State as a background document for the U.S. national paper presented at the 1979 United Nations Conference on Science and Technology for Development.

The following topic is now under study and will be the subject of a future BOSTID report:

- **Jojoba**

For a complete list of publications, including those that are out of print, please write to BOSTID at the address above.